Also by the Author

The Odyssey of Echo Company:
The 1968 Tet Offensive and the Epic Battle
to Survive the Vietnam War

In Harm's Way:
The Sinking of the USS Indianapolis
and the Extraordinary Story of Its Survivors

Praise for Doug Stanton's *12 Strong*

"Stanton packs a huge amount of research into a thrilling action ride of a book. Rousing, uplifting . . . for those who like their military history told through the eyes of heroic grunts, sergeants, and captains. Think of Stephen E. Ambroses's *Band of Brothers* . . . deserves a hallowed place in American military history."

—Bruce Barcott, cover of *The New York Times Book Review*

"Truth is stranger than fiction. And in the hands of . . . Doug Stanton, it's mesmerizing. . . . *12 Strong* is a riveting piece of literary journalism."

—*Detroit Free Press*

"Meticulous and thorough . . . difficult to put down. You find yourself riding along with the soldiers."

—*Northern Express*

"*12 Strong* tells the important story of the Special Forces soldiers who first put American boots on the ground in Afghanistan in 2001. Fighting alongside the Northern Alliance, the troops, often riding on horseback, achieved important victories against the Taliban."

—*Small Wars Journal*

"Spellbinding . . . the book reads more like a novel than a military history. . . . Stanton delivers page after page of harrowing descriptions. . . . The book has unforgettable characters. . . . The Horse Soldiers' secret mission remains the U.S. military's finest moment in what has since arguably been a muddled war."

—*USA Today*

"A great read with amazing detail. I would recommend it to any reader, but particularly those interested in understanding a different aspect of war—the one fought by the U.S. Special Forces . . . [and] the campaign that is considered the most successful in all of America's engagement in Afghanistan."

—*The Christian Science Monitor*

"12 *Strong* reads like a cross between an old-fashioned Western and a modern spy thriller."

—*Parade* magazine

"[An] absolutely riveting account, full of horror and raw courage . . . A seminal battle and, in Stanton's prose, a considerable epic."

—*Publishers Weekly* (starred review)

"A fascinating account . . . This is not just a battle story—it's also about the home front. An important book."

—*The Today Show*

"A great story and great reporting . . . Incredible. It takes you deep inside what the Special Forces do."

—*Morning Joe*, MSNBC

"Superbly crafted."

—*Booklist*

"A lively and exciting battle chronicle that will be popular."

—*Library Journal*

"Stanton is a masterful storyteller."

—*Military Officer* magazine

"Stanton has done excellent work. [He] provides great insight and understanding [and] gives the reader a feeling . . . of how these 'Horse Soldiers' succeeded. . . . [Gives] family, friends, and anyone interested in Special Forces a glimpse of the world of the SF operator living and fighting behind enemy lines."

—J. R. Seeger, *Studies in Intelligence: Journal of the American Intelligence Professional*

"Buy and read *12 Strong*. Stanton tells this story so well that I would find myself nearly 160 pages along before I needed a coffee refill."

—*Mansfield News Journal*

"An amazing story. The pacing is like [that of] Stanton's *In Harm's Way*. We give it the highest recommendation."

—Ward Carroll, Military.com

"Stanton has captured the inner workings of . . . Special Forces operations. A thorough recounting of the events told from the soldiers' point of view."

—*Veritas: Journal of Army Special Operations History*

"*12 Strong* is a great read—a riveting story of the brave and resourceful American warriors who rode into Afghanistan after 9/11 and waged war against Al Qaeda. We're hearing many of these stories for the first time—and from those who waged a war worthy of Rudyard Kipling, James Bond, and Davey Crockett."

—Tom Brokaw

"In the spirit of *Black Hawk Down* and *Flags of Our Fathers*, Doug Stanton plunges into the heart of a single mission and returns with a stark understanding not only of what happened but what was truly at stake. Through precise reportage and hauntingly rendered battle scenes, Stanton shows that we may ignore this 'forgotten' theater only at our own peril."

—Hampton Sides, author of *Ghost Soldiers* and *Blood and Thunder*

"Doug Stanton's *12 Strong* is as gripping as the most intricately plotted thriller. It is a masterwork of thrilling military action, brilliant in-depth journalism, and powerful storytelling. Finally Americans can know how just a few dozen courageous U.S. soldiers beat the Taliban under the most extreme and dangerous conditions imaginable. I could not put this book down."

—Vince Flynn

"Not just an epic war story, *12 Strong* is a beautifully written, intimate portrait of the men and women who lived the battle on the fields of fire—and at home, too. Their secret mission against the Taliban was intelligent, brave, and undertaken with great care for the good people of Afghanistan. Doug Stanton's superb account is an invaluable insight for policy makers and the public for years to come."

—Greg Mortenson, author of *Three Cups of Tea*

"Doug Stanton's *12 Strong* is the story of the first large American unconventional warfare operation since World War II. My Green Berets were launched deep into enemy territory to befriend, recruit, equip, advise, and lead their Afghan counterparts to attack the Taliban. The Horse Soldiers succeeded brilliantly with a highly decentralized campaign, reinforced with modern airpower's precision weapons, forcing the Taliban government's collapse in a few months. Doug Stanton captures the gritty realities of the campaign as no other can."

—Geoffrey C. Lambert, major general (retired), U.S. Army, commanding general of the U.S. Army Special Forces Command (Airborne), 2001–2003

12 STRONG

THE DECLASSIFIED TRUE STORY
OF THE HORSE SOLDIERS

Previously published as
*Horse Soldiers: The Extraordinary Story
of a Band of U.S. Soldiers Who Rode
to Victory in Afghanistan*

The story that inspired the movie

DOUG STANTON

**SIMON &
SCHUSTER**

London · New York · Sydney · Toronto · New Delhi

A CBS COMPANY

This edition first published in the United States by Scribner,
an imprint of Simon & Schuster Inc., 2017
This edition first published in Great Britain by Simon & Schuster UK Ltd, 2017
A CBS COMPANY

1 3 5 7 9 10 8 6 4 2

Simon & Schuster UK Ltd
1st Floor
222 Gray's Inn Road
London WC1X 8HB

www.simonandschuster.co.uk
www.simonandschuster.com.au
www.simonandschuster.co.in

Simon & Schuster Australia, Sydney
Simon & Schuster India, New Delhi

The author and publishers have made all reasonable efforts
to contact copyright-holders for permission, and apologise
for any omissions or errors in the form of credits given.
Corrections may be made to future printings.

A CIP catalogue record for this book
is available from the British Library

Text set in Electra

Paperback ISBN: 978-1-4711-7082-9
eBook ISBN: 978-1-4711-7335-6

Printed and bound by CPI Group (UK) Ltd, Croydon, CR0 4YY

Certain names and identifying characteristics have been changed.
Insert photographs are courtesy of FOB-53 (Forward Operating Base 53) unless otherwise noted.

This book is dedicated to the men and women
of Fifth Special Forces Group and their families.

And to my family,
Anne, John, Kate, and Will;
and my parents,
Bonnie and Derald Stanton;
and Deb, Tony, Genessa, and Wylie Demin.

And, finally, Grant and Paulette Parsons.

I also wish to acknowledge a heartfelt debt of gratitude to
Sloan Harris, Colin Harrison, and Blake Ringsmuth. None finer.

Without their unwavering support,
this book would not have been written.

I am the kit fox,
I live in uncertainty.
If there is anything difficult,
If there is anything dangerous to do,
That is mine.

—Sioux warrior's song

CONTENTS

AUTHOR NOTE

The events recounted in this book are based on more than one hundred interviews of Afghan soldiers, Afghan civilians, U.S. soldiers, and U.S. civilians. These interviews, some of which were in-depth and stretched over a series of days, took place in Afghanistan and in the United States. Most dwelled on the subjects' firsthand recollections of events related in this book. In addition, the author traveled in the region described in these pages and in particular inspected the Qala-i-Janghi Fortress. The author's research also included examination of personal journals, previously published media accounts, contemporaneous photography, and voluminous official U.S. military logs and histories.

Many of the events described in *12 Strong* transpired under extreme circumstances, some of them traumatic to those who experienced them. For these reasons and perhaps because memory is often imperfect, the recollections of some of the participants conflicted at times. While the author has made every attempt to present an accurate portrait of the events involved, he has related the version that seemed most consistent with other accounts.

KEY PLAYERS

AFGHAN GENERALS

Abdul Rashid Dostum
Atta Mohammed Noor
Naji Mohammed Mohaqeq

CIA PARAMILITARY OFFICERS

Mike Spann
Dave Olson
J. J. Sawyer
Garth Rogers

U.S. SPECIAL FORCES COMMANDERS

Major General Geoffrey Lambert, United States Special Forces
Command, Fort Bragg
Colonel John Mulholland, Fifth Special Forces Group, Fort
Campbell and K2, Uzbekistan
Lieutenant Colonel Max Bowers, Third Battalion, Fifth Special
Forces Group, Fort Campbell and Mazar-i-Sharif, Afghanistan

KEY PLAYERS

CAPTAIN MITCH NELSON'S TEAM
(RIDING WITH DOSTUM)

Captain Mitch Nelson, team leader
Chief Warrant Officer Cal Spencer, assistant team leader
Sergeant First Class Sam Diller, intelligence operations
Sergeant First Class Bill Bennett
Sergeant First Class Scott Black
Sergeant First Class Sean Coffers
Sergeant First Class Ben Milo
Master Sergeant Pat Essex
Staff Sergeant Charles Jones
Staff Sergeant Patrick Remington
Sergeant First Class Vern Michaels
Staff Sergeant Fred Falls
Staff Sergeant Sonny Tatum, Air Force combat controller
Staff Sergeant Mick Winehouse, Air Force combat controller

TURKISH SCHOOLHOUSE,
MAZAR-I-SHARIF, AFGHANISTAN

Admiral Bert Calland, III, Special Operations Command Central
Lieutenant Colonel Max Bowers
Major Kurt Sonntag, executive officer
Major Mark Mitchell, ground commander
Major Steve Billings
Captain Paul Syverson
Captain Kevin Leahy
Captain Craig McFarland
Captain Andrew Johnson
Captain Gus Forrest

KEY PLAYERS

Sergeant Major Martin Homer
Master Sergeant Roger Palmer
First Sergeant Dave Betz
Sergeant First Class Pete Bach
Sergeant First Class Bob Roberts
Sergeant First Class Chuck Roberts
Staff Sergeant Jerome Carl
Staff Sergeant Jason Kubanek
Sergeant First Class Ted Barrow
Sergeant First Class Ernest Bates
Staff Sergeant Malcolm Victors, Air Force combat controller
Master Sergeant Burt Docks, Air Force combat controller
Captain Don Winslow

CAPTAIN DEAN NOSOROG'S TEAM
(RIDING WITH ATTA)

Captain Dean Nosorog, team leader
Chief Warrant Officer Stu Mansfield, assistant team leader
Sergeant First Class Darrin Clous, intelligence operations
Master Sergeant Brad Highland
Staff Sergeant Jerry Booker
Sergeant First Class James Gold
Sergeant First Class Mark House
Staff Sergeant Brett Walden
Sergeant First Class Martin Graves
Staff Sergeant Evan Colt
Staff Sergeant Francis McCourt
Sergeant First Class Brian Lyle
Staff Sergeant Donny Boyle, Air Force combat controller

KEY PLAYERS

SBS (SPECIAL BOAT SERVICE)

Chief Petty Officer Steph Bass, U.S. Navy (deployed with SBS)

MEMBERS OF U.S. ARMY TENTH MOUNTAIN DIVISION WHO TOOK PART IN THE RESCUE AT QALA-I-JANGHI BOMBING

Staff Sergeant Thomas Abbott
Private First Class Eric Andreason
Private First Class Thomas Beers
Sergeant Jerry Higley
Private First Class Michael Hoke
First Lieutenant Bradley Maroyka
Specialist Roland Miskimon
Sergeant William Sakisat
Specialist Andrew Scott

HELICOPTER PILOTS AND CREW, 160TH SPECIAL OPERATIONS AVIATION REGIMENT (SOAR), K2, UZBEKISTAN

Aaron Smith
Greg Gibson
John Garfield
Larry Canfield
Dewey Donner
Carl Macy
Tom Dingman
Jerry Edwards
Steve Porter
Vic Boswell
Alex McGee

Jim Zeeland
Carson Millhouse
Ron White
Donald Pleasant
Will Ferguson
Kyle Johnson
Bill Ricks
Ron Mason
Barry Oberlin
Ross Peters

12 STRONG

★ ★ ★

Afghanistan

N

0 miles 50
0 km 50

UZBEKISTAN

KYRGYZSTAN

CHINA

TAJIKISTAN

TURKMENISTAN

Mazar-i-Sharif

Konduz

Herat

AFGHANISTAN

Kabul

Peshawar

Kandahar

PAKISTAN

IRAN

Karshi-Khanabad, "K2" base camp for U.S. forces

UZBEKISTAN

AFGHANISTAN

Konduz

Mazar-i-Sharif
(location of Qala-i-Janghi Fortress)

DARYA BALKH
RIVER

Dehi

0 miles 50
0 km 50

UPRISING

Qala-i-Janghi Fortress
Mazar-i-Sharif, Afghanistan

November 24–25, 2001

Trouble came in the night, riding out of the dust and the darkness. Trouble rolled past the refugee camp, past the tattered tents shuddering in the moonlight, the lone cry of a baby driving high into the sky, like a nail. Sunrise was no better; at sunrise, trouble was still there, bristling with AKs and RPGs, engines idling, waiting to roll into the city. Waiting.

These were the baddest of the bad, the real masters of mayhem, the death dealers with God stamped firmly in their minds. The city groaned and shook to life. Soon everyone knew trouble had arrived at the gates of the city.

Major Mark Mitchell heard the news at headquarters nine miles away and thought, *You're kidding. We got bad guys at the wire?*

He ran downstairs, looking for First Sergeant Dave Betz. Maybe he would know what was happening.

But Betz didn't know anything. He blustered, "One of the Agency guys came down and told us we got six hundred Taliban surrendering. Can you believe that?"

Surrendering? Mitchell couldn't figure out why. He thought the Taliban had fled from the approaching forces of the Northern Alliance to Konduz, miles away. American Special Forces and the Northern Alliance had been beating them back for weeks, in battle after battle, rolling up territory by coordinating airstrikes from the sky and thousands of Northern Alliance soldiers on the ground.

1

They now stood on the verge of total victory. Konduz was where the war was supposed to go next. Not here. Not in Mazar. Not at Club Mez.

Besides, these guys didn't *surrender*. They fought to the death.

Die fighting and you went to paradise.

Mitchell stood at the dirty plate-glass windows and watched. Here they came, a motley crew of the doomed, packed into six big trucks, staring out from the rancid tunnels of their scarves. Mitchell could see their heads over the barricade that ringed his headquarters, a former schoolhouse at the junk-strewn edge of the city. The prisoners—who surely included some Al Qaeda members—were still literally in the drivers' seats, with Northern Alliance soldiers sitting next to them, their AKs pointed at the drivers' heads. The prisoners turned and stared and Mitchell thought it was like looking at hundreds of holes punched in a wall.

"Everybody get away from the windows!" said Betz.

Major Kurt Sonntag, Captain Kevin Leahy, Captain Paul Syverson, and a dozen other Special Forces soldiers knelt behind the black-and-white-checked columns in the room, their M-4 rifles aimed at the street. Behind them, in the kitchen, the local cook was puttering—the air smelled of cooked rice and cucumber—and a radio was playing more of that god-awful Afghan music that sounded to Mitchell like somebody strangling a goose.

He had been looking forward this morning to overseeing the construction of the medical facility in town, and the further blowing up of mines and bombs that littered the area like confetti. Each day, a little bit more of the war seemed to be ending. Mitchell had even started to wonder when he would get to go home. He and a team of about a dozen Special Forces soldiers had moved into the schoolhouse only forty-eight hours earlier. Their former headquarters inside the Qala-i-Janghi Fortress, nine miles off, in Mazar's western quarter, had given them the shits, the croup, and the flu, and Mitchell was glad to have moved out. It seemed a haunted place. Known as the House of War, the fortress rose like a mud golem from the desert, surrounded by struggling plots of wind-whipped corn and sparse cucumber. Its walls towered sixty feet high and measured thirty feet thick under the hard, indifferent sun.

The Taliban had occupied the fortress for seven years and filled it with weapons—grenades, rockets, and firearms, anything made for killing. Even Enfield rifles with dates stamped on the bayonets—1913—from the time that the Brits had occupied the area. Before their hurried flight from the city two weeks earlier, the Taliban had left the weapons and smeared feces on the walls and windows. Every photograph, every painting, every rosebush had been torn up, smashed, stomped, ruined. Nothing beautiful had been left behind.

After three years of Taliban rule, there were old men in Mazar with stumps for hands. There were women who'd been routinely stoned and kicked on street corners. Young men who'd been imprisoned for not wearing beards. Fathers who'd been beaten in front of their sons for the apparent pleasure of those swinging their weapons.

The arrival of Mitchell and his soldiers on horseback had put an end to that. The people of Mazar-i-Sharif, the rugmakers and butchers, the car mechanics and schoolteachers, the bank clerks and masons and farmers, had thrown flowers and kisses and reached up to the Americans on their horses and pulled affectionately at the filthy cuffs of their camo pants. The locals had welcomed the balding, blue-eyed Mitchell and two dozen other Special Forces soldiers in a mile-long parade lining the highway that dropped into town out of the snowy mountains. Mitchell had felt like he was back in World War II, his grandfather's war, riding into Paris after the Nazis fled.

Now thirty-six, Mitchell was the ground commander of the Fifth Special Forces Group/Third Battalion's Forward Operating Base (FOB). It had been a distinguished nearly fifteen-year career headed for the top of the military food chain. His best friend, Major Kurt Sonntag, a thirty-seven-year-old former weekend surfer from Los Angeles, was the FOB's executive officer, which technically meant he was Mitchell's boss. In the tradition of Special Forces, they treated each other as equals. Nobody saluted, including less senior officers like Captain Kevin Leahy and Captain Paul Syverson, members of the support company whose job it was to get the postwar operations up and running, such as providing drinking water, electricity, and medical care to the locals.

Looking at the street now, Mitchell tried to figure out why the Taliban convoy was stopping. If anything went bad, Mitchell knew he was woefully outnumbered. He had maybe a dozen guys he could call on. And those like Leahy and Syverson weren't exactly hardened killers. Like him, these were staff guys, in their mid-thirties, soldiers who had until now been largely warless. He did have a handful of CIA operators living upstairs in the schoolhouse and eight Brits, part of a Special Boat Service unit who'd landed the night before by Chinook helicopter, but they were so new that they didn't have orders for rules of engagement—that is, it wasn't clear to them when they could and could not return fire. Doing the math, Mitchell roughly figured that he had about a dozen guys available to fight. The trained-up fighters, the two Special Forces teams that Mitchell had ridden into town with, had left earlier in the day for Konduz, for the expected fight there. Mitchell had watched them drive away and felt that he was missing out on a chance to make history. He'd been left behind to run the headquarters office and keep the peace. Now, after learning that 600 Taliban soldiers had massed outside his door, he wondered if he'd been dead wrong.

The street bustled with beeping taxis; with donkeys hauling loads of handmade bricks to the city-center bazaar; with aged men gliding by on wobbling bicycles and women ghosting through the rising dust in blue burkhas. *Afghanistan.* Never failed to amaze him.

Still the convoy hadn't moved. Ten minutes had passed.

Without warning, a group of locals piled toward the trucks, angrily grabbing at the prisoners. They got hold of one man and pulled him down—for a moment he was there, gripping the battered wooden side of the truck, and then he was gone, snatched out of sight. Behind the truck, out of sight, they were beating the man to death.

Every ounce of rage, every rape, every public execution, every amputation, humiliation—every ounce of revenge was poured back into this man, slathered on by fist, by foot, by gnarled stick. The trucks lurched ahead and when they moved on, nothing remained of the man. It was as if he'd been eaten.

The radio popped to life. Mitchell listened as a Northern Alliance commander, who was stationed on the highway, announced in broken English: *The prisoners all going to Qala-i-Janghi.*

Remembering the enormous pile of weapons cached at the fortress, Mitchell didn't want to hear this. But his hands were tied. The Afghan commanders of the Northern Alliance were, as a matter of U.S. strategy, calling the shots. No matter the Americans' might, this was the Afghans' show. Mitchell was in Mazar to "assist" the locals in taking down the Taliban. He figured he could get on a radio and suggest to the Afghan commander presiding over the surrender that the huge fortress would not be an ideal place to house six hundred angry Taliban and Al Qaeda soldiers. But maybe there was a good reason to send them there. As long as the prisoners were searched and guarded closely, maybe they could be held securely within the fort's towering mud walls.

And then Mitchell thought again of the weapons stockpiled at Qala-i-Janghi, the piles and piles of rockets, rifles, crates of ammo— tons of violence ready to be put to use.

Not the fort, he thought. *Not the damn fort!*

Belching smoke, grinding gears, the convoy of prisoners rumbled past the fortress's dry moat and through the tall, arched entrance. The prisoners in the trucks craned around like blackbirds on a wire, scanning the walls, looking for guards, looking for an easy way out.

In deference to the Muslim prohibition against men touching other men intimately, few of the prisoners had been thoroughly searched. No hand had reached deep inside the folds of their thin gray gowns, the mismatched suit coats, the dirty khaki vests, searching for a knife, a grenade, a garrote. Killer had smiled at captor and captor had waved him on, *Tashakur. Thank you. Tashakur.*

The line of six trucks halted inside the fort, and the prisoners stepped down under the watchful eye of a dozen or so Northern Alliance guards. Suddenly one prisoner pulled a grenade from the belly-band of his blouse and blew himself up, taking a Northern Alliance officer with him. The guards fired their rifles in the air and regained control. Then they immediately herded the prisoners to a rose-colored, plaster-sided building aptly nicknamed "the Pink House," which squatted nearby in the rocks and thorns. The structure had been built by the Soviets in the 1980s as a hospital within the bomb-hardened walls of the fortress.

The fort was immense, a walled city divided equally into southern and northern courtyards. Inside was a gold-domed mosque, some horse stables, irrigation ditches encircling plots of corn and wheat, and shady groves of tall, fragrant pine trees whipping in the stiff winds. The thick walls held secret hallways and compartments, and led to numerous storage rooms for grain and other valuables. The Taliban had cached an enormous pile of weapons in the southern compound in a dozen mud-walled horse stables, each as big as a one-car garage and topped with a dome-shaped roof. The stables were crammed to the rafters with rockets, RPGs, machine guns, and mortars. But there were more weapons. Six metal Conex trailers, like the kind semitrucks haul down interstates in the United States, also sat nearby, stuffed with even more guns and explosives.

The fortress had been built in 1889 by Afghans, taking some eighteen thousand workers twelve years to complete, during an era of British incursions. It was a place built to be easily defended, a place to weather a siege.

At each of the corners rose a mud parapet, a towerlike structure, some 80 feet high and 150 feet across, and built strong enough to support the weight of 10-ton tanks, which could be driven onto the parapet up long, gradual mud ramps rising from the fortress floor. Along the parapet walls, rectangular gunports, about twelve inches tall, were cut into the three-foot-thick mud—large enough to accommodate the swing of a rifle barrel at any advancing hordes below.

In all, the fort measured some 600 yards long—about one-third of a mile—and 300 yards wide.

At the north end, a red-carpeted balcony stretched high above the courtyard. Wide and sunlit, it resembled a promenade, overlooking a swift stream bordered by a black wrought-iron fence and rose gardens that had been destroyed by the Taliban. Behind the balcony, double doors opened onto long hallways, offices, and living quarters.

At each end of the fort's central wall, which divided the interior into the two large courtyards, sat two more tall parapets, equally fitted for observation and defense with firing ports. A narrow, packed foot trail, about three feet wide, ran around the entire rim along the protective, outer wall. In places, a thick mud wall, waist-high, partially shielded the walker from the interior of the courtyard, making it possible to move along the top of the wall and pop up and shoot either down into the fort, or up over the outer wall at attackers coming from the outside.

In the middle of the southern courtyard, which was identical to the northern one (except for the balcony and offices overlooking it), sat the square-shaped Pink House. It was small, measuring about 75 feet on each side, too small a space for the six hundred prisoners who were ordered by Northern Alliance soldiers down the stairs and into its dark basement, where they were packed tight like matchsticks, one against another.

There, down in a dank corner, on a dirt floor that smelled of worms and sweat, brooded a young American. His friends knew him by the name of Abdul Hamid. He had walked for several days to get to this moment of surrender, which he hoped would finally lead him home to California. He was tired, hungry, his chest pounding, skipping a beat, like a washing machine out of balance. He worried that he was going to have a heart attack, a scary thought at age twenty-one.

Around him, he could hear men praying as they unfolded hidden weapons from the long, damp wings of their clothing.

The following morning, November 25, two CIA paramilitary officers, Dave Olson and Mike Spann, kitted up at headquarters in Mazar and prepared to drive across town to the fort. Both men hoped to interrogate as many prisoners as possible.

Mitchell was in the school cafeteria, drinking chai and eating nan, a delicious, chewy flat bread, when Spann and Olson walked up. Mitchell knew Olson the better of the two. Spann, a former Marine artillery officer, had joined the Agency three years earlier. He wore blue jeans and a black sweater, and was of medium height, with severe cheekbones and a crooked smile, his blond hair cut close. Olson was tall and burly, with a thin salt-and-pepper beard over an old case of acne. He spoke excellent Dari, the glottal, hissing language of the local Northern Alliance fighters, and he was dressed in a black, knee-length blouse, called a *shalwar kameez*, over beige pants.

Mitchell noticed immediately that the two CIA guys weren't carrying enough ammunition. For whatever reasons, they had about four ammo magazines between them. Mitchell preferred the standard operating procedure of bringing four magazines apiece on a mission. Olson and Spann carried folding-stock AK-47s slung over their shoulders and 9mm pistols strapped in holsters on their legs. Spann carried another pistol tucked at the small of his back in his pants' waistband. Neither man had a radio, which Mitchell also thought was strange. But then again, these CIA guys had always brought their own party with them. He figured that whatever Olson and Spann were doing this morning, it was their own educated business.

Olson announced, "We're going out to Qala to talk to these guys, see what we can find out."

The previous night, there had been a brief gunfight outside the schoolhouse, and Mitchell, sensing that the situation in the city was increasingly tense, had asked Olson if he himself and a couple of his men could go and provide security while the two CIA officers conducted their interrogations at Qala. Mitchell knew that interrogating prisoners was officially the CIA's job, but he was worried about his friends' safety. No, said Olson, you guys need to stay away. To Mitchell's thinking, he was a bit nonchalant about the whole thing.

All three men knew that the prisoners included many hard cases: Chechens, Pakistanis, Saudis—the epicenter of Al Qaeda. The men who had surrendered were the heart of Osama bin Laden's most skilled army. Maybe—just maybe—one of them knew where bin Laden was.

Watch your back, thought Mitchell.

Olson and Spann started out the lobby's front door to a truck parked in the circular drive. Beyond the wall, the busy midmorning traffic buzzed by. The vehicle slipped into the stream of cars, trucks, and donkey carts, and was gone.

Sergeant Betz walked up and stood beside Mitchell, watching them go. He said, "I don't like the looks of that."

Mitchell asked him why.

"I dunno," said Betz. "I like a guy to carry a lot of ammo when he leaves."

About a half hour later, Olson and Spann entered Qala.

At the fort, Abdul Hamid climbed the steps from the basement of the Pink House and blinked in the morning sun, his arms tied behind him with a turban. The stairway resembled a collapsed brick chimney as it emerged from the dark hole that reeked of piss and shit.

Abdul was led past the Pink House, the walls of the fort soaring around him. About a hundred other prisoners had already been led into the courtyard, also trussed with their own clothing, arms behind their backs, sitting cross-legged on an apron of trampled weeds twisting up from hardpan mud.

Mike Spann bent down and peered at Abdul.

For the life of him, he couldn't figure out where the kid was from or who he was. Arab? Pakistani? *Canadian?* He studied Abdul's tattered British commando sweater, sensing that the prisoner—*what was he, twenty, twenty-three?*—could speak at least some passable English.

"Where are you from?" Spann demanded. "You believe in what you're doing here that much, you're willing to be killed here?"

No answer came.

"What's your name? Who brought you here to Afghanistan?"

The kid on the carpet dropped his head, stared at the *shalwar kameez* bunched around his knees.

"Put your head up!" Spann yelled.

The young man's face was sunburned, his eyes the color of cold tea.

Spann let his gaze linger, and then raised a digital camera and framed a shot. The photo would be sent by encrypted satellite communications back to headquarters, where the image would be cross-referenced against a digital lineup of terrorists and known Al Qaeda soldiers.

"Mike!"

It was Olson, lumbering across the dusty courtyard. He'd spent the last five minutes talking with another group of prisoners. Olson towered over the young man on the ground.

"Yeah," said Spann, "he won't talk to me . . . I was explaining to the guy we just want to talk to him, find out what his story is."

"Well, he's a Muslim, you know," mused Olson. "The problem is, he's got to decide if he wants to live or die. . . . We can only help the guys who want to talk to us."

It was Spann's turn: "Do you know the people here you're working with are terrorists, and killed other Muslims? There were several hundred Muslims killed in the bombing in New York City. Is that what the Koran teaches? Are you going to talk to us?"

Then it was back to Olson: "That's all right, man. Gotta give him a chance. He got his chance."

Olson scuffed the dirt with his boot; Spann, exasperated, hands on hips, looked at the prisoner.

Finally, Spann said, "Did you get a chance to look at any of the passports?"

"There's a couple of Saudis, and I didn't see the others."

They agreed that the young man wasn't going to tell them anything, and the two CIA officers started walking away along a gravel path lined by pine trees toward the gate in the middle of a tall mud wall that divided the fort into its separate courtyards. They were headed to the former headquarters to regroup.

At one point, Olson turned to see Spann stopped on the path, joking with a group of Northern Alliance soldiers. He turned back and kept walking.

By the time Olson reached the middle gate he heard the explosion of a grenade, followed by a burst of gunfire. He turned.

Spann was frantically attempting to fight off a gang of prisoners who were beating at him with their fists and screaming, *Allah Akbar!*—God is Great!

Olson started running toward Spann, and as he did so, Spann emptied his pistol into the crowd, then reached behind to the other gun, hidden in his waistband. He fired and fell to the ground under the storm of flesh.

Seeing that Spann was down and thinking he was already dead, Olson spun around to see a Taliban soldier running at him, firing from the hip with an AK-47.

Olson could hear the snap of the rounds passing and was amazed he hadn't been hit. The guy kept coming, and finally Olson, momentarily frozen on the spot, raised his pistol and shot him.

The man skidded to a stop at Olson's feet, so close Olson could almost touch him with his boot.

He next turned and fired at the crowd of people beating on Spann. He was pretty sure he killed a few of them. He sensed he was being rushed again and spun to shoot another man running at him. By now, he was out of bullets.

And so he ran. He ran down the path and into the northern courtyard, past the ruined rose garden fronting the grand balcony. He ran up the steps and into the inner courtyard, where he made a phone call, alerting Mitchell and Sonntag back at the schoolhouse.

"I think Mike's dead," Olson said over the phone. "I think he's dead! We are under attack. I repeat, I am receiving heavy fire!" RPGs were hitting the balcony wall, rocking the place.

Back in the southern courtyard, Abdul Hamid had been shot in the leg and lay in the dirt. He tried crawling back to the basement steps, but it was too far. He wondered if he'd ever see his mother again, in California. He wondered who the strange men were who had been asking him questions. He wondered if they knew his real name: John Walker Lindh.

Meanwhile, one of the prisoners walked up and fired twice, point-blank, at Mike Spann.

By the hundreds, the Taliban prisoners jumped up from the ground where they'd been ordered to sit by Spann and Olson.

They shook off the turbans binding their wrists and looked wildly around, not sure what to do next.

Up on the fortress walls, a dozen or so Northern Alliance guards were pouring fire into the courtyard, raking the hard ground, raising divots of mud, mowing men down.

Several minutes later, the prisoners found the weapons cache.

They swung open the metal doors of the long Conex trailers and beheld hundreds of rifles, grenades, and mortars, spilled at their feet.

They scooped up the weapons and scattered around the court-yard, crouching behind mud buildings, in bushes, inside storerooms built into the walls. They started returning fire. The air roared.

Wounded horses soon littered the courtyard, twitching and braying in the dust, as the hot sun beat down.

Mitchell arrived with a ground force half an hour after Olson's call. He pulled up outside the fortress gate, got out of the truck, and gazed up at the walls. He couldn't believe the intensity of the fight. Several hundred guns must've been firing at once. Mortars started arcing over the walls and exploding around his truck.

He and his men ran to the base of the fort and started climbing.

The wall pitched skyward at about a 45-degree angle. They scut-tled up hand-over-hand. At the top, out of breath, Mitchell peered at the mayhem below.

Dead men were scattered up in the grove of pine trees, blown there by grenade blasts. They hung from the tree limbs, heavy and still, like blackened ornaments.

He saw prisoners running among the trees, turning to fire up at the walls. There were six hundred of them down there, Mitchell knew. And they wanted out.

He again counted the number of his own force: fifteen men. *Fifteen.*

Before leaving for Afghanistan, Mitchell had been asked by his commander, *"How will you die?"* It was a blunt way of asking how he planned to stay alive. Until now, he hadn't given the answer much thought.

Massive explosions punched the sky. He figured the prisoners had finally found the mortars. It was only a matter of time before they zeroed in on the guards on the walls.

The gunfire was filled with pops, fizzles, and cracks, like the snapping of enormous bones. Mitchell worried that the fighters inside were breaking out. He expected them to pile over the top wall at any moment.

In a matter of minutes, something had gone terribly wrong. *We fought so hard. And we won. But now we're losing so damn quickly . . .*

He thought of his wife, then his two daughters. He had been worried that they were growing up without him. And now he thought: *They'll never know me at all.*

Mitchell took out his pistol and prepared to be overrun.

PART ONE

GOING TO WAR

Comfort Inn Airport Motel
South Portland, Maine

September 11, 2001

Little Bird woke.

Beside the motel bed was a small nightstand, a radio alarm clock, a Bible. Beyond that, beyond the dirty window, a parking lot filled with the cars of commuters, of people on vacation, heading to work, to their families, to the rest of their lives. It had been a peaceful night for Little Bird at the end of a long, restless year. This morning, he was going to heaven.

There is no god but one God and Muhammad is his messenger.

His real name was Mohammed Atta, the nickname given to him by his father, a stern and dour lawyer who saw Atta as soft and overly sensitive, as too frail, too lazy, too fretful. Hadn't he even timed the young Atta's three-minute walks home through the streets of Cairo, after school, finding fault if he were seconds late? *Little Bird, you take too long! Why?*

Arrivals, departures. *There is no god but one God and Muhammad is his messenger. . . .*

He rose.

Wearing jeans and a blue polo shirt, the floral bedspread smooth and shiny beneath his delicate hands, he looked like another harried tourist eager to start his day. Overhead, jet airliners roared away, taking off.

•   •   •

Ten minutes later Atta stood at the South Portland, Maine, airport, holding a ticket he had purchased on the Internet two weeks earlier in Las Vegas. He'd gone there for a final organizational meeting. Four days earlier, he'd celebrated his birthday in a place called Shuckum's Oyster House, in Hollywood, Florida. Through the night, he played pinball and drank cranberry juice and watched his fellow assassin, Marwan al-Shehhi, drink liquor and look at the women and nod to the music. He hated their touch, women. Their smell. Their sex. He had five days to live. He was thirty-three years old and he'd just planned the largest attack on American soil in world history. Still, everything bored him. Eating bored him. Sleeping. Breathing. The only thing worth living for was dying.

As Atta walked through the metal detector at the Portland airport, he carried a four-page note in his pocket that read:

"When you get on the aircraft . . . think of it [your mission] as a battle for the sake of God. . . . Do not forget . . . the true promise is near and the zero-hour had arrived. Always remember to pray if possible before reaching the target or say something like, *There is no god but one God and Muhammad is his messenger*."

By 5:45 a.m., he was through security, along with another assassin, Abdul Aziz al-Omari. Fifteen minutes later, Atta's plane took off, and swung out over the Atlantic headed for Boston. There, Atta was to switch planes for a flight to Los Angeles on American Airlines Flight 11.

At 6:52 a.m., inside Boston's Logan International Airport, seven minutes after he landed, his cell phone rang. It was Marwan, also in the airport, calling from a nearby terminal. The two men must've spoken quickly—*Is everything ready? Yes, brother, everything is ready. There is no god but one God and we are his messengers.*

And then they hung up.

By 7:40 a.m., Atta and his team of four others were seated comfortably, the jet pushing back from the gate. Thirty-four minutes later, United Flight 175, with Marwan al-Shehhi and his four comrades on board, also took off from Logan Airport in Boston.

At the same time—8:14 a.m.—Atta proceeded to take control of American Airlines Flight 11. He and his compatriots sprayed Mace and yelled they had a bomb on board in order to move the passengers to the rear of the plane. At 8:25, air traffic control in Boston heard a voice say: "Nobody move. Everything will be okay. If you try to make any moves, you'll endanger yourself and the airplane. Just stay quiet."

In first class, where the hijackers had been sitting, one man lay slumped with his throat cut. Two flight attendants had been stabbed. They were still alive, one with an oxygen mask pressed to her face, the other with minor wounds.

At 8:44 a.m., the plane dipped low over New York City.

A flight attendant on board, Madeline "Amy" Sweeney, was talking on a cell phone to an air traffic controller when she looked out the window and said, "Something is wrong. We are in a rapid descent. . . . We are all over the place." And then, before the line went dead: "We are flying low. . . . *We are flying way too low!*" A few seconds passed. "Oh my God, we are way too low!"

At 8:46 a.m., American Flight 11, racing at the speed of nearly 500 miles per hour, rammed into the north tower of the World Trade Center. Ten thousand gallons of aviation fuel exploded, with the force of 7 million sticks of dynamite.

Fourteen minutes later, on board United Flight 175, which was now under Marwan's control, a young man named Peter Hanson was making a phone call to his father, Lee, back in Easton, Connecticut.

"It's getting bad, Dad," he said. "A stewardess was stabbed. They seem to have knives and Mace. They said they have a bomb. It's getting very bad on the plane. Passengers are throwing up and getting sick. The plane is making jerky movements. I don't think the pilot is flying the plane. I think we are going down. I think they intend to go to Chicago or someplace else and fly into a building. Don't worry, Dad. If it happens, it'll be very fast. My God, my God."

At 9:03 a.m., United Flight 175 hit the south tower.

Thirty-four minutes later, American Flight 77 dove into the Pentagon.

At 10:03 a.m., United Flight 93 exploded in a field near Shanksville, Pennsylvania.

Cal Spencer had just gotten out of the Cumberland River and was pulling the Zodiacs onto the boat trailers when somebody in the truck said, "Jesus, can't anybody turn up the heat?" Spencer shook off the morning's cold, grumbled, reached over, and punched a few buttons on the dash. Then he wheeled the truck for home through the lush Tennessee countryside, the trees along the interstate painted with the first, faint brush of autumn.

Spencer ticked off the day's tasks: hurry back to the team room, finish up the avalanche of paperwork that always fell on your desk when you finished a training mission, get home, help Marcha make dinner, check in with Jake about his homework, go to bed. Get up. Repeat.

Sandy-haired, wry, with the lanky build of a baseball player, Spencer was usually ready with a sly quip to liven up any dreary moment. But not this morning. The night had been miserable. He'd led eight Special Forces soldiers up the Cumberland River in the dark, using night vision goggles and a GPS. The work wasn't hard—Spencer had done this kind of thing hundreds of times, and he was bored. They had delivered their "package," a second Special Forces team traveling with them, to the predetermined infiltration point on the river, shut down the outboard motors, and waited.

They didn't have long. Seconds later, the large shadow of a Chinook helicopter appeared over the trees, floated downward, and hovered just feet above the water, a metal insect descended from the heavens, louder than hell, the twin rotors sending mountains of cold spray into the air. The helo was flown by the top-notch pilots of the 160th Special Operations Aviation Regiment (SOAR). SOAR's guarantee was that they would arrive thirty seconds on either side of their scheduled time, no matter what, and they had kept their promise. Created in the wake of the 1980 rescue attempt of U.S. Embassy workers in Iran (the Army had determined that poor air support had caused the mission's failure), the SOAR pilots lived a secretive existence flying exquisitely planned missions in the world's hairiest places. They operated from a base situated behind acres of barbed wire in a remote corner of Fort Campbell, a twenty-minute drive down cracked, two-lane blacktop from Spencer's own team room. Spencer never knew any of the pilots' names and they didn't know his. It was better that way: in case the real thing ever happened—if they ever went to war—nobody could compromise anyone's identities. Spencer's team completed the linkup and he and his men yanked the outboards to life and roared back upriver. Their job was over. Time to get back to the boat landing.

And then they ran into fog. They had to stop the boats because some in the group were blindly bumping into the riverbank. Suddenly, a barge came steaming out of the dark—the big vessel was headed straight for them.

They all sped for the riverbank, where—better safe than sorry—they tied up to some low-hanging tree branches and prepared to spend the night.

Spencer hadn't dressed for the cold weather. He sat in the bottom of the boat wrapped in a poncho, teeth chattering, trying to pretend the chill didn't matter. Sergeant First Class Sam Diller found a bunch of life jackets and piled those on top of himself, eight in all, yet still he lay on the boat bottom shivering.

When they pulled up at the boat ramp the next morning, Cal Spencer wanted nothing more than to be curled up at home, martini in hand, watching TV.

He thought that maybe at age forty he was getting too old for this kind of work. As a chief warrant officer, Spencer was the senior man on the team. He was a father figure to the younger guys, and a brother to Diller and Master Sergeant Pat Essex, who had served with him in Desert Storm. Essex was thin, and stern, of good Minnesota stock (he had grown up in California), and wanted to spend his retirement years bird-watching. Sam Diller was from a holler in West Virginia that Cal thought probably didn't even exist anymore. He was also, Cal thought, one of the smartest guys on the team.

They were all good men, and someday, Spencer mused, they might just get a real mission. As he pondered all of this, news of the first plane hitting the World Trade Center broke over the truck's radio.

Marcha Spencer was still at home in bed at Fort Campbell when the phone rang. It was her best friend, Lisa, Diller's wife.

"Turn on the TV!" she said. Lisa sounded freaked out, and that wasn't like her. She was one of the most poised people Marcha knew, tough as nails.

Marcha flipped on the set and couldn't believe what she was seeing. The first plane had hit the north tower and the building was in flames. Marcha stared at the screen, not comprehending the picture.

"Those people," she murmured. And then she thought: *Cal. He'll be leaving soon.*

"They're going," said Lisa, on the other end, reading her mind.

The two women immediately tried to figure out where their husbands would be deployed. What country? Who had done this? Cal had a deployment scheduled in a few weeks to Jordan, a training mission with the Jordanian army. Marcha knew that'd be canceled now.

As they were talking, the second plane hit.

"Oh my God!" the two friends screamed at each other into the phone. "Oh my God!"

Looking at the TV, Marcha said to her friend, "This one really scares me, Lisa. This one feels different." Something, they knew, had just ended, and something had just begun.

Hearing the news of the attacks, Spencer floored the big five-ton truck, double-timing it to Fort Campbell, headquarters of the U.S. Army's Fifth Special Forces Group. The huge post consists of over 100,000 scrubby acres of buckled hills, scorched firing ranges, and humid tangles of kudzu, sixty-one miles northwest of Nashville. The third largest post in the United States, it actually sprawls between two states, the majority of it located in Tennessee, near Clarksville. Fort Campbell's post office sits in Kentucky, outside the farming burg of Hopkinsville. The entire area is ringed and quilted by strip malls, franchise steakhouses, discount furniture stores, and cornfields. Spencer imagined Marcha at home watching the news on TV. He knew she was tough, but he didn't know how she'd react to the terrible images.

As he drove, he felt sick to his stomach, confused. He was pretty sure that as soon he got back to base, they'd begin packing to leave. Beyond that, everything was a question mark.

The banter in the truck was nonstop.

Can you believe this?

Who the hell did this?

We're going to war, right?

Spencer could only imagine what was going on in those planes. He'd been in firefights, he had seen people killed and torn up and blown away, but this was different. These were civilians.

•   •   •

On September 11, 2001, Dean Nosorog had been a married man for exactly four days, a fact that continued to surprise him. He, Dean Nosorog, from a hardscrabble farm in Minnesota, married to the prettiest girl on the planet!

At the moment the planes hit, he and Kelly were in Tahiti, asleep in their hotel room overlooking a black crescent of beach dotted by palm trees. They got up, had breakfast on their balcony, and decided to go mountain-biking. Dean, madly in love, felt a world away from his real life. The best part was that he had two weeks of this easeful living left. They walked out of the hotel hand-in-hand, oblivious to the first news reports of the attacks playing out on a TV in a corner of the lobby.

As they pedaled up a mountain road, no one passing by would have guessed Dean was a secret soldier in a part of the Army most Americans knew little about, the U.S. Army Special Forces. Sun-burned, freckled, with unruly red hair, dressed in cargo shorts and T-shirt, Dean resembled a young pharmaceutical salesman on holi-day. He kicked off and sped the bike toward town, yelling to Kelly to catch up.

At the bottom of the hill, they pulled into a tiny French pizzeria and ordered. An American woman quickly walked up out of breath and said, "Have you heard?" She was in tears.

"A plane," she blubbered. "A plane just ran into a building in New York."

Dean looked at Kelly, cocking an eyebrow. *What?*

"And there was a second plane," she went on. "A second plane hit another building."

Dean's face dropped. "Hurry," he told Kelly. The two of them raced back to the hotel.

The lobby was now filled with Americans, all of them, it seemed, also on their honeymoon. Dean pushed up to the TV: everybody huddled close, in shock. Dean watched for several minutes and then he turned to Kelly.

"I've got to make a phone call," he said.

He started back toward their room and along the way paused at the front desk, picking up a copy of the *International Herald Tribune*.

The newspaper's headline stopped him cold.

Massoud was dead. The leader of the Afghan people in their fight against the Taliban. The great man had been at war for more than twenty years—the ultimate survivor. Now he was dead.

Not Massoud. Assassinated: September 9, 2001. In Afghanistan. Dean felt that the timing could not be an accident.

All week, the Lion had been close to dying—he just hadn't known it.

Handsome, graying at the temples, with a sharp smile and eyes like black enamel, the Lion had been restless. He had decided he would attack the Taliban that night, September 9. In the camp, men had been loading AK-47 magazines with ammo, sorting and counting RPGs, feeding their exhausted, shaggy horses, whose neighing and snorting caromed off rock walls scoured by cold mountain winds.

Massoud had been fighting the Taliban for seven years and—he had to admit—they were about to win. He was boxed into a tiny slice of the Panjshir Valley, his boyhood home, a verdant swath of hills and violet escarpments, facing death. Taliban and Al Qaeda fighters, with the help of foreign Arabs, were pummeling his resistance forces, their last obstacle to taking total control. Still, he vowed to press on. He would never give up the fight. He would bite and claw at the enemy. He would kill, wound, harass. Ahmed Shah Massoud, the Lion of the Panjshir, an ally of the United States and the CIA during the 1979–89 Soviet invasion of the country, was the last chance Afghan citizens had to defeat the Taliban.

He had fought the Soviets for ten years, until they retreated in defeat. And then he had fought the men with whom he had fought the Soviets. Thousands and thousands had died on all sides, and Massoud, although revered, had plenty of skeletons in his own closet. Massoud had been at war for twenty-two years.

In April 2001, Massoud had traveled to Strasbourg to ask for international assistance. There, he had said to the press, "If President Bush doesn't help us, then these terrorists will damage the United States and Europe very soon—and it will be too late." Massoud had been describing the Taliban and a man named Osama bin Laden, the billionaire son of a Saudi construction mogul. No one listened.

That neglect had given bin Laden the summer to organize a secret plan to kill Massoud. The mastermind was a man named

Ayman al-Zawahiri, an Al Qaeda commander. Bin Laden's army, totaling some 3,000 trained soldiers, had merged with the Taliban, consisting of 15,000 farmers and butchers, teachers and lawyers. Together, they wanted to return the Middle East to the fourteenth century, into a golden age ruled by Islamic law. A big step toward that great step back would be the elimination of Ahmed Shah Massoud.

In his hotel room, Dean dialed the Fifth Special Forces Group Headquarters at Fort Campbell. He got the answering machine of his battalion commander. Frantic, he left a message. "Sir," he said, "I'm in Tahiti, I just saw what happened on TV." He found himself talking way too fast. "I want to talk to you, I want to know what is going on. What do you need from me?"

Dean glanced again at the newspaper in his hand. "I'm looking at the *International Herald Tribune*. The leader of the Northern Alliance was assassinated, according to this article, by bin Laden."

Dean knew bin Laden and he knew Massoud. The famous guerrilla had been the leader of something called the Northern Alliance, a shaky collection of three Afghan tribes who'd fought the Taliban. With Massoud dead, Dean knew that the Alliance would be in danger of falling apart. Massoud's assassination and the attacks he'd seen on TV this morning must, Dean guessed, be a coordinated assault. He hung up, redialed, and started looking for flights to Fort Campbell.

At his base camp in the Panjshir Valley, Massoud had been on the phone when the assassins arrived, two Arabs traveling on stolen Belgian passports, and posing as television journalists who had cajoled to meet with Massoud for several weeks.

He welcomed the visitors and they seated themselves opposite him. Massoud requested tea for his guests. He had asked to see the list of interview questions while chatting with the cameraman as he set up his equipment. He told them they could begin the interview.

The Arab aimed the eye of the lens at Massoud's waist. He switched it on. Inexplicably, the camera flashed with blue fire, choking the room with smoke. The fire leaped straight at Massoud.

As the smoke lifted, Massoud lay bleeding in the charred hulk of his chair, a hole blown through the chair's back.

The cameraman was dead, cut in half by the force of the bomb that had been strapped to the battery pack on his waist.

Massoud had whispered to a bodyguard, "Pick me up." Fingers on his right hand had been blown away; his face was a bloody mess. Someone quickly stuffed cotton into his eye sockets. He had been shot through the heart by shrapnel. There was no hope of saving him.

His aides stashed him in a refrigerator at a Tajikistan morgue, and vowed to stay mum. They feared the resistance fighters would lose heart and run.

Within a day of Massoud's death, the hills around Konduz echoed with rumors of his passing, carried through canyon and river-wash by handheld Motorola radios. In a trench far north of the city, a young man from California had settled down in the beige, talcum earth and claimed it as Allah's last battleground.

Abdul Hamid knew the final battle was near. He had just marched north with 130 other fighters for 100 miles from the city of Konduz, stepping over rock and thorn in thin sandals, going without food or water, a bandolier of ammunition slung over his shoulder, two grenades stuffed in a bag at his waist, to reach this place of war, the desolate village of Chichkeh. With Massoud rumored dead, they would finish off what was left of the corrupt enemies of Islam.

Abdul was a member of bin Laden's crack unit called the 055 Brigade, part of the larger Al Qaeda army. He had met the great man a month earlier near Kandahar, the birthplace of the Taliban and their spiritual home of study—the word *talib* is Arabic for "student." The training camp, Al Farooq, was frequented by Saudis, by Chechens and Pakistanis, the very best of the most serious martyrs drawn to the searing light of bin Laden's message. Some of the camp followers had achieved martyrdom in the previous several years by bombing embassies in Nairobi and Saudi Arabia, even an American warship named the USS *Cole* in Yemen. These righteous attacks had killed many people, Americans among them.

His comrades, Abdul knew, had died in order to drive the dirty infidel, the Jew, the Christian, the Buddhist, the atheist from the land of Muhammad, which was Saudi Arabia, the hard-rock cradle of Islam, the holy cities of Mecca and Medina. After the bombing of the USS *Cole*, which killed seventeen sailors and injured thirty-nine more, Abdul had e-mailed his mother and father in California that the ship's presence in the Yemenese harbor had been an "act of war." Abdul's father was deeply disappointed that his son felt this way. But he also knew that there was nothing he could do to change his mind. He felt that his son was long past the point of being influenced by a father's opinions.

Bin Laden had made it clear in his *fatwas*—his edicts—that it was an insult for the infidel to have based his soldiers in Muhammad's land after the first Gulf War, what the Americans called Desert Storm. As a result, every American, bin Laden claimed, must be killed wherever he was found.

Abdul Hamid knew that bin Laden dreamed of an ancient world of Islam coming back to life. Violent jihad would be his time machine. Abdul Hamid had come to Afghanistan eleven months earlier to help the Taliban in his fight.

At Al Farooq, soldiers-in-training were taught to shoot a rifle, throw a grenade, use a compass, and poison water, food, and people. Abdul had heard bin Laden speak at the camp. One of the men at Al Farooq had approached Abdul and asked if he wanted to carry the struggle to Israel or the United States. It would be easier for him to serve as a "sleeper" agent in such places than it would a dark-skinned Saudi or Pakistani. Part of Abdul Hamid was still a boy from Marin County, a young man who had sold his CD collection of rap music to pay for his ticket to the land of jihad. Part of him still wanted to go home and see his father, a lawyer, who had named him after John Lennon, and his mother, who had home-schooled him and loved him when other kids made fun of him. What he heard today, though, troubled him. There had been an attack, a big one, in the United States. Martyrs had flown jet airplanes into some buildings. They had killed thousands and thousands of Americans. He had told the bearded man, No. His fight was here, in Afghanistan, with the Taliban. He did not want to kill Americans.

The tiny speakers of the Motorola radios in the trenches around Abdul squawked the news: *The world in flames, America tumbling down.*

Major Mark Mitchell walked into the packed mess hall at Fort Campbell just as the first scenes of the attack were playing out on a

TV mounted in the big room. Mitchell nearly dropped his fork as the news sunk in. There were about seventy-five guys inside, and they'd fallen silent as they stared at the monitor. Mitchell had just come from morning physical training out on the parade field, playing Ultimate Frisbee with the men in the support company. He was standing there holding his mess tray with an omelet, biscuit, and salsa on top, when he heard the reporter on TV explain that a Cessna, or some kind of other small plane, had flown into the World Trade Center.

"That was not a Cessna," Mitchell said. "And no commercial pilot in the world would fly into a building. He'd be ditching in the Hudson River."

Some of the younger guys at one of the tables started snickering, announcing that the footage had to be a prank.

"It's a clear, blue-sky day," Mitchell went on, "there's not a cloud in the sky."

"It's an accident!" somebody else said.

Mitchell turned and shouted: "Shut up! This is not a joke. This is not a laughing matter!" The depth of emotion startled even him.

The younger soldiers shut up. Seconds later, the second plane hit.

Mitchell hurried out the door and across the parking lot, under the scraggly oak trees, across the burnt grass, and up the concrete steps to his office.

As the operations officer for Third Battalion, Fifth Special Forces Group, it was Mitchell's job to help make sure that the group was ready to mobilize to anywhere in the world within ninety-six hours. The group's main theater of operations was the Middle East, under the command of Central Command (CENTCOM) and General Tommy Franks, down in Tampa at MacDill Air Force Base, and he knew the folks in Florida would already be making plans.

When Mitchell turned down the hall, he saw his fellow staff officers standing around looking stunned. He could tell already that everybody was running on all cylinders, but with nowhere to go. He wondered what they'd look like by the end of the day.

Mitchell found another group of them standing around a TV in the conference room upstairs, watching the towers burn. A staff officer walked up and told him the Pentagon had just been hit.

Mitchell couldn't believe it.

The towers had been a matched pair, but the attack on the Pentagon, in a different city, nearly simultaneously—this raised the game to a different level.

By the time Greg Gibson, a helicopter pilot, passed through the guard gate at the 160th SOAR headquarters, he had heard enough on the radio to realize the crash was not an accident. Walking into the hangar, he told his crew to get ready to break down the Black Hawks and twin-rotored Chinooks for travel.

Gibson knew they were going to war.

Mitchell rushed back down the stairs to his boss's office, Third Battalion commander Lieutenant Colonel Max Bowers, to update him on what he knew.

Bowers had been at home finishing a shower when he heard news of the attacks delivered by his five-year-old son standing in the bathroom.

"Dad," the young boy said, "a plane just flew into a big building in New York! Really!" His son had been watching TV in another part of the house and Bowers hadn't heard any of the reports.

Bowers, wrapped in a towel, smiled, ruffled his son's hair, and told him not to joke about something like that, then he hurried to get dressed for work. On the way, he turned on the radio in the car.

Bowers's short salt-and-pepper hair was still wet from the shower as he gravely listened to Mitchell report the strike on the Pentagon. Mitchell could tell Bowers was upset, and he knew Bowers was not one to scare easily. A powerfully built, forty-two-year-old career officer, he was smart and articulate. In Bosnia in 1999, Bowers had snuck into the war-torn country by commercial flight, a risky move in contravention of U.S. policy at the time (Congress would eventually authorize U.S. troops on the ground). Once there, Bowers had called back to Fort Bragg on a pay phone at the airport to say, "I'm in," and then hung up. He proceeded to help guide the air war by spotting targets from the ground. The mission was a success: nobody had gotten hurt but the bad guys. Instead of a demotion, upon his return Bowers was secretly applauded.

Bowers told Mitchell that there were reports coming in from the Department of Defense in Washington on "the red side"—classified e-mail—and that, for the moment, the plan was: there was no plan.

Mitchell hurriedly left Bowers's office to call Maggie.

"Where are you?" he demanded.

"I'm driving," said Maggie. "I've left the post."

"Get back here. Now!"

She could tell Mark was worried, and that had her worried.

Maggie, a mother of two, had left to do some errands at Target in the mall. She'd left their older girl, age three, with a nanny, and had taken the younger, age two, with her.

"Did you hear what happened?" Mark asked.

"Oh, Mark, I was listening to the radio . . ."

"You need to get home and get in the house."

Without further thought, she wheeled the Ford Explorer around and headed back in the other lane. It was nine thirty in the morning.

Because security had been tightened at Fort Campbell, each car was now painstakingly searched at the front gate. Normally, Maggie would have been able to flash her driver's license at an armed soldier at the gate and drive on. Traffic was now backed up for at least a mile on Highway 41A, all the way from a place called Sho-West, a strip club, to the Army surplus store and pawnshops outside Gate 4, the post's main entrance. There were nearly four thousand families living on base, and it seemed every one of them was trying to get back home.

Cal Spencer and his team got back from the Cumberland River around midmorning, then promptly ran smack into the same traffic jam that had snarled Maggie Mitchell. Sam Diller got out of the truck, stood bowlegged in the middle of the road, and looked up and down the long line of cars coming and going from the fort. The snarl was particularly bad because of a concrete barrier cutting off a lane.

"We can get through," said Sam. "We just need to move one of these barricades."

Spencer agreed. He and Sam and a couple other guys on the teams bent down and heaved ho.

"I don't think you should be doing that," said an approaching guard from the 101st Airborne, the regular Army. He had an M-16 slung across his chest. He looked scared by the morning's events.

Guys like Spencer and Sam had spent their Special Forces careers avoiding the regular Army. And now here was this kid, afraid to break the rules when it made sense to break them.

They ignored him and kept grunting until the cement divider was moved.

Spencer ran back to the truck, drove it through, and then helped the others push the big chunk back into place. Then they sped past the Fort Campbell bank, the PX, the Kentucky Fried Chicken and Taco Bell, to Third Battalion Headquarters. Spencer felt he was seeing all of these ordinary sights for the very first time.

As the morning wore on, from his office at Fort Bragg, North Carolina, Major General Geoffrey Lambert was working the odds.

The son of Kansas Mennonite farmers, Lambert, at fifty-four, was the commanding general of United States Army Special Forces Command (USASFC Airborne), which was composed of 9,500 men spread across the world in different geographic zones. Fifth Group ran the Middle East and Africa. Seventh and Third Groups, based at Bragg, operated in Africa, interdicting narcotraffickers and insurgencies. Tenth Group was based in Colorado and operated in Europe. (Lieutenant Colonel Bowers had been part of Tenth Group when he went into Kosovo.) The Pacific Rim, Indonesia, and the Philippines fell under First Group's oversight, based in Fort Lewis, Washington. The other groups, Nineteenth and Twentieth, which were made up of National Guard soldiers, went wherever they were needed, but also had battalions committed to regions around the world. These groups formed a wide net cast by highly secretive men.

Lambert knew that Fifth Group's bell had just been rung.

It had taken him about ten seconds to figure out who had masterminded the attacks, and who had carried them out. For the past several years, he had observed a top-secret intelligence program called data mining that had identified one man, an Egyptian by the name of Mohammed Atta, as a serious terrorist with links to a Saudi named bin Laden, who was a financier of terrorist training camps for men like the Egyptian. Months earlier, the people involved in the program had tried telling the FBI what they had discovered, but Army lawyers had discouraged the disclosure, even though the project had identified the hijackers. Lambert figured they knew everything there was to know about Osama bin Laden and his military training camps in Afghanistan, but none of the legal minds could decide if the surveillance was lawful. Now Lambert felt sick that more effort had not been made to warn someone. (Lambert, extremely upset, later agreed with lawyers that the information not be shared with the FBI.)

Usually, there was a contingency plan on the shelf at the Department of Defense for invading a country. But when it came to Afghanistan, Lambert knew that no such plan existed—nothing, not one scrap of paper describing how you mobilized men and weapons to take down the place. Since the end of the Cold War, U.S. military planners had been floundering in the backwash of old conflicts, unsure of how to prepare for imagined threats from unfamiliar enemies.

The attacks were, to Lambert's mind, perfect examples of the kind of violence that the future held for Americans. They had come quickly and cheaply, carried out by a small number of men communicating by mobile phones and the Internet. The damage had equaled that which might've been achieved by an entire army, yet it had cost only a half million dollars to inflict—small change, all things considered.

Lambert had been a hard-charging U.S. Army Ranger, serving ten years in the jungles of Latin America, fighting all kinds of insurgencies, great and small, secret and public, when he'd left to become a Special Forces officer. Being a Ranger, he said, was a fine thing, but being in SF was a lot finer.

Lambert was resolved to get his guys in this new fight, though he figured that General Tommy Franks, the CENTCOM commander, would never consider them a viable option. Franks didn't understand Special Forces soldiers, nor did he really like them. Few officers in the big, regular Army were fond of this secretive branch of independent-minded warriors. Over the past several years, funding for Special Forces had been cut, and few people outside the community realized that many of the teams lacked the equipment they needed. Hell, a lot of the teams didn't have a full contingent of men—short a medic here, a weapons specialist there.

Vietnam had done them in. During that war, the men of Fifth Group grew their hair, slept in hammocks, took native women as girlfriends, and lived and fought in the jungle far beyond the reach of anyone's official control. They had also committed some of the conflict's worst atrocities.

In the late 1980s, the unit was shunted into deeper twilight, providing what was called FID, or Foreign Internal Defense—a euphemism for training the armies of foreign governments. Other missions, such as serving as policemen on Native American reservations, had been more routine. Desert Storm had provided the first use of Special Forces in combat since Vietnam, much to the initial chagrin of General Norman Schwarzkopf, who'd seen their dirty work while commanding in Southeast Asia. The SF guys weren't particularly happy with their assignment; hunting SCUD sites was, they figured, the kind of soldiering better suited to gung-ho Marines, the publicity-savvy Navy SEALs, or the Army's secretive counterterrorism unit, Delta Force. They were masters at the snatch and grab—"door-kicking," Lambert called it. Let these other troops handle hardware hunts and scavenger forays.

Special Forces trained to do something different from everyone else. They fought guerrilla wars. This fighting was divided into phases: combat, diplomacy, and nation-building. They were trained to make war and provide humanitarian aid after the body count. They were both soldier and diplomat. The medics worked as dentists, fixing the teeth of local villagers; the engineers, experts at the orchestrated mayhem of explosives and demolition, were trained to rebuild a village's bridges and government offices. They spoke the locals' language and assiduously studied their customs concerning religion, sex, health, and politics. Their minds lived in the dark corners of the world. Often, they were the senior-ranking American officials in a country, hunkered down in the dirt drawing out a water treatment plan with some warlord and acting as America's de facto State Department.

The SEALs and the regular Army generally studied none of a country's languages, customs, or nuances. Special Forces thought first and shot last. They were the velvet hammer. Lambert knew that in the history of the United States, Special Forces soldiers had never been given the chance to fight as a lead element.

To get them off the bench and on the field was going to take considerable back-channel maneuvering. The U.S. Army had zero interest in deploying a bunch of cowboys in Afghanistan who had served as inspiration for the best scenes in *Apocalypse Now*.

Lambert was banking on the cooperation of the CIA and its covert ties in the country. During the 1979–89 Soviet occupation of Afghanistan, the Agency had funded the *mujahideen*—"freedom fighters." Of all the diplomatic and military branches of the U.S. government, the CIA had kept the closest watch on Afghanistan. It was the go-to authority. It was also the progenitor of U.S. Special Forces.

After World War II, the OSS (Office of Strategic Services) had disbanded and its members migrated into either what eventually became the Central Intelligence Agency or into Special Forces. The unit itself wasn't officially formed until 1942, as the First Special Service Force, comprised of American and Canadian soldiers. It adopted a red arrowhead with a dagger down the middle as its emblem. The symbol was not a haphazard choice. The World War II Army soldiers who had fought as guerrillas had drawn their inspiration from the Apache Scouts of the nineteenth century. They had survived behind German and Japanese lines by relying on the goodwill of the friendly people they encountered. Often outgunned and outnumbered, they ambushed instead of employing frontal assaults. They harassed supply trains. They attacked in several places at once and vanished into the woods. They followed none of the regular rules of warfare. Relishing their lethal craft of stealth and surprise, these World War II soldiers even nicknamed themselves "the Devil's Brigade," sneaking into German trenches in the night and slitting the throats of a surprised enemy. Daylight would reveal a chilling scene: paper arrowheads left on the foreheads of the dead men.

Earlier, in the American Revolution, this homegrown brand of war had been practiced by Ethan Allen in the northeastern states and Francis Marion, aka the Swamp Fox, in the South. A particularly rough group of marauders named Rogers' Rangers were known for their extraordinary exploits during the French and Indian War on the British side. The credo of the band was simple: "Let the enemy come till he's almost close enough to touch, then let him have it and jump out and finish him up with your hatchet."

Lambert relished the prospect that his soldiers would be called to fight this kind of war. He was also worried. The men were untested. Once he launched them, quick rescue would be impossible.

By midday, when Mitchell checked with Bowers, there still was no plan. The order stayed the same: *Be ready to leave at a moment's notice.* As Mitchell walked the halls, he heard the crackle and burble of the same news playing on the radios and TVs. Each minute seemed trapped in the endless looping of one moment: the towers falling, the towers falling. By late afternoon, the day felt like it had lasted for a year.

When it was clear nothing more was going to happen, that no plan for action was forthcoming from Washington, Mitchell decided to go home. He pulled into his driveway well after dark, staring at the lights burning in the windows of his trim brick ranch house. He and Maggie lived in a far corner of Fort Campbell, in a place called Werner Park that was filled with deer, bordered by woods, and populated by officers and their wives. The day before, September 10, he had returned from a trip out west, ten days in a rental car spent pulling into ranchers' driveways, asking whether the U.S. Army could use their property for a war game. *A game.* He shook his head at how much had changed in one morning.

Mitchell got out and walked across the lawn, flipping and reflipping the interior switch that all men reach for who are not violent but who live in a violent world.

Warrior/Father.

Warrior/Father.

Father.

He pushed open the door and kissed Maggie. *I will do anything to protect this place*, he thought.

Cal Spencer arrived home at the end of the day in his ratty second-hand Mercedes and swept Marcha up in a hug when he walked through the door.

"I can't believe it," he said quietly.

She felt him tense as he said this. She stepped back and looked at him.

"You're going away," she said.

He nodded. "Yeah."

"When?"

"I have no idea."

Marcha knew that soon Cal would walk around the house look-ing for things that usually he had no trouble finding, like his life insurance policy. He'd grow short-tempered, and he'd talk about what would happen to her and the boys if he didn't come home, if he got killed on this deployment. She never liked having this con-versation, and her refusal always led to a fight, usually a bad one.

Cal kissed her and went into the garage and started packing, pulling stuff off of the metal storage shelves—his sleeping bag, his CamelBak for drinking water, his headlamp. He was worried how their sons were going to take the day's attack. Their oldest was living in Mississippi and working at a job he liked. He was going to be fine. Luke, the middle child and a high school junior, kept his feelings inside, like Cal did. Jake, the youngest, a sophomore, was easygoing. He wanted to be an actor or a comedian. Cal and Marcha worried about him the least.

So when Jake walked through the door after school and Cal saw the quizzical, sad look on his face, he knew that this leave-taking was going to be hard. He knew that Jake had spent the day at school hearing about the attacks in New York, and that he'd imagined the worst for his dad.

"Are you leaving, Dad?"

"Yeah, I am."

Jake nodded and kept walking down the hall to his room.

Cal started to follow him, but stopped. He'd leave him alone for now.

He imagined Jake in his room in front of his Nintendo. He was pretty sure he wouldn't be playing a war game, where your bullets rarely ran out and a wound to the chest was just a momentary inconvenience. The boy had had enough of war, with his dad gone almost half his life. *How in the world,* Cal wondered, *can a man love his family and want to go to war at the same time?*

President George W. Bush appeared on TV the next day, September 12, and declared war on Al Qaeda. Over the next twenty-four hours, a military response began to emerge. Tommy Franks proposed to

Secretary of Defense Rumsfeld and President Bush that America invade Afghanistan with 60,000 troops. He explained that such a massive migration would take six months.

Donald Rumsfeld hated the plan. "I want men on the ground now!" he said.

In response, CIA director George Tenet proposed sending in CIA operators with Special Forces soldiers. At Special Operations Command, this alternate plan was refined and sent back up the chain to Rumsfeld. At Fort Campbell, Mitchell and Spencer followed the developments closely—in the news and in the battalion hallways. Dean, still in his hotel in Tahiti, felt cut off. He was spending a good part of each day on the phone trying to book a flight back to the States. On September 14, he and Kelly finally arrived back in Clarksville. When he took her to a team picnic to meet everyone, his battalion commander walked up and shook her hand and said, "Welcome to Special Forces. Your husband will be leaving soon."

Several days later, for the first time in American history, President Bush approved a plan using Special Forces as the lead element in the war in Afghanistan.

The plan involved using massive American airpower—cruise missiles and laser-guided bombs—to blast the Taliban out of the country. U.S. Special Forces on the ground would spot targets, build alliances among the locals, and whip them into fighting shape. The Afghan Northern Alliance—Massoud's old fighting force of several tribes led by different warlords—would make up the bulk of the ground power. The CIA would grease the wheels, many of them unturned in years, with money and intel, and help the SF soldiers link up with the Afghans.

General Franks had spent part of the week holed up in a room at the Sheraton Hotel in Tashkent, Uzbekistan, convincing the country's president to let the United States base its troops in the former Soviet Republic. Beset by Islamic terrorists of their own called the IMU, the Uzbekis had not been an easy sell, but Franks had succeeded.

On September 18, to a packed White House cabinet room, President Bush announced: "The war starts today."

Naturally, the plan was classified, top secret.

Mitchell, Spencer, and Dean started arriving at work at 4 a.m. and stayed until midnight. Birthday parties, anniversaries, normal life were canceled. Nothing mattered except the war—getting ready for the war, planning to survive the war, and coming home alive. All the men believed Fifth Group could be deployed at any *hour*. Nobody went anywhere—the mall, the dentist, the movies—without leaving a cell phone number with the staff person in charge. Teams went to the firing range and shot thousands of rounds of ammo at pop-up targets. They practiced patrolling, ambushes, winter survival techniques. They marched, lifted weights, and, in the absence of real intel (they were told CIA analysts at Langley were working on it), read anything they could find about Afghanistan on the Internet. They cleaned weapons, took inventory of broken equipment, and made lists of the necessary gear.

They needed a lot, so much so that it was embarrassing. Their group commander, forty-five-year-old Colonel John Mulholland, had taken charge of the soldiers just two months earlier, after a staff posting in Washington, D.C., and a stint as a student at the National War College. He was working around the clock to meet the needs of his men. Tall, massively built, with the intense, pinched gaze of someone who did not suffer fools, Mulholland had served during the 1980s under General Lambert as a Special Forces lieutenant in Latin America, and he'd worked as a Delta Force operator in the mid-1990s. He assured each team that they could have whatever new equipment they needed. After much cajoling and pressuring on his part and Lambert's, the Pentagon had agreed to pull out the golden credit card.

There was no time to requisition supplies the old-fashioned way, so new methods were created. Sergeant Dave Betz and his staff called camping stores like REI and Campmor and bought all the socks and tents they had on the shelves—literally everything in stock. Same with the clothes, and when the dealers ran out—as they did with a particular black fleece jacket everyone wanted—the guys called North Face headquarters and bought direct. There were soldiers perusing back issues of *Shotgun News* magazine and ordering pistol holsters and ammo magazines for AK-47s. They bought CamelBak water hydration systems, thermoses, water filters, tan winter boots made by a company called Rocky's, duffel bags, Iridium satellite phones, generators, tool kits, compressors, electric conversion kits to convert 12-volt DC to 110-volt AC, camp stoves, fuel, and headlamps. Staff guys carried new radios and laptops and PDAs into team rooms, gizmos the men had never seen before. The guys liked the lightweight Garmin Etrex GPSs—the military GPS being heavy and the size of a writing tablet—and couldn't purchase enough, ordering them all over the country, three hundred to four hundred at a time. A supply sergeant would e-mail a supplier, "I want all your GPSs. *Hold them.*" And they bought batteries. One of Betz's men personally drove over to an enormous store near Fort Campbell called Batteries Plus, bought every double-A they had, and drove off with them in the trunk of his car. As he departed, the sales clerks stared slack-jawed at their empty racks.

Everything not in car trunks and backseats was shipped overnight by FedEx, the delivery trucks pulling up to a gray two-story building that resembled a grain elevator balanced atop a warehouse. The place was called the Isolation Facility, or ISOFAC. There, the equipment was stacked along walls on aluminum pallets—"palletized"—and each bunch was covered with a waterproof material nicknamed an elephant rubber. The covering was meant to protect the gear during the long plane journey to a place called Karshi-Khanabad, in Uzbekistan, a staging base, also top secret. Everybody called it K2.

On September 18, Lieutenant Colonel Bowers announced: "I need a man for a secret mission."

Stepping forward was a seasoned master sergeant from New Hampshire named John Bolduc, nicknamed "the Skeletor," after the He-Man comic book superhero. Shy, quiet, thin as a whip, Bolduc held the record for finishing a grueling eighteen-mile rucksack march required to pass Special Forces selection in just three hours. It took most people eight to cover the distance. At the time, Bolduc had actually had to wake up the guys in the lead van; they hadn't expected anybody to show up until dawn. They thought Bolduc was kidding when he told them he was done. Bolduc was not a kidder.

At Fort Campbell, the master sergeant's men would follow him anywhere, even off the edge. He would emerge from the woods during the team's "fun runs," sprint off a high rock ledge, and drop fifty feet to the quarry water below, still pedaling in the air, looking over his shoulder to make sure his team was following him. If they were, he knew he was doing a good job.

The irony, given his dedication, was that Bolduc had recently put in for retirement after eighteen years in the Army. In fact, he'd received some of the official paperwork on September 11. His fellow soldiers tried to talk him out of leaving, but he would have none of it.

"I only have one reason for retiring, " he told them then. "I have a teenage daughter who doesn't know me and I don't know her." Bolduc had added up the years to discover he'd spent half of his daughter's life away from home, deployed to godforsaken snatches of sand; winning, but losing, too.

Still, he was torn. He'd jumped into Panama as a Ranger in the takedown of Noriega, but this was the big time. As a soldier, you lived for wars like this.

As he stood before Bowers's desk, the battalion commander asked Bolduc, "Are you retired yet?"

"No, sir, not yet." For once, the Army's glacial pace of dealing with paperwork had opened the door instead of closing it.

"Are you ready for a mission?"

"I am."

"I can't tell you where you're going. But plan as if you're going into combat."

"When do I leave?"

"Tonight."

Several hours later, Bolduc was standing on the airstrip at Fort Campbell, clutching a rucksack loaded with Claymore mines, hand grenades, radios, and ammo magazines. Most intriguing was the getup he'd been instructed to wear during the mission. The disguise, consisting of paisley bell-bottom pants, a blue nylon shirt, and a funky slouch hat, was designed to make him look like an American hipster on parade in a former satellite of the Soviet Union. But after trying on the clothes, he decided that what he really looked like was a seventies porn star—and he stashed the duds in a closet at home. He now stood nervously on the tarmac in Levi's, flannel shirt, and hiking boots, waiting for his adventure to begin. A white Citation jet with no tail numbers swooped down out of the sky, rolled to a quick stop, and picked him up.

Twenty-four hours later, Bolduc was in Uzbekistan, at K2, a desolate piece of real estate oozing chemical waste, misery, and the rank bloom of failure. The Russians had used the place to stage their failed ten-year war in Afghanistan against the likes of Massoud and his men, the fabled *mujahideen* fighters. The Russians had gotten their heads handed to them, losing 50,000 men in the war. Historians credited their defeat as one of the causes of the collapse of the Soviet Union, one final and fatal proxy battle between the United States and the USSR in the Cold War.

Bolduc's job was to help redesign this flat, muddy ground as the new secret home for a group of soldiers determined not to repeat the Soviets' mistakes, a place from which the entire fury of America's military could be unleashed on the Taliban. Bolduc had never done anything like this in his life—he wasn't an engineer—but he fully intended to try. This was the Special Forces' way: you improvised like mad.

The CIA was doing its part, too. On September 19, paramilitary officer Gary Schroen loaded three cardboard boxes, each packed with $3 million in hundred-dollar bills, into an unmarked Suburban, and headed off to see his boss, Cofer Black, at CIA Headquarters in Langley, Virginia. The money was meant to bribe the Afghan warlords that the Special Forces troops were to work with—rough, mercurial characters with names like Abdul Rashid Dostum, Atta Mohammed Noor, and Naji Mohammed Mohaqeq. For the past decade or more, they had fought each other for control of their country. Schroen would be flying to Afghanistan with the money that day, aiming to convince them to work together to kick the Taliban's ass. He was doubtful of the outcome. You can't buy an Afghan's loyalty, he thought, but you sure can try to rent it.

Cofer Black was very specific about another aspect of Schroen's mission. Writing later about the conversation, Schroen would recount Black saying, "I have discussed this with the President," he said, "and he is in full agreement. You are to convince the Northern Alliance to work with us to accept U.S. military forces. . . . But beyond that, your mission is to exert all efforts to find Osama bin Laden and his senior lieutenants and to kill them.

"I don't want bin Laden and his thugs captured," Black explained. "I want them dead. Alive and in prison here in the United States, they'll become a symbol, a rallying point for other terrorists."

And then Black shocked Schroen: "I want to see photos of their heads on pikes. I want bin Laden's head shipped in a box filled with dry ice.

"I want to able to show bin Laden's head to the President. I promised him I would do that. Have I made myself clear?"

Schroen believed that he understood the mission.

The following day, Cal Spencer got the call. As the CIA built up alliances with different Afghan tribes, the Air Force would conduct its air war. Spencer's team, the Pentagon had decided, would provide combat search and rescue (CSAR) for bomber pilots who'd been shot down by the Taliban. During the Soviet invasion of Afghanistan, the United States had supplied Stinger missiles to the anti-Soviet *mujahideen* fighters. After the Soviets' retreat, the Taliban had emerged from the ranks of these highly trained soldiers. The United States had turned a blind eye to the radical religious beliefs of some of its favored *mujahideen*, and now it would pay the price. The extremists had become terrorists, anxious to use the Stinger missiles against U.S. soldiers.

CSAR was a heady mission, and Spencer had spent years training for it as a young sergeant at Nellis Air Force Base in Las Vegas. He and his team would be working behind enemy lines on a thin wire of support. If anything went wrong, they were on their own. They welcomed the challenge, but there was one problem: their team didn't have a captain. A week earlier, Captain Mitch Nelson had been kicked upstairs to Fifth Group Headquarters, promoted out of the job he loved, following the team's return from a training mission.

Nelson, thirty-two, the son of a Kansas rancher, was miserable in his new staff position. He hated office work, but if he expected to move up in rank, he had to punch his ticket and slay the beast that was administrative paper shuffling. The worst part was that one of his best friends, Dean, still had a year of team time left (each captain got two years), and now Dean for sure was going to get into the fight while Nelson would be wielding staplers and paper clips.

Nelson wanted in, too. He spoke Russian; he'd recently been in Uzbekistan with Spencer and the rest of the team. He was itching for this.

He paid a visit to Lieutenant Colonel Bowers's office.

"Sir," he said. "I need to be back on my team."

Bowers looked at him and said, flatly, "No."

Like the other men he had served with on the team, Nelson was stubborn and independent, and he had chafed under Bowers's command. He figured that his request was denied because the lieutenant colonel did not like him. In fact, when it came time to select the teams for the CSAR mission, Bowers had nominated another group of men in his battalion. Spencer's team had gotten the CSAR mission only because other teams had recommended them so strongly to Colonel Mulholland. Bowers had grudgingly obliged.

Feeling the urgency of the moment, Spencer and Master Sergeant Essex also appealed to Bowers on Nelson's behalf: "We really need him," they said. Bowers, either convinced or resigned, gave Nelson his job back.

"You've got six hours to leave," Bowers told the two men. They had an airplane inbound that very minute to take them to K2, in Uzbekistan.

Essex had no idea where Nelson was, and it was his job to make the trains run on time. He called Nelson on his cell phone, getting the voice mail. The young captain and his wife were in Nashville at an obstetrician's office. His wife was expecting a baby in two months, and Nelson, an excited father-to-be, had turned the phone off before going into the appointment.

"Hey, man!" Essex yelled into the phone. "You need to get back here, like right away!"

Essex was unsure of how to proceed over a nonsecure phone line, so he said, "Because we're . . . *leaving* . . . *now!*"

He hoped like hell Nelson would make it back to the team room in time.

He need not have worried. Several hours after it was ordered, the mission was canceled. There was no explanation for the change in plans (and there never would be). The team was ordered not to travel more than an hour's distance from the post. Deflated and dismayed, the men stood down. They could conquer any enemy but one: the men who pulled the levers at the Pentagon. For now, all they could do was wait.

•   •   •

But not for long. On October 4, they got the call for real: the mission was on. As before, no explanation was given.

Before they left, Major General Lambert flew down from Fort Bragg to address the crowd. Digging into the pocket of his crisp camouflage pants, he removed a piece of jewelry. Most of the soldiers had only heard of this special item: the gold ring of war. A chipped ruby blinked back at them as Lambert held it aloft.

"This ring has been through hell and back," said Lambert, "worn on the hands of men who are dead or retired, men whose work won't be talked about for years, if ever."

Lambert knew the ring's history well. Back in 1989, as a young commanding officer, he'd asked one of his sergeants to take part in something called the Expert Infantryman's Badge Test: five days of running, shooting, and shitting in the woods.

Lambert had asked the sergeant to take the test as a show of his leadership among the men under his command, and the man had passed, but just barely. Lambert, impressed that the sergeant hadn't quit, had given the guy his own Expert Infantryman's Badge, which he'd had displayed in a frame on his office wall.

Without pausing, the sergeant had taken off a ruby ring he was wearing and thrust it out to Lambert.

"Here, we'll trade. This damn thing's been in Bolivia, Panama, Vietnam, Thailand, Pakistan, Belgian Congo, Bosnia, you name it," he said. Then he smiled. "Now it's yours."

Lambert decided that henceforth the ring would be carried into battle or on a mission during each SF deployment. The only caveat was that the man chosen to wear the ring of war had to bring it home safely. It was a bit of voodoo, military-style.

As he looked out at the conference room, Lambert remembered one man in particular who had worn the ring five years earlier. The man had appeared to be in excellent health, but then surprisingly failed a routine physical fitness test. Within a week, doctors diagnosed him with Lou Gehrig's disease. Within days, he couldn't even use his hands.

Lambert had arranged for an immediate medical retirement and his fellow soldiers painted the ring wearer's house, helped him sell it, and moved his family to a new city to be closer to relatives. The ring stayed behind.

Several months later, Lambert received a phone call from the guy's wife. She had said he was depressed.

"Is it okay if I put you on speakerphone?" she asked.

Lambert, choked up, said sure.

He hadn't known what to say at first. He imagined his old friend sitting in his wheelchair, frozen, unable to move or speak. They taught you a lot in Special Forces, but nothing had prepared him for this.

"You're a great man," Lambert had said. "And I admire you."

His wife narrated her husband's reactions. "He's smiling!" she said. "He seems happier."

Lambert spoke for a bit longer, then he hung up, thinking, *I'm going to fix this.* He packaged up the ring and sent it to her. "Tell him to put this on," he wrote. "I want him to have it."

But there was no returning from this mission. The man had put the ring on and sat in his wheelchair and he stared at it, mute, trapped. When he died, he still had it on. His wife gave the ring back to Lambert, telling him, "He would want you to have this."

Now, gazing at the men before him, Lambert asked them, *"How will you die?* I want you to think about that."

It was not a rhetorical or even moral question. Part of each team's mission was to consider the ways in which it could be annihilated, in order to avoid such a fate.

He handed the ring to the officer in charge of the meeting, and said, "Give this to your best man. Make sure he brings it home."

John Bolduc had come home from the prewar planning trip to Uzbekistan on September 22, and Sharon, his wife of seventeen years, and his teenage daughter, Hannah, sat down together and talked about his retirement. Bolduc told Hannah that he felt they didn't know each other, and he said he was still hoping to change that. But now that chance would have to wait. He was going away, he said. But he promised to come back. It was Bolduc who was given the ruby ring.

They went to the video store and rented some movies for his last night at home. Husband and wife lay on their bed, with Hannah between them, and stared at the screen. Sharon told him, "You know, I could just jump on your legs and break them, and you wouldn't have to go." She was only half kidding.

•   •   •

41

Lying in bed, Marcha Spencer looked over and asked her husband if he was awake. Then she asked where he was going tomorrow. Could he tell her? she wondered. She already knew the answer.

"You know I can't tell you that," said Spencer.

In truth, Marcha didn't want to know any geographic specifics; all she really wanted to know was whether or not Cal would survive.

Lying there, Cal next to her, Jake asleep in his room down the hall, it had been a good life, but still . . . she never said anything, not out loud, you couldn't. She burned at the lack of recognition—nobody, and she meant *nobody*—knew exactly what dear Cal Spencer did when he went away on his missions. He came home sunburned, sand in his pants, ready to get back in the swing of family life, walking around with a screwdriver in his hand, looking for things to fix. My God, during their twenty years of marriage, he'd been absent for at least half of it. The sum never failed to stun her. Ten years of No Cal! She and the boys would get along, she knew that. They would grow old together, she and the boys, and Cal would be gone. But who would know what he had done with his life? The Army never let them say anything about their missions—when Cal came home, there was no welcoming party at some sunny airport, with a brass band playing and the TV cameras rolling. What Cal Spencer did when he was away would forever be a mystery. What she and the boys endured remained hidden from everyone.

In the past, Cal had had dreams of dying in the desert—some desert, any desert—but not tonight. He steered his mind to the sound of the crickets outside. They sang above the whir of air conditioners up and down the block.

He figured that Death was out there, a black dot on the horizon. Either you would pass it by, or not. Until that time, there was little use worrying.

He tried not to think about being killed, but he had the feeling that if it was going to happen, it would happen with this deployment. He had read enough intel to know how the Taliban treated their prisoners. The biggest thing was not to get captured. He'd decided he would not be taken alive.

Cal Spencer fell asleep, knowing he might never again sleep in his own bed.

• • •

The following day, Marcha dropped him off at Fifth Group Headquarters. Cal leaned in the car window and looked at her.

"I love you," he said. "And I'll be back." He smiled.

"Okay."

Marcha was crying as she drove off.

The parking lot was the scene of all kinds of turmoil. Ben Milo told his oldest teenage son, "You're the man of the house. You make sure you make things as easy as possible." The boy started sobbing.

Lisa Diller drove away dry-eyed. She'd been cleaning the house the night before, and all day today, right up until it was time to drive Sam to headquarters. Whenever Sam deployed, Lisa cleaned. It was her way of dealing with the stress of her husband's leave-taking, her way of saying goodbye.

They had been through just about everything together. Sam had come home from the Battle of Mogadishu in Somalia in 1993, dropped his duffel bag in the living room, and said, shakily, "I'm home." Lisa took one look and she could tell he was falling apart. She nursed him back. But at least he had returned. As she drove home now, her stoicism melted and she broke down. Coming into the kitchen of the ranch house, she poured a beer and got into the hot tub and soaked in the thick, new silence of the house. It was going to be a long, long winter.

Late that night, October 5, Spencer crawled on board a converted school bus. The vehicle rolled across the post, the windows blackened so no one could see inside. There were few people to look; the neon of the Kentucky Fried Chicken restaurant was dark, the bank sign blinked for no one except the enormous eye of the waning moon. Cold wind blew trash across the street. Few of the guys talked, the bus's heavy springs squeaking beneath them. Steadily, it rumbled onto the airfield.

Shouldering his daypack, Spencer stepped off the bus into the roar of the C-130's four huge engines.

The plane's rear ramp was down, its belly palely lit. Together, the men leaned forward through the prop blast, clutched their black watch caps, and trudged up the ramp. They buckled themselves in and the plane shuddered beneath them, its green nose hunting in the dark over the tarmac, hunting speed, lift, release.

Spencer felt the deck beneath him lighten and drop and knew he was airborne. Aloft, with the sides of the plane cooling in the darkened heavens, he swallowed an Ambien, reached into his pack, and unrolled a foam sleeping mat on the frozen aluminum floor. There in his pack he found the letters Marcha had written him, each one numbered and dated, indicating when he should open them.

He opened them all at once and started reading hungrily: *Dear Cal, I have never told you how much I love you . . .*

Once done, he placed each letter carefully back in his pack. He lay on the humming deck of the plane and fell asleep, floating over the Atlantic, heading east, into the dawn, to the secret base in Uzbekistan.

Before Dean took off, he and Kelly went for a drive on a perfect, sunny Sunday afternoon. Winter—and Dean's departure—seemed an impossible thought. The blue autumn sky looked as if it had been polished with a hank of velvet. Back at their tiny apartment, moving boxes were still stacked to the ceiling.

They hardly talked as they drove country roads for several hours. At the parking lot next to the church, the scene of so many departures over the years, Kelly finally dropped Dean off. He dragged his rucksack upstairs to the team room, then ran back downstairs and out the door to Kelly. He had to see her again. There she was, still standing beside the car, looking so beautiful, and scared.

When he'd set his sights on her two years earlier, he'd been a student in the Special Forces Qualification Course at Fort Bragg, and because he couldn't get enough language study there (he spoke fluent Russian), he decided that once a week he'd drive to Chapel Hill, an hour away, and take a conversational course in Russian at the University of North Carolina.

One night, a pretty girl with curly red hair had walked into the classroom and taken a seat next to him. Taking one look at her, Dean felt himself turn to jelly. "So," he stammered, "uh, why are you taking this class?"

"I'm planning a career in international business," she said. The way she said this, so confidently, made him think, *I am going to marry this woman*.

Their courtship was a whirlwind. Early on, Dean had to leave

Fort Bragg briefly and he spent night and day fretting that Kelly would forget him while he was gone. He ordered a florist to send her flowers in his absence. He wrote her letters in advance and arranged for them to be delivered.

He proposed marriage six months later during a visit to her family's home in New Hampshire, standing beside her father's grave. Dean firmly believed that if the old man were alive, he—Dean— would be honor-bound to ask permission to marry his daughter.

"I wish you could've met him," Kelly said.

Dean suddenly got down on one knee and turned to her father's headstone: "Sir, with your permission," he said, and then he pivoted and looked up at Kelly.

"Kelly," he asked, "will you marry me?"

Now, saying goodbye to her after being married less than a month, he lifted her hands and said, "I love you." He walked away backward, looking at her. He turned and climbed the steps to his team room, his new combat boots ringing on the metal stairs.

The entrance door slammed shut behind him.

Dean paused on the steps to collect himself. Part of him wished that he was good with words so he could have written something for her, letting her know what a lucky guy he was to have met her.

He jogged to a hallway window on the second floor and looked out, but she was gone.

He shouldered his gear and walked into Fort Campbell's prison-like Isolation Facility.

Dean disliked ISOFAC for its mind-numbing seclusion but also loved it for its promise: to enter ISOFAC was to sit in waiting for your violent birth into war.

ISOFAC was a closed world within the tightly held universe of the post itself. Guards manned the gate in the chain-link fence, tipped with concertina wire. From the outside, it was a windowless jumble of metal-gray blocks. The ugly architecture suggested that everything important was going on in the inside, which was no doubt true.

The building was divided into two levels, with the sleeping quarters in the top floor. On the first floor, Dean walked through the main planning room, which was sparsely furnished with desks, chairs, and several dry-erase boards mounted on cinder-block walls. The walls pulsed a dazzling white under banks of fluorescent lights. Each team was allotted two rooms—one for planning its mission and another for sleeping. Each planning room contained chairs and desks, a fifteen-foot-long chalkboard, and a continuous roll of four-foot-wide butcher paper that fed through an easel. The Spartan sleeping room consisted solely of twelve bunks—topped with thin, brown plastic mattresses—squared smartly along the walls.

Down the hall was another room for hand-to-hand combat training, with padded walls and wrestling mats on the floor. At the end of the hall was the arms room, where weapons were locked up, and a gym with weights, and the cafeteria. It was as spare as a new prison. The odd-looking tower that rose from the clutter of metal, lending the air of a defunct chimney, was used to rig and pack parachutes.

No talking was allowed in the hallways.

There were six other teams in the Isolation Facility, including John Bolduc's. (Among various teams, ODA 555, "Triple Nickel," would land near the Panjshir Valley shortly after Nelson's touchdown in Dehi, making Nelson and his men the first U.S. soldiers to enter Afghanistan.) Because the teams were not supposed to talk with each other, they sat elbow to elbow at cafeteria tables during meals, drinking the gritty Kool-Aid (the cooks could never get the stuff mixed right), pretending the others were not in the room. They were forbidden to speak in order to maintain a separation among the teams' missions. If one of the teams was captured and tortured, they would have little to tell the enemy.

CIA analysts appeared and visited some team rooms and not others, though they had little to share this early in the war. A few days earlier, on October 7, the U.S. Air Force had started bombing Taliban soldiers dug in throughout Afghanistan. As Dean studied maps of the country, measuring roughly the size of Texas, he saw that it was a surreal contradiction of 17,000-foot mountain peaks, vast desert, and squiggles of green rivers flowing through forested valleys. He learned from intel reports that the bombing campaign was proving to be a challenge. Flying at 20,000 feet over snowcapped mountains and expansive khaki-colored plains, it was difficult for the pilots to designate their targets (the threat of antiaircraft fire prevented lower-altitude flights). It was clear to Pentagon officials that the pilots needed boots on the ground to guide the way—sooner rather than later. They needed guys like Dean.

Dean would hear one of the CIA guys pad down the hall, knock on a team's planning room door, and close it gently behind him with a soft click. Dean burned with the desire to hear that knock on his team's door. He studied whatever he could get his hands on about Afghanistan, including Ahmed Rashid's book *Taliban*, and scraps of classified intel about a warlord named Abdul Rashid Dostum, who reputedly was tolerant of both prostitution and opium production in his camp, and with whom the CIA was hoping to do business. Another warlord was named Atta Mohammed (unrelated to the hijacker, with whom Atta Mohammed, because of name similarity, was sometimes confused). In comparison to Dostum, Atta Mohammed was a pious Muslim.

Using his laptop in the team's planning room, Dean searched for more information about these two shadowy men, as well as whatever he could find about bin Laden and Al Qaeda. But there wasn't much hard intel immediately available. One day, an ISOFAC staffer dropped off an armload of tattered *National Geographic* magazines and a few Discovery Channel television shows on VHS tape about the history of Afghanistan, and when Dean asked, "What's that stuff?" the guy replied, "Consider it more intel."

It dawned on Dean that the U.S. government was woefully underprepared to send him and his men into Afghanistan. The country had not been lately in anyone's intelligence bull's-eye. When Dean finally did hear the CIA analyst knock on his door, the briefing was anticlimactic and he learned little that he didn't already know. Thankfully, someone had gotten the idea of phoning the publisher of a book called *The Bear Went Over the Mountain*, about the Soviet experience in Afghanistan, and asked that they send 600 copies. The publisher no longer even stocked the book and had to scramble and send an electronic version to a printing plant, from which fresh copies were express-shipped to the ISOFAC. Dean was pleased when it arrived. He went to sleep each night plugged into his earphones listening to a Barry White CD and poring over the tome.

After a week in isolation, he felt he was as ready as he ever would be. On October 13, he was asleep in his bunk when the lights in the room snapped on. Dean sat up, swearing and rubbing his eyes.

"Hurry up!" yelled an ISOFAC staff member. "You guys are moving out! The plane is here!"

"Well, that plane didn't just fall out of the sky," groused Dean. "Who's the genius who didn't call ahead and say, 'I am about to land'?"

An hour later, he was airborne.

They left in quick succession. The day before his departure, Mark Mitchell had stayed up until 2 a.m. writing a list of things for Maggie to do in his absence—"Pay $300 to this credit card every month until it's paid. If it gets paid off and I'm still gone, add the $300 a month to the savings account." He'd spent the last few days getting the carpets cleaned, fixing the garage door, tightening the loose faucets in the bathroom. Nothing was left to chance. Maggie freely admitted she couldn't fix anything and Mark was happy to oblige. "I may not be handsome," he joked with her, "but I sure am handy." She thought he was the handsomest, most honest man in the world.

She'd met him at Fort Stewart in Georgia, where she was teaching grade school and he was a lieutenant in the Army. They dated for a month, and Mark asked her to accompany him to a family wedding in Milwaukee. While they were dancing, he proposed. Maggie, surprised, said, "You're drunk, aren't you?"

"Yes, but will you marry me?"

"Ask me again in the morning when you're sober."

He did, and she accepted.

Married now for ten years, she loved it that he still had something interesting to say. He'd graduated from the Jesuit Marquette University in Milwaukee, with a degree in engineering. As a teenager, he'd gotten up at four in the morning to deliver his paper route and then clean the bar of an Irish pub in downtown Milwaukee. By eight o'clock he was in school.

In high school he ran cross country, played football, and wrestled. He had a quirky sense of humor. He loved to play the piano and knew most of Ray Charles's numbers. He liked Sting and the Talking Heads. He loved Coen brothers movies. He had an artistic side that Maggie rarely saw except when he was taking photographs.

His humble demeanor disguised brute strength. Mark could lift himself hand-over-hand forty feet up in the air, with sixty-five pounds of field gear on, while down below, as he clung on the rope, other guys were bent over their knees, vomiting into the cinders. Mitchell's training had taught him to ignore pain and mental exhaustion. It taught him to pay attention to nuances, to what a man says and what he doesn't say. It taught him to think about using his brain first, his weapon next. During his deployments, he had tried to set aside his ideas of how he thought Pakistanis, Jordanis, Saudis would act, and to listen and observe what they actually said and did. He hoped to do the same with the Afghans he would encounter.

The son of a federal prosecutor, Mitchell had grown up listening to stories of his father's exploits. Back in the 1970s, Milwaukee was at the center of the burgeoning U.S. heroin market. He had admired his dad's courage, and never forgotten that he lived in a world where there were bad people, and it was a just thing to punish them. He'd entered the military to prove himself in such a world, what Maggie called his "Hoo-ah" macho stuff.

After fours years as an infantry officer, he found Special Forces and it changed him. He loved who they were, what they did. They were the best kept secret in the world. It seemed the Navy SEALs got all the publicity, with a new movie every summer about some cool shit they had blown up in a fictional South American country. Even the Delta Force, which the U.S. Army didn't officially acknowledge as existing, got more press than Special Forces. None of the guys in SF had ever written a book, which was fine with them. They knew the public spotlight only helped the enemy target you more easily. There was a Special Forces bumper sticker around Fort Campbell that read: THE QUIET PROFESSIONALS. They joked about it—"Hey, man, we're the *quiet* professionals," delivered with their best DJ voice—but they meant it, too.

Around four o'clock in the afternoon on October 24, Maggie and the girls dropped Mark off at the church parking lot. He picked up his daughters, one in each arm, and told them he loved them. He'd spent the day in the house watching the Disney Channel with the girls. They'd played tag in the yard. For lunch, they'd gone to Burger King. All the while, he'd been anxious.

The key was to keep life "normal." More than anything, he wanted to get the leaving over with. He expected to be gone six months, maybe a year. All his gear, his rucksack with his food, sleeping bag, a change of brown desert camo fatigues, his ammo and weapons, was packed neatly in a pallet on the plane waiting at the airfield. In a small knapsack slung over his shoulder he carried a Nalgene bottle with water, a toothbrush, a razor, and some old issues of *U.S. News & World Report,* the only newsmagazine he could ever find time to read. All that was left to do was to walk on the plane.

Mark boarded the bus and rode out to the airfield to the C-17 grumbling against its anchors on the runway. After less than six weeks of preparation, he and the rest of the Fifth Group were ready to take down a country.

Twenty-four hours later, he stepped off in the dark at K2, in Uzbekistan, the sky engorged, oozing the milk of a new starlight.

He was in. *Game on.*

PART TWO

HORSEMEN, RIDE

Dehi, Afghanistan

October 16, 2001

Finally, thought General Mohammed Mohaqeq, the Americans were coming. *Finally . . . finally.*

When the two towers that had seemed to stretch all the way to heaven had swallowed themselves and lay in heaps like strange, smoking beasts, the Americans finally heard the cry that Mohaqeq believed all people hear at the hour of their dying. They had awakened to the soundtrack of life in Afghanistan.

He had heard that sound for decades now, in Kabul, Mazar-i-Sharif, and today in a place called Safid Kotah—the White Mountain—where he had been fighting several thousand Taliban soldiers at the very hour America was attacked.

Mohaqeq had been hunkered in his mud hut near the mountain, his field headquarters, wearily poring over a map, when an aide rushed in with a radio and handed it to him. "Sir, there is news on the Motorola." Mohaqeq, leader of the Hazara people, with 2,500 men under his command, had listened and then set the radio down. He was speechless. He had been to the United States as a young man. New York. So big. A giant. He knew immediately that the Americans would be coming. For seven years, he had been fighting the Taliban, and losing. He was overjoyed.

He and his people had endured unbelievable hardship in that struggle. Along the streets of Mazar-i-Sharif, past the butchered goats draped in shimmering veils of flies, past the vegetable hawkers,

51

past old men beating spoons out of hubcaps, past the war orphans staring mute at the sun, hands out, begging for even a crumb, past all of this the Taliban had come, to the doors of the Hazara. The crazed soldiers had kicked in the doors.

They had dragged the men—old men, young men, boys—any Hazara male unfortunate enough to be caught cowering under these rickety, dry eaves, and hauled them into the street, slit their throats and castrated them, and left them to rot in the road, dark eyes frozen wide as the killer's dagger had snickered across their stretched necks. So much damage and sorrow that it had seemed to General Mohammed Mohaqeq that it would take years for any man to ever live normally again.

But he would try.

In early October, Mohaqeq had received a mysterious visit from a most dangerous man, General Abdul Rashid Dostum, who had told him, "Brother, we are going to be visited by some special friends. What is your feeling about this?"

Mohaqeq explained that he had been writing letters for the past year, three hundred in all, to the United Nations in New York City, in America, beseeching: "You must do something to help us. The Taliban are killing us." Mohaqeq said he would take help anywhere he found it.

"Our friends," Dostum continued, "need our help. I want you to lay fourteen lightbulbs in front of your home, and light them."

"But how will I light them?" asked Mohaqeq.

Mohaqeq did not know where he would find electricity in this cold, rocky place along the Darya Suf River. He was several days' ride from any electricity in any direction. The nearest large city was Mazar-i-Sharif, sixty miles upriver by horse.

The man held up his hand. "You will figure it out. If they see fourteen lights, they will know it is safe to land."

And then he left.

Now, one week later, in the dark of the autumn midnight, Mohaqeq bent down and connected an electric cable strung with fourteen bulbs to a gas-powered generator he'd been able to procure. The bulbs flared and lay glowing in the fine dust that puffed underfoot with the lightest step.

Mohaqeq stepped back to admire his handiwork. There was no sound except the *thwap* of the approaching helicopter.

The helicopter landed and several strange men got out, wearing American dungarees and flannel shirts, carrying guns and computers and heavy black duffels. Mohaqeq made them tea in his headquarters and fed them bread, and they seemed to sleep with their eyes open—they were that vigilant—and in the morning Mohaqeq drove them up the river several miles to the village of Dehi, the pickup jouncing over rocks and ruts and straining through the iron-dark shadows of the valley.

At a bend in the river, facing a broad yellow plain, they came to a mud building—a paddock surrounded by high mud walls, braced by heavy timber. The Americans named the place upon arrival: "the Alamo." They anxiously set about sweeping and unpacking.

Also in the camp was Abdul Rashid Dostum, who seemed to Mohaqeq to already have a close relationship with the CIA, and Atta Mohammed Noor, a fierce lieutenant who was commanded by Mullah Fahim Khan. The owly, gray-bearded mullah had succeeded Ahmed Shah Massoud, who had been assassinated September 9, and now lay buried on an Afghanistan mountaintop.

These fierce men formed the triumvirate of the Northern Alliance, the Taliban's mortal enemy, which had been organized and commanded by Massoud.

It was Fahim Khan who had spirited Massoud's torn body out of the country into Tajikistan for safe burial after his assassination, and it was Khan who had held the Northern Alliance together in the aftermath of the great warrior's death in September. Massoud had spent years forging the front with Dostum's Uzbeks and Mohaqeq's Hazaras, alongside his own Tajik soldiers, and the Alliance had come close to crumbling.

Some 7,000 men had been fighting all summer against the Taliban in the Darya Suf River Valley, rife with thousands of land mines, dotted by villages that stood mute in the mountain light, the broken bric-a-brac of buildings strewn on cold ground in the aftermath of months of Taliban attacks. In some places, the Taliban had barred the inhabitants inside their homes and burned the villages to the ground.

Through the summer, Mohaqeq, Dostum, and Atta's men had

been pushed farther and farther south, deeper into the valley. At this same time, Massoud was battling in the north, about seventy-five miles away, in an ever-thinning slice of the verdant Panjshir Valley, his supply lines in threat of being cut. Holding this northern front, if he could, Massoud would wait for Mohaqeq, Atta, and Dostum to ride up from the south, attacking the outnumbering Taliban from the rear.

If the Alliance could do this, the Taliban front, which stretched eighty miles east and west, from Konduz near the Pakistan border to Mazar-i-Sharif, would be divided.

The Alliance's goal was to capture Mazar-i-Sharif. If a man held Mazar, then he could hold the north. And if he held the north, he could capture the capital, Kabul. From there, he could attack the desert wastes in the south stretching from Kandahar to the border with Pakistan. His army would rule Afghanistan.

But capturing Mazar had so far proven impossible. In 1998, the Taliban had marched into the city, laid waste, killing an estimated 4,000 to 5,000 people, and held it ever since. They were encamped in a mammoth mud fortress, complete with a moat and gunports cut into its high walls. The place was called Qala-i-Janghi, and it happened to have been Dostum's former headquarters when he commanded the city in 1997 with a 20,000-man militia. Dostum was eager to return to the fort and reclaim what was his.

Under his protection, Mazar had been a quasi-cosmopolitan city, by relative standards, escaping the plague of devastating urban warfare and aerial bombings that had leveled other parts of the country. As the Taliban captured parts of the eastern and southern sections of the country in the mid-1990s, Dostum snorted, "I refuse to live in a country where a man can't drink vodka, and where women can't wear skirts and go to school." He could be a benevolent dictator.

There had been much for him to protect. Rich oil and gas deposits lay nearby. The city's airport boasted the country's longest paved runway, capable of landing transport and supply aircraft; the bridge north of the city over the ancient Oxus River (which Alexander the Great had forded during his conquest of the area) could be used to move men and matériel from Uzbekistan. This was why it was such an important piece of ground to hold in the defeat of the Taliban. But increasingly, Mazar's capture seemed unrealistic.

The men under Dostum, Mohaqeq, and Atta's command had marched and ridden horseback against the Taliban in countless fierce gunfights. Their loyalty to their leaders remained unwavering. However, now their supplies were running low, and winter was coming. The mountain passes of the Hindu Kush Mountains, catapulting 25,000 feet from the desert floor, would soon freeze; large parts of the country would be locked in the white, hoary iron of winter.

For the men, breakfast was often a dusty rind of flat bread. At night, the exhausted soldiers cloaked their horses with warm blankets and slept uncovered in the open under piercing starlight. In the mornings, they drank from the cold, rank buckets only after the horses had taken their fill and lipped from their masters' hands a morning's breakfast of weevily oats. Then the men saddled up, battered rifles across their saddles, and rode back into the withering fire of another day.

Their supplies and matériel had been seriously depleted by the fighting in nearby Safid Kotah. There, the Taliban had dug in with some two hundred bunkers along the flinty rock face of the mountain, including emplacements for tanks. The White Mountain, already dusted by the early snows of autumn, had to be taken. Mohaqeq and Dostum attacked it for a month, starting in mid-September.

Some 2,000 men on horses tried riding up the sloped face, rising 7,000 feet, but they were cut down by walls of small-arms and tank fire. They had no choice but to skirt the back side and picket the horses and begin climbing the rock by hand, their battered AK-47 rifles flopped over their shoulders on ragged slings made from twined duct tape. On the way up, step by step, this was hard, vicious, hand-to-hand fighting, often in knee-deep snow. Mohaqeq's men did not have combat boots; they wore scuffed men's dress shoes, or scrabbled along barefoot. Several thousand heavily armed Taliban soldiers stood at the top, firing down at them as they climbed. Their bullets punched the snow around the climbing men with sickening thuds.

Supplies dropped so low among Mohaqeq's men that each was given just five bullets before a gunfight. To compensate, the under-supplied fighters started setting ambushes at night. When they captured a Taliban bunker, they scooped up precious grenades and stuffed their suit coat pockets full of stray rounds. They drove the captured Taliban tanks off the mountainside and cheered when they crashed at the bottom. They did not want the Taliban to recapture them, and they did not have the expertise to keep them running. A Taliban general on the mountaintop was so convinced of his invincibility that he told Mohaqeq over his walkie-talkie: "If you take this mountain from us, I will give you my wife."

After thirty days of fighting, in mid-October, the men took the mountain.

Mohaqeq's aide called out on his radio, "What do you think now, brother? We have come for your wife."

"We have been ordered to retreat!" came the harried reply.

The Taliban soldiers fled the White Mountain en masse, an exodus of tanks, old Russian armored personnel carriers, and fleets of battered black Toyota pickups, sending a plume of dust and diesel smoke across the horizon, the army appearing and disappearing over rolling hills as it ran.

About fifteen miles north of Safid Kotah, the Taliban army stopped and turned, swinging its turrets and rifle barrels back down the valley, resetting for a next battle with Mohaqeq and his men.

Mohaqeq had halted the Northern Alliance's rout near the village of Dehi, a lonesome, windswept settlement of low adobe storefronts and hitching rails for horses lining a muddy main street. Dehi was the end of the road, so to speak, the farthest south Taliban tanks had been able to travel in pursuit of Alliance fighters. Mohaqeq and his cohorts had lurked in the shadows of the granite cliffs, amid the sound of the jade shallows of the Darya Suf River, just out of range of the Taliban guns.

But now Mohaqeq faced a dilemma that had plagued him for seven years. Whenever the Northern Alliance soldiers took new ground, a Taliban tank would roll into view over the next rocky ridge and charge down the slope at the ragged horsemen, who were bent over their horses' necks, firing their AKs wildly over the tops of the steady animals' ears, before wheeling suddenly in retreat and hightailing out of harm's way.

Mohaqeq knew they would need something to stop those tanks. Something the Taliban had never expected.

•　　•　　•

At Dostum's camp in Dehi, the Americans introduced themselves as "Baba Daoud," or Brother Dave (actually Dave Olson), who was tall, broad-shouldered, and sported a sparse, black beard; "Baba J.J." (J. J. Sawyer), who seemed the oldest, with his striped beard and haggard, concerned squint; and "Baba Mike" (Mike Spann), who appeared to be the youngest—thin, pale, muscular, with short sandy hair, an intense man.

These men, Mohaqeq learned, worked for the American CIA. Mohaqeq was disappointed there weren't soldiers among them; these men had guns, but they weren't military fighters, Mohaqeq knew that much. He had seen that they spent a lot of time typing on computers, which they connected to small black foldable antennas that looked like spiderwebs spun from strange plastic. They also carried brick-sized stacks of American bills in nylon duffel bags; this money, Mohaqeq guessed, was part of the bounty to be paid to Atta Mohammed Noor, Fahim Khan's subcommander. With it, Atta was to buy food and ammunition for his men. Khan himself was encamped at his headquarters 210 miles south, near Kabul, in the village of Barak.

Two days earlier, a man named Gary Schroen, the leader of another CIA team recently arrived in Khan's village, had handed over $1.3 million in cash to the recalcitrant warlord. (There were two separate CIA teams in Afghanistan; the team lead by Baba J.J. and Baba David in Dehi reported to Schroen's headquarters.) Schroen had plopped the cash down on a table in a nylon bag, and none of Khan's men had immediately moved to take possession, as if the money did not matter.

When one of the men did pick it up, the minion's eyes widened in surprise, and he gave an extra tug on the straps to lift it. Schroen looked on, amused.

The wily Khan did not have high regard for American soldiers; after several decades of fighting with Ahmed Shah Massoud, he did not believe there was anything an American could teach an Afghan about killing and repelling invaders on his soil. The money, however, would go a long way toward changing his mind. Atta, on the other hand, grudgingly wanted American bullets, blankets, and bombs. But most of all, the fiery guerrilla fighter wanted the respect of the United States military.

In early October, when the American bombs had started hitting around the country—hitting at nothing, really, but sand and the occasional Taliban bunker—Atta had become so angry that he announced he was immediately ending the war until the Americans discussed their plan with him.

Atta was sure that the errant bombing had only boosted the Taliban's morale. He could hear them laughing about it on their radios. At the same time, it had made his soldiers question the seriousness of the Americans.

Some of his soldiers, part of the Tajik force formerly under Massoud's commmand in the north, had been poised in a village twenty miles from Mazar-i-Sharif. Atta was convinced they could capture the city if only the Americans would drop bombs on targets he had identified. The Pentagon ignored Atta Mohammed Noor and his men.

In early October, the Americans were strictly fighting an air war. Without U.S. troops on the ground to spot and identify Taliban positions as bona fide targets, the Pentagon was not going to take the word of a lone Tajik fighter. At one point, without any air support, Atta's troops were overrun and five men were captured by the Taliban. The men were beaten and nooses placed around their necks. They were dragged from the bumper of a pickup truck until dead. Atta, a pious man who prayed to Allah five times a day, was troubled by these deaths, which he felt could have been avoided with the Americans' assistance. He was embarrassed that he had failed to convince America to turn its might in support of his struggling men.

But he had learned something several days earlier that had brightened his mood. The American government had decided to bring to Afghanistan highly trained men who were used to operating behind enemy lines. He believed they would know how to spot Taliban targets on the ground. Atta was anxious to meet this special breed of men.

What Atta learned next infuriated him: he would not be working with any of these men in his camp. Instead, they were going to work for Dostum in his. For over twenty years, since the Soviets' invasion in 1979, he and Dostum had been perennial enemies. Atta believed the Uzbeki was a rank opportunist.

During the Soviet occupation, Dostum had actually fought for the Russians by guarding the oil and gas fields in his birthplace of

Sheberghan from guerrilla attack. Dostum loved drink, women, and song; he was not pious. The atheism of the Communists did not bother him. He followed power. From it flowed safety and prosperity in an uncertain, violent land.

But when Dostum saw that the Soviets' puppet government was going to fold, he had abandoned his patron and taken up arms with Massoud against the collapsing presidency in Kabul. Now, when Atta heard that he himself would not be receiving American assistance in his fight against the Taliban, he threatened to attack an imminent air drop of supplies meant for Dostum's troops. Atta meant this threat with every ounce of his tall, wiry frame, although he knew it could bring about his death.

This brewing contretemps between the two warlords had caused no end of consternation for Baba J.J. at the Alamo and his boss, Gary Schroen. Trying to salvage the situation, Dostum, ever the diplomat, assured Atta that he would share fifty-fifty any assistance he received from the Americans.

He seemed to be telling the truth, Atta thought. His feelings were further soothed when J.J. presented him with a package. A grateful Atta opened it and discovered inside $250,000 in American bills. He announced that the money would be useful in feeding, clothing, and arming his men.

He added that he would call off the attack on the supply drop. He and Dostum would fight together, against the Taliban.

About 250 miles to the north, in Uzbekistan, at K2, Colonel John Mulholland gathered Mitch Nelson's team—twelve somber, burly guys in tan camo and black watch caps—at the helicopter's side and told them, "You might not come back from this mission alive."

Nelson was just glad to be leaving K2, which he and the team had come to see as a cesspool. Puddles of multicolored goo, like ruby-tinted antifreeze with cold coffee splashed in, lay glistening outside his tent, which had flooded in the recent rains. The ground was literally regurgitating its polluted past, left there by the Soviets when they invaded Afghanistan. (Somebody in command headquarters would later guess that some of the liquid was "nuclear" in nature, which meant the place was possibly radioactive.) The camp needed to be lifted about six inches off the ground and drained.

Colonel Mulholland looked the men in the eye when he told them they might not come back to this camp alive, and everyone on Nelson's team appreciated that. *At least he's got the balls to tell us this might be suicide,* thought medic Scott Black. *No bullshit and no kidding here.* Black's grandfather had been a paratrooper in World War II, a member of the 101st Airborne, made famous by *The Band of Brothers* television show, which had been on TV back at Fort Campbell just as Black was leaving for Afghanistan. He wondered if Granddad back in southern Michigan had ever done anything as daring as this, landing in enemy country with just a handful of guys, ready to gut it out over a long winter's campaign. . . . He knew the answer was yes. Talking to his grandfather, Black felt that he had a bridge to the past. He also felt like he was walking a tightrope into the future.

Standing at the side of the helo, Mulholland asked that the men bow their heads and pray. It was October 19, 2001. Go time.

The Army chaplain, a tall, taciturn kid, a recent graduate of Wake Forest, drawled a few words asking God's protection, and that they should vanquish their enemies, and be fine men in this hour of war, and come back home. The reverend had carted boxes of Bibles, Korans, the Talmud, even Buddhist texts, into the camp to better serve all the faiths of the men in his charge. So far, the makeshift chapel, a mud-spattered tent with a plywood pulpit, had remained empty. But now the men prayed.

Cal Spencer reached down the front of his black fleece jacket and drew up the medallion hanging there on its silver chain, St. Michael, the patron saint of paratroopers. He recited his favorite verse from the Bible, "A greater love hath no man than this, that he lay down his life for his friends," and dropped the medallion back down his jacket front. When the chaplain was done, Spencer said to all of the guys, "All right then, brother," and with his M-4 rifle in one hand, he reached up with the other to the helo's side and climbed the rear ramp, where inside it was dark and smelled of old canvas, rubber hoses, and aviation fuel; he could feel the thrum of the helo's rotors spinning overhead, coming at him through the thin skin of the airframe in cold, metallic waves.

Spencer stood up in the cramped quarters. The interior was dimly lit by several bare lightbulbs in their own cages of wire. Overhead ran a maze of silver hydraulic lines and red and green electrical wires stretching the length of the ship, with the interior measuring about thirty feet by almost eight feet across. The interior walls were padded with gray quilting for insulation. Web seats, six on a side, lined each wall. The insulation did nothing to deaden the thump of the rising engines. The helo rattled and hummed under Spencer's feet.

He started moving around the gear stacked in the middle of the metal deck—thick plastic crates, filled with ammo, extra food, radios, winter clothing, each weighing 120 pounds. These were stacked chest-high, provisions to get them started on the ground until Spencer could coordinate resupply drops in Afghanistan.

He and the team would be kept fed, clothed, and armed by matériel depots located in Germany, where the gear and food would be packed by Air Force personnel in bundles about the size of VW Beetles, and then flown to Incerlik, Turkey, where they would be unloaded and wheeled onto MC-130 Combat Talon turbo props. From Turkey they would be flown another seven hours over mountains into Afghanistan and kicked out the door at 20,000 feet over drop zones often no bigger than a football field. Twenty thousand feet seemed awfully high to Spencer to kick anything out of an airplane with the hope it would hit the bull's-eye, but he filed this worry among many he had at the moment. He was most worried about surviving the helicopter flight into Afghanistan.

Intel was scant on what to expect. The pace of the U.S. bombing, which had begun October 7, had proceeded so quickly that the 160th Nightstalker pilots hadn't received any briefing from the CIA on the effectiveness of the campaign. Spencer knew the Taliban had in their possession Stinger missiles, which locked onto the heat trail thrown by an aircraft. One of the team's missions, once they were on the ground, was to look for these missiles and buy them back from friendly Afghans or from easily bribed warlords looking to make a buck.

The war effort was so new that the CIA had little else to tell Spencer and the pilots. Military planners at the Pentagon believed that Spencer and his team would spend the bitter winter training the ragtag Northern Alliance soldiers for a spring offensive, which wouldn't commence until seven months after they had landed.

Mulholland had little faith that more than two towns could be captured before winter's thaw. Time and weather were on the Taliban's side. If the Taliban could kill enough soldiers through the winter, the Americans would lose heart in the effort. Spencer would be locked in a siege scenario, eating boiled goat and keeping his head down around a dung fire, battered by winter blizzards. The operating idea was that the entire war would take a year and a half before Kabul fell.

Spencer did not look forward to spending the winter in such austere conditions, but he knew he probably would have to. The team's planning tent had sat across the muddy road from the CIA's own well-kept hovel, but even if the CIA officers possessed a treasure of information, they didn't have the authority to pass any of it on to Spencer. In essence, he and the Special Forces team were about to fly into Afghanistan blind. Medic Scott Black thought it was like the guys in Normandy having to hit the beach without any idea of the German gun emplacements in the surrounding hills.

Spencer reasoned that if they were going to be shot at by Stinger missiles, there was nothing he or any of the guys could do about that. Worrying wouldn't stop a rocket from coming up from the ground and hunting the aircraft like a dog chasing a rabbit.

Soon he was sweating under layers of polypropylene underwear as he grabbed more gear from the rest of the guys as they walked up the ramp. Nobody was saying a word. Just the shuffle and grunt of men moving in synchronized grimness.

On top of the pile, packed in their own duffel, were six bottles of vodka for General Dostum. Back in Special Forces training ten years earlier, Spencer and the men had learned the value of ingratiating oneself with the warlord at hand. Also stacked up were about a dozen white nylon bags of oats, apparently for the general's animals. *Horses?* Spencer thought. *I'll be damned. Horses.*

He looked at the bags of feed and gave them no more thought.

Master Sergeant Pat Essex, Spencer's teammate, fully expected to die when he landed, the victim of gunshot or mortar attack. He'd taken part in dangerous infils before. On a training mission, he'd watched as Nightstalker pilots cut their own landing zone using the rotors of the helicopter as giant hedge clippers. They'd been landing in a pine forest and he marveled as the helo dropped into the hole of its own making—pine boughs, pinecones, pine bark flying all over the cabin. Essex now accepted the fact of his upcoming death with a breathtaking resignation that seemed to imply that he would, in fact, survive the helicopter landing. He had become like one of those Kung Fu warriors on late night TV, men who can't be killed because they are already dead.

Even more so, the very thought of being shot at when he landed pissed Essex off. If he was dead, he'd never get the chance to become a park ranger for the National Park Service, one of his life-long ambitions, beyond soldiering. His attitude to the business of war was simple. "We got food on the table, a house to live in, clothes to wear," he liked to say, explaining the meaning of his military career to himself and his family.

Beyond that, he didn't know what his career meant. He said history would have to take care of that. He knew one thing: he would not tell his own kids to find a life in the Army. He wanted them to be able to spend more time at home, and to see the world from more places than behind a gun.

This devil-may-care attitude hid his tremendous drive to achieve competence. Essex had a near-photographic mind, able to absorb details about a warlord's past battles, family life, feuds, triumphs, curlicues of personality. With the jut of his squared chin, straight blond hair, and gold wire-rimmed spectacles, Essex generally glowed with a can-do attitude without coming off as a know-it-all. He didn't consider himself an authoritarian, but he got people to work for him and he got people to believe in him. He was part news reporter, part diplomat, part angel of death. He could look at a topographic map of a chunk of ground and pretty quickly intuit where the strongpoints were, where the ambushes were likely to occur. He didn't have to look at a map twice.

In fact, that the team had this mission, their raison d'être for sitting in this bird at this very moment of liftoff, had largely to do with Essex's stubbornness. He had heard that another team at K2 might be getting this very job, and that Essex's team was in danger once again, as it had been back at Fort Campbell, of being broken up and every one of them—Nelson, Spencer, Diller, all of them—being turned into liaison officers, desk jockeys, stapler apes.

Essex and the team had been put to work with shovels and picks digging ditches, erecting tents, and hauling gravel for the base's roads. Shortly after they arrived at K2, their combat search and rescue mission was canceled. One night, the worst sandstorm in fifty years blew up and pelted the tents with biblical gusts. When it wasn't blowing bitter dust, it was raining. Ankle-deep water coursed over the tents' plywood floors, which bowed under the sagging weight of the men and their gear. Computers, printers, radios, all had to be stashed up on tables to keep them from shorting out in the flood. For a while, it had looked like they wouldn't get to do anything but build roads and tents while U.S. bombers, flying at 15,000 feet, threw shrapnel around the country and failed to hit Taliban troops.

When Essex heard that the first team had blown an opportunity to prove their pre-battle worthiness during a briefing session with Colonel Mulholland, all by complaining about the mission's uncertainty and danger, Essex had walked into the Old Man's office practically strutting. He had been building tents all day, pounding metal rods into the ground, and untangling miles of heavy rope. He was sick of that crap, for sure.

The first team chosen for the job had complained to Mulholland that "the communications plan for this mission stinks." This was true: there weren't enough radios to go around for an entire team. Communications were going to be dicey—at times they might be nonexistent. The implication was that if you got into trouble, you might not be able to call anybody to help get you out. "Well, how can we go in-country with a commo plan like that?" the other team had whined. Mulholland had quickly shown them the tent door.

When Essex walked in, he said, "Communications are a problem, sir. We'll make it work."

Essex knew that the other team had also been asked, "Have you ever called in close air support with a B-52?" Essex had thought to himself, *Well, shit. Nobody has ever called in close air support with a B-52, not even the U.S. Air Force, because it's a strategic bomber.*

Essex answered Mulholland, "Well, sure, I can call close air with a B-52. Have I ever done it? No. But could I do it? Yeah."

The colonel appreciated this attitude. The air campaign had so far been ineffective and various U.S. press reports were already questioning the prospect of its success.

With each passing day that American planes bombed but did not kill any Taliban—not enough of them, at least—the locals grew angry and dismissive of the Americans, who had been cowed in places like Vietnam and Somalia when dead and wounded U.S. soldiers began coming home.

The town of Karshi-Khanabad sat about two miles from the camp, on the other side of a thirty-foot-high berm meant to keep the place secure from enemy attack. And there were enemy around, terrorist cells of the IMU—the Islamic Militant Union—operating within the country in hopes of returning the former Soviet republic to a fundamentalist state. The Americans were forbidden to go beyond the berm, or to fraternize with any of the local workers and suppliers who streamed in and out of the place in cargo vans. Fear of mortar attack loomed in the camp.

In a matter of a few weeks, the air control tower had been swept and rewired, a hospital was built, and food, clothing, ammunition were cached in newly built warehouses, and the communications tents quickly sprouted bouquets of satellite antennas. (In less than a month, sixty C-17 transport planes would offload several thousand tons of supplies.) The place, in the words of one Army historian, "looked like a gold-rush boom town."

The entire cantonment was approximately a half mile long and a quarter mile wide. It took five minutes to walk down a dirt service road to reach the Chinook helicopters on the taxiway. Next to them sat the Black Hawks, the Chinooks' security escorts, bristling with rockets and door guns, and looking in the bronze light of morning and evening like an ancient army of insects rearing on ash-black legs.

Colonel Mulholland's Joint Operations Center, or JOC (pronounced "Jock"), consisted of twenty tents made of heavy green vinyl, each about the size of a two-car garage, and joined end to end, making a long, fluorescent-lit tunnel. The soldiers in camp nicknamed the place "the Snake." It was filled with the latest space-age gear—satellite communications, computers, display screens, miles of wire, and hundreds of blinking lights. All of this was powered by tall banks of diesel-run generators that roared day and night throughout the camp.

The JOC sat next to an aircraft taxiway and the concrete bunkers used as living quarters by the 160th SOAR pilots, even grimmer accommodations indeed, to Spencer's mind, than his own team's.

The concrete Quonsets had been used by the Soviets as aircraft hangars during their 1979 invasion of Afghanistan, and they had sloping, bell-curved roofs, which the men, pilots and soldiers alike, used as a makeshift gym. By loading their rucksacks with sixty pounds of rock and strapping them on, they could hike up and down the slope of the roofs until their leg muscles screamed for rest. They made dumbbells out of five-gallon plastic buckets of water balanced on each end of a broomstick, and pulled off countless reps in between mission planning sessions. When they weren't lifting the buckets, they were using them once a week to shower with. The buckets were painted black and by setting them in the sun, the water inside would heat ever so slightly. They fashioned a hose to run off the bucket that delivered a thin, tepid stream down on their filthy heads. The common latrine was a dark, deep hole in the middle of a barren field, over which you squatted *en plein air*, unadorned by screening trees or even a scrap-lumber wall.

When Nightstalker mission commander John Garfield had landed at K2 and stepped off the airplane, the first thing he saw was a soldier squatting in the field and casually reading a magazine. *I'm in the Stone Age now,* he thought. *Just one click away from clubbing your own meat for dinner.*

The preflight briefing, which had taken place earlier that morning, had been about as freewheeling as Essex had ever seen. Mulholland got right to the point. Their mission: link up with the warlord General Rashid Dostum at a village called Dehi. Three CIA officers had arrived there the week before to prepare the way by familiarizing themselves with the warlord's capabilities and intel network. If they survived this, they were to capture the city of Mazar-i-Sharif, about sixty miles north of Dehi. How they accomplished any of this was up to them. Trust no one, warned Mulholland. Not Dostum—not any Afghan. *No one in this country is clean,* he said.

Nelson had next stood up in the cramped tent and launched into their mission plan. Gaunt, bow-legged, with short blond hair, Nelson spoke in a high, reedy voice and looked very much like a windswept son of the American High Plains, which he was. "Sir," he began, "we are a-goin' to win."

He had drawn arrows with a red Sharpie pen on a wide roll of paper fed through an easel, marking the team's sure, solid advance. The arrows bulldozed through miles of sand, scree, rock, pomegranate, blue chicory and sweet acacia, pine tree and poplar, over riverbeds, up cliffsides, across forlorn, twilit plateaus, and kept pushing for the horizon, where at night the stars would come up out of the singed cauldron of the autumn night and wheel overhead and smear the sky with ancient phosphorescence.

The briefing took about five minutes. When it was over, Mulholland looked at all of them—he was deeply moved by their earnestness, their belief that they could win. He really didn't know what would happen to them. He really didn't know if they would be killed as soon as they landed.

This mission, this war, had the air of a lark, a deadly one—it all sounded so simple on paper as Nelson explained it, but they all knew the first rule: that the plans for war were the first things to be thrown out once the war started. From that moment, you lived minute by minute.

"You guys got the job," said Mulholland. "Good luck."

It had been a long day for the hard-as-nails, forty-six-year-old Mulholland. Towering at six foot five, with hands as large as oven mitts, he was not a man easily cowed. But Mulholland had been on the phone for most of the morning getting his ass chewed by Secretary of Defense Donald Rumsfeld. "Why in the hell aren't there American soldiers on the ground in Afghanistan?" the SecDef had asked.

The weather, was all Mulholland could say, deeply exasperated.

And this was true. The weather *was* spooky, something right out of a child's dark fairy tale. For the past three nights, Mulholland had been trying to put Essex, Spencer, Nelson, and the rest of the team into Dostum's camp, about 250 miles to the south of K2, but the helicopters had been blown back by freak sand- and snowstorms that hadn't shown up in any of the U.S. Army's weather forecasts.

The Nightstalker pilots ferrying the guys over the mountains were tearing their hair. On the first night's attempt, they had returned to base after nine hours of flying in total whiteout, unable to discern any difference between the ground and sky. They made up a name for this: *flying into the Ping-Pong ball*. Afterward, the 160th Nightstalker pilots walked back from the flight line to the dank, bat-infested Quonset bunker they called home, completely white-faced. They sat on their cots and stared at the floor, like terrified mimes.

"You couldn't have shoved a hot buttered pin up my ass," said one of them, the otherwise surly and battle-hardened Greg Gibson, who just weeks earlier had been driving his jeep to work at Nightstalker headquarters when he heard the news of the attacks in New York. Back then, he thought he was the man for the job. A veteran of the world's worst wars for the past thirty years, he had been ready to kick some Al Qaeda ass. Now he was finding this flying over Afghanistan's mountains the scariest flying he'd ever done.

And that was the problem: the mountains. They soared from what seemed the loneliest-looking valley floor on the face of the earth, a place that more resembled the bottom of a drained ocean at the end of time, straight up to 16,000, 18,000, 20,000 feet, draped in frozen garlands of snow.

Back home, the highest Gibson and the guys usually flew in the Chinook, the Special Operations workhorse of a helicopter, was around 3,000 feet, in Colorado on training missions. On the occasion they hit 10,000 feet, they felt like they'd flown to the top of the world. In Afghanistan, low altitude *started* at 10,000 feet. Gibson and his men were literally flying their helicopters into no-man's-sky. Nobody, and he meant nobody, in the history of U.S. Army aviation, had ever flown a helicopter this high, this far.

And they were trying to do it at night. But that wasn't the part that bothered them. Flying "blacked out," as they called it, with no lights on inside the aircraft or on the exterior, and using state-of-the-art avionics, was done for security's sake. These pilots feared daylight flights the way vampires fear a noon sun. *Hard to see, harder to hit. The Nightstalkers own the night!* The problem was that they couldn't get the goddamn weather forecast right.

On the second night, they returned to base equally demoralized and freaked out by their inability to punch through the cloud layer (*It just wouldn't quit. . . . There was no cloud layer. . . . It was all a white fuzz!*), and Gibson, along with his mission commander, John

Garfield, practically grabbed the poor weatherman by the ear and dragged him to the Nightstalkers' bunker.

"All right," said Garfield, a normally affable fellow, a former Delta soldier with over three hundred combat jumps to his name. "Tell us what we're finding up there at ten thousand feet. If we run into it again, we're taking you up with us!"

The weatherman hurried back to his table of charts, graphs, and computer models and started rethinking the problem. And the problem was that he was relying too much on technology. When he broke the satellite photos out of their video display and examined each image individually, he discovered something that shocked him. Hidden except in this static view was a mass of . . . sand, snow, and God knew what else. Amazingly, the blob measured hundreds of miles across, and materialized not from the ground up but in midair, at around 10,000 feet. And it was completely undetectable except under the naked eye, after thorough and anxious study. He gave this freakish weather formation a new name, "the Black Stratus."

Now, on the third night, as the Nightstalkers prepared for liftoff, they were ready for the Black Stratus. Just knowing what this thing was comforted them.

Essex settled into one of the Chinook's web seats in the back, on the right. There was just enough room for his legs if he sat with them bent to his chest, hiking boots touching the gear in the middle. Beside him was his rucksack, his new home away from home, for what he figured would be a year's trip into the void. Fighting and shooting. Killing strangers and making new friends. His tan shirt pockets were stuffed with the fussy lagniappe of a neatnik: one of the Garmin GPSs that the guys back at Fort Campbell had FedExed to him at the last minute; five unsharpened pencils; a writing notebook that could double as a journal; reference cards called "TA [tactical air] cards" with directions on how to call in close air support, calculate distances, recon ambush sites, and fire mortars; dog tags; a couple of hundred bucks in American currency; and something called a "blood chit," which was a kind of Get-Out-of-Jail-Free card to be presented to any Afghans he met should he be trapped behind enemy lines and need assistance. Essex drolly reflected that there would be little time in his journey during which he would *not* be behind enemy lines, seeing that he was a guerrilla fighter and all. The blood chit offered a cash reward to anyone who gave assistance to the American soldier in distress.

In his rucksack he carried ammo for his long rifle, the M-4 carbine; clothes; and enough food for five days—one meal-ready-to-eat (MRE) a day—plus purification tablets for drinking water. His mesh, many-pocketed load-bearing vest resembled in cut and style a fly fisherman's; only Essex's was filled with grenades, more ammo for his 9mm pistol, compass, water bottle, an emergency MRE, and batteries for an interteam, handheld radio. On top of his rucksack, in its own duffel, sat the gift of vodka for his warlord-to-be, General Dostum.

Because he fully expected to be killed in the first five minutes after he hit the ground, he kept busy alongside Spencer, helping the rest of the team load its gear. *There's nothing I can do now*, he reasoned. *I'm just a chunk of cargo. If something goes bad, I'll just have to deal with it then.*

He watched as the guys marched up the ramp and slung their heavy rucksacks on the pile in the center of the deck.

Captain Nelson was squeezed in his seat up front, on the left side, behind lead pilot Alex McGee. Nelson was already wired into the ship's headphones, pulled down over his ears atop his black cap, nervously listening to the cross-talk and chatter between McGee and his copilot, Jim Zeeland, going through their preflight checks.

Sitting next to him, imperious and wry, was Sam Diller, the intel sergeant on the team, and at forty, the oldest. Earlier in the day, he had been sitting in his tent drinking coffee when a staff officer poked his head inside and said, "Hey, it's going down tonight. The weather is fair, and you're going." They'd loaded their gear onto a truck and rode the quarter mile out to the helo, plopped it down on the ground, and waited. The aircrew had come out about the same time as they did. Diller was wedged in his seat next to his buddy, the jovial, bearded senior medic Bill Bennett. He and Bennett had been on the team in 1991 to help oust Saddam from Kuwait, and Bennett had rejoined the crew several days after September 11. Back home, Bennett—a quiet, handsome, unassuming thirty-three-year-old with fifteen years of service—had a wife and young son with whom he liked to kayak and hike on the weekends in the Tennessee foothills. But all that was now a million miles away.

Like most everybody on the team, Bennett spoke Arabic and had been trained in the art of sniper fire, mortar launch, and high-altitude parachute jumps from 25,000 feet, during which he would breathe from a minitank of oxygen strapped to his waist. He and Diller were keeping an eye on an eager group of younger gunfighters, average age thirty-two, each with about eight years in SF. "This is your first rodeo," Diller had told junior weapons specialist Sean Coffers, who had joined the unit from the 101st Airborne. "You're sticking with me." Nearly everybody was married and had children; there were more than a few broken marriages in the rearview mirror. Sitting next to Bennett was Vern Michaels, the good-humored senior communications sergeant, and junior engineer Patrick Remington. These six guys were the Alpha cell—a split team—of the A-team, the twelve-man detachment.

On the other side of the aircraft, the right side, Chief Warrant Officer Cal Spencer was running herd on the six-man Bravo cell. Spencer was sitting near the ramp, next to Pat Essex, his coleader. Crammed in beside them were junior engineer Charles Jones and weapons specialist Ben Milo, a burly, affable thirty-year-old weapons sergeant from the suburbs of Chicago.

Milo was usually the motormouth of the team, the one guy they always had to keep away from the press (not that the press ever came around, but still they worried). But now Milo was sitting there in his tiny web seat not saying anything, and Spencer was trying to think of something to say, some joke to crack, to perk him up. Milo was a devout Catholic, an amateur artist (he loved drawing 1970s vintage record album covers, like those of Pink Floyd and The Grateful Dead), whose beefy appearance disguised the restless high-speed spirit of an overachiever. Several weeks earlier, the announcement that he would be deploying on this mission had gone off like a bomb in his family and caused Milo's wife, Karla, to start crying. The mother of four, including two teenagers, and herself a nursing student at night school, Karla had not been prepared for Ben's departure. They went to their youngest son's grade school to pick him up and spend time with him before his father left. Ben was angry. He wanted to kill those involved with his bare hands. He couldn't understand why they had hurt civilians on American soil. If they'd hit a military target, he felt he could've grasped the reasoning. A few days before leaving, Milo was driving with a friend, another weapons sergeant on the team, and finally lost his temper. *They piss me off, and they pissed my wife off!* he started yelling. He ranted so long about the jihadis and what he was going to do to them that his teammate had to pull over and stop laughing before driving on.

"Oh, I'm serious now," Ben was muttering. "Those motherfuckers are gonna feel my pain, 'cause I'm one pissed person. Oh, I'm very serious now!" But sitting in the helo, he was wondering what it would mean, what it would *feel* like, to kill a man. What would his priest back home think? He figured he hadn't the slightest idea. But he did know that he'd just have to keep pulling the trigger when he got in a fight. Milo crossed himself and prepared for liftoff.

Next to Milo was Scott Black, who as a junior medic was more worried about saving lives than ending them. The intel he'd gotten on General Dostum was that the warlord was overweight, a heavy drinker (hence the housewarming gift of vodka), that he had diabetes and didn't have use of his right arm; that he tired easy and his eyesight was failing; that the ruthless codger was a footstep from the grave. This man, this raconteur, this bon vivant, a former plumber and peasant's son, would be Black's new best friend, and clearly he was a walking medical emergency. It was Black's thankless job to make sure the weakened, forty-seven-year-old Uzbeki didn't die on his watch. How embarrassing would that be? Black figured Dostum's own militia would probably try to kill him if Dostum died under his care. Black was ready to hit the ground and begin CPR on the guy immediately.

Sitting next to Black was the team's comedian, a tall, gangly soul named Fred Falls, a junior communications sergeant. Like Spencer, Falls delighted in dry, corny humor. The previous night the two of them had sat in their tent watching the 1985 Chevy Chase/Dan Aykroyd movie *Spies Like Us* (using a laptop propped open on a table made from sawhorses), a screwball comedy about two hapless Americans who get lost in Afghanistan and are captured by *mujahideen* fighters. At one point, Chevy Chase and Dan Aykroyd are strung up by their ankles, ready to be tortured, and they try cajoling their captors into letting them go. Spencer and Falls thought the movie was hysterical, while the rest of the team just rolled their eyes.

In reality, everyone had already decided that they would not be taken alive, if a gun battle came to that. They'd sat on their cots and written what they called their "death letters"—last missives home to wives and family about last thoughts. One Special Forces soldier had poured his heart out. He truly expected not to come home at all. "If you are reading this letter," he wrote to his family, "things are not well for me. And I [had] so many things I wanted to do with you both. I love you and think of you as often as possible. You made me the happiest man in the world." He had told his fellow soldiers, "Look, we're in this together. And we need to know that coming back isn't really an option for us. If we get killed in the process, we get killed. I don't want [us] to shy away from what we have to do."

After writing their letters, the men removed wedding rings and emptied wallets of any possibly incriminating photos of family and friends (images and information that could be used against them in a torture session), and dropped these tokens of identity in large manila envelopes provided for the occasion. These were sealed and handed for safekeeping to the chaplain.

On Nelson's team, the Bravo cell's job in Afghanistan would be supporting the Alpha cell by running the "log train" (short for logistics), which meant making sure that everybody had enough "beans, bullets, and blankets" to fight another day. This job was not as glamorous as being on the Alpha cell, but the distinction went unspoken here. As Spencer reminded everyone, there were no "small" jobs in the business of war—just dead egomaniacs.

Tonight, they were going "in single ship." This meant that the entire group was flying into Afghanistan on one helicopter. Usually, half the twelve-man team flew on one bird, while the other half followed behind on a second. This way, if either one of the helicopters was shot down, the entire group would not be wiped out. Within Special Forces doctrine, the team was to resemble an amoeba, dividing and thriving in even the most austere environments.

This was why the group had two of everything: two medics (Bennett and Black); two commo sergeants (Michaels and Falls), whose job it was to run the radios and communications within the team and with headquarters back at K2; and two weapons specialists (Coffers and Milo), tasked with the mind-boggling challenge of memorizing weapons and ammo used around the world and training the teams in their use. Engineers Pat Essex and Charles Jones kept the teams supplied, organized, and operating swiftly like a mini-corporation whose business was handcrafted violence. Sam Diller, as the intel sergeant, was point man for information flowing among the CIA, the Pentagon, or headquarters in K2, and the two teams.

The leanest level Nelson had taken the group to in training was three guys to a cell. This meant that from an organism of twelve guys, he could split the team into four pieces. "We train like the terrorists do," Nelson was always reminding everybody. "We fight like terrorists." Sam Diller believed that "each SF trooper ought to be able to shoot all the bad guys, bomb them, know how to use the radio to report what happened, and do the necessary medical care afterward."

"You know," Diller liked to say, "the whole man concept."

In his shirt pocket, alongside his toothbrush, Diller carried a dog-eared edition of the poet and warrior Sun Tzu's favorite aphorisms. Diller had memorized many of them and quoted at will in his West Virginia drawl: "When on offense, strike like lightnin' from the clouds. All war is deception." Diller further liked to paraphrase Steve McQueen in *The Magnificent Seven* when describing the team's esprit de corps: "Mister, we deal in lead."

Diller was traveling so light that he had even cut down his pencils to three inches each, to shave off ounces. He carried a pair of heavy-duty mittens bought at Walmart in a back pants pocket. He was lean, broad-shouldered, with a face like carved oak.

He had come a long way from the battered house trailer on the West Virginia hillside that he'd shared with Lisa, his wife, before joining the Army in 1986. Diller had played football in college and gotten a degree to become a history teacher. He loved teaching and he loved working with kids. He was a study in contrasts, with his high-holler voice and quick mind. He spoke deliberately, slow and methodical as lug nuts dropping in an oil pan. He was a man of few outside interests, except gun collecting. He'd spent at least half his life sleeping outdoors as a soldier, yet he hated camping. He kept his own counsel. During Desert Storm, his sergeant had asked him to maintain the team's diary, a daily record of events. At the end of a month, Diller had written one terse sentence: "We trained, it was hot, and I got sick." His sergeant told him to quit keeping the diary.

But Diller was crazy like a fox. His father had been the editor of a West Virginia daily newspaper, and an aficionado of musical theater, an erudite man who wore bow ties and loved his rough son unconditionally. Diller was headed for a career in academia when one night, shortly after getting his degree, his life took an unexpected turn.

He was at a bar and he got into a fight, a bad one, which is to say the other guy, the poor, unlucky sonofabitch, didn't win. Even as the cops were hauling him away, Diller realized his teaching career was finished before it began. He pled to a misdemeanor, and was sentenced to three years' probation. "Schools don't hire teachers who tend to wrap other guys around their fist," he realized.

He and Lisa moved into the trailer up on the hillside crowded around by kudzu, chickens, and car parts, and Sam got the only job he could find, mixing cement for a construction company twelve hours a day. It was backbreaking, god-awful work, but there was a part of Diller that felt he deserved this as penance. He had a brother who was in the Army, and one weekend the man came to visit Diller while he was on leave. His brother told him that he was actually getting paid to be on vacation. Diller couldn't believe it. He'd never thought of joining the Army. Within a matter of weeks, he'd enlisted. He and Lisa left the house-trailer and never looked back.

He went to work in the First Cavalry Division as a scout, a man trained to sneak into enemy lines as an observer. And then another unexpected thing happened: Diller got bored. The Army had so many rules and regulations, it seemed no one could find their own ass with both hands unless somebody told them how. He grew miserable. He brooded. He spoke even less, which heretofore hadn't seemed possible. Lisa could barely stand to be around him. One day he met a Special Forces soldier who told him, "Man, you'll dig what I'm doing, it's everything you want." Diller said, "Tell me more."

He was a voracious reader and he already knew enough about the history of warfare that he figured he might like being a guerrilla fighter. He liked the odds of being outnumbered, outgunned. He liked the idea of fighting for his life with his back against a wall. He liked being able to think for himself. Within a matter of weeks, he had volunteered for Special Forces selection. When his two years of training were over, when he was handed his Green Beret on graduation and could proudly call himself a Special Forces soldier, he discovered that he felt more likable in SF. He had been sure Lisa was going to divorce him. Now he wasn't so sure.

Diller chalked up this life change to something he called "the Big Boy Rules." The Rules were hard to define, harder to embody, even harder to follow. But they went like this: When you followed the Rules, you agreed to do what you said you were going to do, no questions asked, no excuses given. You agreed to pay for your own mistakes. When you lived by the Rules, though, every man was working for himself; and every man was devoted to the brother on his left and right. The Rules were righteous, they were real, and they seemed to resemble how the universe really worked. Diller grasped them immediately.

Take the way he helped train the younger guys on the team. Diller always led the way. Every dawn he pushed them through a four-mile run, followed by enough sit-ups and push-ups to make an Olympian god throw up, and then after that came thirty minutes of hand-to-hand combat training, based on the street-fighting techniques of the famed Gracie brothers, the Brazilian brawlers who'd gained notoriety in the 1990s for their bloody *Ultimate Fighter* matches on pay per view cable TV.

The Special Forces soldiers called this mode of fighting "grappling." It was meant to disable and do great bodily harm with an economy of movement—eye-gouging, groin-stomping, arm-breaking, and choking were the order of the day. In training, you could "tap out" when the pain got too great, or when you felt yourself slipping into unconsciousness as your buddy gripped your windpipe. Grappling was what you did when you were out of bullets at last and you'd even thrown the gun itself at your attacker, before diving in and tearing him to bits. It was mean, vicious work, and it made you into a walking human-demolition team. Diller excelled.

He believed that the everyday SF soldier could survive circumstances that would fry the circuitry of normal men. He was treated like a pro and he trained like one, too. And he did this away from the limelight of reporters and politicians. He got paid decent money (about $4,000 a month for a fifteen-year veteran) to "take on the worst of the bad guys" and to "train the lamest guys on the planet" to be soldiers. He was proud of saying: "We ain't got 'hero' stamped on us, none of us does."

Going in single ship tonight was risky. But because Colonel Mulholland had been on the phone all day with Rumsfeld, he felt it was a risk he had to take. Simultaneously, he was launching another team in the east of the country, and this had left him with only one helicopter with which to deliver Nelson's team to the United States' primary initiative in the south. Diller oddly saw this bold chance-taking by Mulholland as evidence of the Old Man's chutzpah. If they were shot down, Diller figured he could fight his way out of the jam, living in the weeds, surviving. But Mulholland? He'd be tarred and feathered. His career possibly might end.

The Chinook rocked and shifted on its six balloon tires. The blunt nose dipped, and the entire length of the craft swung in the air, and rose.

At the helicopter landing zone in the Darya Suf River Valley, 250 miles to the south, Ali Sarwar awaited the arrival of Diller and his team.

He didn't know these men, but he was hanging his future on them. He closed his eyes and strained to hear the approaching clatter of the huge American machine. Nothing. Not yet. Patience . . .

For most of his life, all he had known was war. He had fought back with Massoud, whom he knew as the greatest warrior of the twentieth century; and with Dostum, the opportunist. Ali Sarwar was now a lieutenant in his tribe's army, the Hazaras, under General Mohaqeq's command. Fair-skinned and green-eyed, his head wrapped in a green scarf, Ali stood at the landing zone with a dozen other men, all eager as he to attack the Taliban. He was portly, dressed in billowing linen pants and heavy oiled boots. On his left shoulder he carried a leather bullet box he had taken from a Chechen he'd killed a week earlier in the fierce fighting at Safid Kotah. He was tired, hungry, and nearly out of bullets for his battered AK-47 after the month-long battle. He lit a cigarette and exhaled slowly at the sky. His father had been a shopkeeper and provided well for his family. Ali was a good student in school. But as a teenager, his house had been under siege—first by the Soviets, and then by factions of Afghan soldiers fighting for control of the city. Walking to school, the mornings were often traced by the crack of interlocking lines of gunfire. All of the men and boys in his neighborhood eventually picked up guns to protect their homes and families. Ali was shot in the foot and hit by bomb shrapnel in the leg, causing him to limp in cold weather.

Always the Taliban had more bullets, more jet fighters, more men. He shifted nervously in the gray talc of the landing zone and hoped those odds were finally changing.

He looked over at the secretive CIA men, Baba Spann, Baba Olson, and the elder, gray-bearded Baba J.J., whose grizzled beard was a sign of wisdom in Ali's culture. He wondered why the other Americans didn't have full hair on their own faces. He figured it was not their way. He didn't care. He was not a fanatic. He liked to smoke, he liked to drink. He only had one wife, even though the Koran told him he could have four. But what kind of man could have time and patience for four wives? No, one woman was enough for Ali Sarwar.

Ali and his men had swept the area, but it was impossible to say if they were safe in this stony silence. Who knew if Taliban were lurking in the rocks and cliffs surrounding them? The moon hung overhead, a bleached horn driven into the flank of the night. Ali could hear the gentle pattering of the horses' hooves in the damp mud of the evening and the burble of men talking quietly in Dari. Everyone was straining to hear the approach of the American machine. One of the men, Baba Spann, his tight, angular face smudged by a sketchy shadow of beard, walked out to the edge of the landing pad, about as big as an American baseball field (Ali had seen pictures on TV, the Yankees of New York playing under bright lights), and the wiry American set something small—about the size of a deck of cards—on the dirt.

Ali waited for the object to do something, but he saw nothing. It just sat there. He figured it was a gadget whose use he would understand in the future, when the future came. He watched as Baba Spann flipped down heavy rubber glasses over his eyes and stared at the gadget. He seemed to nod, pleased, and then he flipped the glasses back. Ali knew the glasses helped men see at night, which seemed another fantastic thing he did not understand about the Americans. He did understand that the men arriving by helicopter would help him win this war against the Taliban. He hated the Taliban with every drop of his blood.

When the Taliban captured Mazar-i-Sharif in 1998, Ali had sent his family—his wife, two sons, and three daughters—to the central highlands, to Bamian, the homeland of the Hazaras, for safekeeping. (Ali was horrified and enraged when in March of 2001 the Taliban dynamited the stone Buddhas that had stood watch over the town for centuries. What man had the right to write the future by blowing up the past?) With his family in Bamian, Ali remained behind in Mazar-i-Sharif to fight.

The city was littered with bodies. The Taliban cut the heads off prisoners and set them on pikes along the streets. Ali was a wanted man. The Taliban announced that if anyone could capture Ali Sarwar, they would set free one Northern Alliance prisoner. Someone—some poor, unlucky man, thought Ali—was snatched up by the villagers and turned in as the criminal Ali Sarwar. The Taliban released a prisoner and then tortured the man they supposed to be Ali. They cut off his head and put it in his lap, and set him by the side of the road, so he could stare at the 10,000 ankles of men, women, and children sleepwalking through the filth and nightmare of their lives.

That was how Ali Sarwar came to be standing at the landing zone waiting for the American bird to roar in.

At liftoff, Captain Mitch Nelson was thinking, *You could fight your heart out and still lose.* That sunk in for Mitch. That really got to him. He steeled himself against the coming blast of cold air as the helo corkscrewed up into the night sky—*one thousand feet, two thousand, four thousand, five.* It was the most amazing thing, the most amazing feeling. Flying into battle was all Mitch Nelson, from Kansas, had ever dreamed about back on the open range amid the winter wheat, sorghum, and vetch. He had ridden bulls in college on the amateur rodeo circuit, he had stared down bodily harm, but this bucking helicopter ride (even higher now—*six thousand, seven thousand, eight thousand*) was a new thing. It completed Mitch. Made him new.

The stench of aviation fuel in the cockpit was overpowering, like burnt caramel. Facing the pilots, he was lined up on the left side, as the long rotors overhead went to work against the frozen air—*wham thrum, wham thrum, wham*. Nearby sat Ben Milo, who was now stone-cold quiet, brooding. And there was Spencer, with his long shafts of pine-straight legs folded up beneath him, Spencer in his black polyester jacket with the collar shot up and his black watch cap pulled down over his ears, the very picture of cool. Nelson knew that Spencer, Essex, and Diller were keeping an eye on him. He knew he had to walk sharp.

He was the captain of the team, it was true, but these three, pushing forty and beyond, were damn near senior citizens. Freaking old. Yet they walked tall. Nelson was going to have to hit one out of the park as the team's leader. He was confident he could do this. Nelson believed there should be two kinds of soldiers: the kind you saw on television news and the kind you didn't see. Nelson was the man you didn't see. The Taliban's goal was to lay waste to them, make them suffer, make them gut it out through a long, painful winter, and then attack them in the spring. Time and shitty weather were on the Taliban's side. And the only thing Mitch Nelson had was the backing of the entire U.S. Army. And that was a scary thing, to have all that American might at your back . . . yet if you died, if you got killed, captured, tortured, no one would ever know you had been here except the people back in Kansas standing around the rodeo ring in the twilight, saying between their teeth, "Remember that Nelson kid? Huh, what happened to him? Got killed, I guess." And so on, until no one remembered that you had been living aboard this very helicopter at this very minute, with this hard wind freezing like enamel at the back of your throat, with the rear ramp down and the snow blowing in, and the side doors open and the wind clawing at the helo's quilted walls, as you sat there shivering and leaning forward on your M-4 rifle while the metal barrel, about as slim as a pool cue, turned to ice in your gloved hands.

Nelson looked around. These twelve guys on this helicopter comprised the entire American fighting force striking back at Osama bin Laden. It was just them, he realized. Them alone. It was enough to make you want to crawl under the poncho liner until the crosshairs of history had passed you by. Nelson strained forward to see out the helicopter's windscreen.

Piercing reefs of black rock loomed out of the night and passed below the broad, froglike belly of the craft. And then as if shooting over shoal water into the deep, the earth dropped away and there was nothing but darkness—a chasm so black it snapped into focus under Nelson as the essence of eternity. It was one of the most exciting things that had ever happened to him.

Flying east toward Tajikistan, they were passing over a mountain the pilots had told him they'd nicknamed "the Bear." As they rose— *nine thousand, ten thousand, eleven thousand*—the temperature plummeted.

The ship's heaters weren't cranking, for security reasons. Back at K2, mission commander John Garfield had decided that if the pilots ran the heaters, this would change the "thermal signature" of the aircraft. This meant that if they were attacked by a heat-seeking missile, the warhead would lock onto the cockpit and kill the pilots, instead of going after the two turbine engines, which were mounted up and aft of the cockpit.

Either way, Garfield pointed out, life would suck, but at least if they survived the initial strike, there might be the chance that someone still alive could crash-land the aircraft.

The doors were open so that the pilots and soldiers could exit in just such an emergency without fumbling with latches and handles. The rear ramp hung down in the night, a ribbed metal tongue measuring about fifteen feet long. As they climbed, Spencer watched frost spider over the exposed metal in the craft, across the deck and up the walls.

He threw his poncho liner over his head, trying to keep warm. He thought, *I'm the old guy here, the young ones are looking to me—am I doing anything weird, anything that would break their confidence?* He threw the poncho off. Shredded bits of snow and micro grit swirled in the air. The cabin howled. The cold air hit him immediately. He figured it was time to—*Jesus Mother Crisco Oil it's COLD in this helicopter!*—he figured it was time to crack a joke—*It is SO freaking cold in this goddamned helicopter!*—and try to lighten everybody's mood, especially Milo's. Spencer called out, "Hey, Milo?"

"Yeah, sir?"

"It's chilly in here!"

"What's that, sir!"

"I said, it's pretty goddamn chilly!"

Milo looked at him like he was crazy.

"Fuckin'-A, sir! You got that right!"

And then Milo closed his eyes and tried to go to sleep.

Ice grew on the barrels of the machine guns mounted in the doors. The Army bastards standing in the doors manning the guns (Spencer didn't know their names) stood still as ceramic gnomes, bundled in layers of parkas, snowpants, gloves, and heavy boots, their helmeted heads bobbing back and forth as they scanned the far, far distant ground for muzzle flashes, missile fires. . . . They knew that a Stinger missile lifted off the ground sparkly and twirling; rockets flew up like the lit end of a glowing cigarette in somebody's hand as they lifted it. . . . It was a helluva ride.

One of those guys manning the guns was Flight Engineer Carson Millhouse, a ten-year career man from Southern California. With his straight black hair and old-timey, gold-rimmed spectacles, Millhouse looked like someone you'd see on the back of a Lynyrd Skynyrd album. The pilots had nicknamed him "Hippy." Growing up in a family of eight kids, he'd spent his youth "despising the military." And then it dawned on him that the easiest way out of the house was to join the Army. Hippy loved the men he worked with and he lived in fear of being promoted out of the unit. If he went up in rank to master sergeant (he was a sergeant first class), he'd find that there were fewer slots available and he'd have to leave and enter the regular Army, where his new pay grade would make life more comfortable, but where the daytime flying and regular routine would bore him to death. This was a terrible thought.

Now, after about an hour of flying, Hippy could barely move his arms. His gloved hands, clenched on the gun handles, felt like they'd hardened in that position. As he looked down at the countryside, he wondered how in hell anybody ever managed to live in this country. Just rows of mud houses and mud fences. Not one damn light burning on the face of the earth. Midnight in Armageddon. It sure sucked being here in Afghanistan, he thought, but at least he didn't have to get off the helo like the soldiers in back and fight in this terrain. Hippy didn't know any of their names, but he guessed he'd probably bumped into some of them on Friday nights with their families back home, going into a Chili's or the Outback Steakhouse on Exit 4 off the interstate. He wished them well and hunkered down in his parka, scanning the ground for gunfire.

The other soldiers on Spencer's team were looking at him nervously, and for a second time he made a show of fluffing up his poncho liner and getting comfy atop the frozen nylon ribs of his seat and going to sleep. Spencer figured it'd take them about two or more hours to fly the several hundred miles to Dehi. Essex was already asleep—he could sleep through anything. Diller, too, was passed out, snoring. Spencer figured racking out for shut-eye was the only sensible thing to do. He lay out in the seat listening as the rotors overhead bit into ever-thinning air and the aircraft strained upward with a terrible clatter.

From his seat up front, Nelson looked down and saw why the pilots had named the mountain the Bear.

Below him was an eye, what looked like a deep, cold lake, winking at him. The broad, stony snout was lifted west, into the wind. The high, steep forehead was sloping down from the north . . . Nelson said it again, *The Bear. Amen.*

Up in the cockpit, lead pilot Alex McGee was straining at the controls as they entered, or he feared they were entering—the hypoxic zone. Going hypoxic was a possibility when you entered airspace above 9,000 feet and your brain started screaming for oxygen. The effect was like shotgunning Champagne and hitting yourself in the head with a ball-peen hammer. Only the results could be a lot more dangerous.

The Nightstalker pilots had been going hypoxic regularly on early missions ferrying supplies and CIA officers to General Dostum's camp. It had become a routine part of flying. The term in use was that lately the missions had become awfully "sporty." As in: "Sporty means you come back to base and you get off the ramp and you kiss the ground for a while. You just take off your helmet, and stare. Whew. That's sporty."

The pilots were a hale and hearty lot—instead of growing glum thinking that they would be killed on these missions, they became world-class goofs. One of the crew chiefs, a thirty-six-year-old fellow named Will Ferguson, had brought his golf clubs along, a sand wedge, an eight-iron, and a three-iron, and a gym bag of balls. One night, bored out of his mind after having watched *Joe Dirt* one too many times back at the hangar—the unit's favorite after-hours flick—Will had gone out to the berm lining the camp and stood in the moonlight and surveyed the wasteland. Piles of toxic dirt oozing chemical waste. Pine trees gently swaying in the breeze.

Will set the ball on the berm, ready to send it into the gloom on the other side, where supposedly Islamic militants were lurking, sniper rifles in hand. *To hell with them,* thought Will.

He fired.

The ball soared, a white orb sinking in the dirty pond of the night sky. *Well, I came, I fought, I golfed,* and he slid the club back down in the bag and returned to the Quonset.

The pilots' quirky sense of humor was second only to their regard for the danger of their missions, especially when it came to going hypoxic. The problem was with the oxygen system on board the Chinook. It was in need of repairs, but the war had spun up so quickly that there hadn't been time to make them.

The crew had landed at K2 on October 6 with the Chinooks and Black Hawk helicopters broken down neatly inside the dimly lit belly of the C-17 transport planes, and they'd had just forty-eight hours to reassemble them. One Army observer described this chaotic, anxious task as resembling "ants attacking a Twinkie." No one had slept on the journey overseas, and when they finished tightening the last bolt and nut, they literally passed out on the oil-stained concrete next to the newly born behemoths. By then, they'd not slept in a week. Garfield walked into the hangar and barked, "Wake Up! We have a mission to do!" Within eight hours, they were ferrying supplies and CIA officers into scattered camps around the country, in preparation for the arrival of Nelson and his team.

The on-board breathing system was made up of a series of oxygen bottles, feeding a series of tubes leading to the kind of black rubber breathing masks you see in jet fighter pilot movies. As the oxygen thinned, you placed the rubber mask over your mouth and took a hit.

Garfield discovered the break in the breathing system when he looked over during one flight and saw his copilot acting goofy, making strange faces and pointing at the helo's windscreen at imaginary shapes in the air. Then the guy took the helo's stick and tried steering, threatening to crash them all. Garfield looked in back and saw that the rest of the crew were acting similarly strangely, and one by one they started passing out, murdered shapes lying on the deck of the helicopter. Garfield tapped the lead pilot on the shoulder and told him that he was shutting off the air to all of the masks except his, and that he, Garfield, expected to pass out any minute. The pilot would be flying alone.

While he waited to be knocked out, Garfield found a disc-shaped gadget in his pocket he called a "Whiz Wheel"—it worked like a slide rule—which the pilots used to make flight computations. He handed the Whiz Wheel to the goofy copilot and told him that the gadget was actually the controls to the aircraft, and that if he wanted to fly, he would have to use the Whiz Wheel.

The guy sat in his seat whapping away at the gadget, making strange, childlike noises. Every once in a while, he would come to, sit bolt upright, throw the Wheel at Garfield, and lunge for the helo's stick. Garfield would have to slap his hands away and hand him back the Whiz Wheel. This went on until the guy finally passed out and Garfield went unconscious himself. The lead pilot successfully completed the mission alone, sucking all the available air through the only workable oxygen mask on board. It was a terrifying experience. As they descended below 9,000 feet, the crew awoke as if summoned by a hypnotist's hand clap. They were groggy for a few hours and their heads rang with the most excruciating headaches. They didn't do this just once—the discovery of the breathing problem was only the beginning. There was no easy, quick fix of the air leak for reasons having to do with the complicated layout of the aircraft. They had to live with it. They went hypoxic on every long-range, high-altitude cruise. Every flight had started to feel like an execution.

As Alex McGee sat at the controls, enclosed in the small booth of the cockpit, it was absolutely dark, except for the faint, sealike glow of the computer screens in front of him, an arm's length away. He was wearing a heavy helmet, Kevlar vest, gray gloves, and night vision goggles, the optic light of which leaked around his eyes, dusting his cheeks an electric lime. Every few minutes he reached up with gloved fingers and methodically paddled over gardenlike rows of switches.

Other than that—the *tick tick tick* of McGee turning switches—there was no sound in the cockpit (if you discounted the turbine's choppy roar). The environment inside was what the pilots called "sterile." Completely quiet. No talking allowed.

The physical strain of flying was constant. McGee's equipment, in total, including the insulated, fire-retardant garment called a Mustang suit, weighed sixty pounds. The pressure of just sitting in the chair and flying the ship was enormous. It was like having a bag of cement sitting at the back of your skull, atop your brain stem. Even folding yourself into the chair was a gymnastic matter. This meant that if you had to piss, you did so in your chair. There was no way to fly the machine and urinate at the same time. The ship's maintenance custodians were always complaining that the pilots were pissing all over the chairs. Jim Zeeland, the copilot, had life a little better. He could urinate into a screw-top liter soda bottle. But the problem was that at high altitude the bottle froze. Now you had a rock-hard bottle of urine rolling around the deck, crashing into people. So, as a rule, they threw these out the open doors of the helicopter. They called them "piss bombs."

Mounted in the console, about three feet from McGee's face, sat a six-inch video screen. Nested deep in the helicopter lay something called the multimode radar. When you flew using the MMR, which McGee was now doing, it was usually because you were flying at twenty feet above the face of the earth at 160 mph, skimming the sand, which was also very sporty flying. Tonight, McGee was flying MMR to make sure that nothing unexpected rose up to meet them in the night. They were nearly 12,000 feet above sea level.

As the radar sniffed the air, it sent signals back to the ship that were expressed on the MMR's video screen as two tiny white triangles, one inverted atop the other.

The position of the triangles on the screen told McGee whether or not the helo could fly over the next rock spire, the next ridge, the next mountaintop. That was always the question: at this speed, at this altitude, at this level of pitch in the rotors, do we have the ability to go higher if a hunk of rock suddenly steps out of the night?

When the triangles touched and formed an hourglass shape on the screen, they were said to be "satisfied." If they were satisfied, you knew that you automatically had enough power, lift, and speed to make it over the next hill. The key to not crashing was in keeping the cues satisfied.

McGee was accomplishing this by nudging the rudder stick between his legs in micromovements, or by pushing at the silver pedals beneath his feet, which helped steer the aircraft, or by reaching down with his gloved right hand and pulling more power in the engine by raising, maybe only one-quarter of an inch, the lever located on the floor in the muted darkness of the cockpit. Every once in a while he would also have to reach over and finger a black dial to adjust the pitch of the rotors. At the same time, Zeeland, the copilot, was looking at something called the E2 Page, which was another video screen that relayed what the terrain looked like ten miles in front of the helicopter.

The mood was complicated by the fact that, in the words of John Garfield, the ground beneath them resembled "flying over dinner plates set in a drainer." At one moment, you had 15,000 feet of nothing beneath the ship; in the next, the helo was suddenly 100 feet above the flinty edge of a mountaintop. And because so very few pilots had ever flown a helicopter this high, it appeared that no one had considered the inherent problem in the multi-mode radar system. Namely, that it shut down whenever the helo rose above 5,000 feet.

The device shut down under the assumption that above 5,000 feet, the sailing would be clear. The engineers had never imagined this flight in Afghanistan over this frozen dish drainer. So, as you flew over one ridge and another, with the radar system shutting on and off, you got this annoying red blinking BAD DATA signal on the console. And after the system shut down, it took several minutes to reboot. In those moments you were flying truly blind, brother, which was very sporty. The pilots found this more unsettling than getting shot at it in combat. In combat, at least you could shoot back.

They'd flown about thirty minutes when McGee aimed the frozen nose of the helo at a notch up the mountaintop, at 12,000 feet. Along each side of the notch, the mountain face soared another 5,000, 6,000 feet into the night sky. Threading through this piece of rock was the only way McGee was going to get over. To go any higher would be to trespass even more deeply in the hypoxic zone, the temperature plummeting further, well below zero. Luckily, tonight, if they made it through this notch, they wouldn't be loitering in the danger zone—maybe twenty minutes, tops, before dropping altitude.

McGee could look up in the rearview mirror mounted in the overhead console and barely make out the shapes of the soldiers in the back, black lumps of clothing, lying motionless. Zeeland told him that the soldiers in back were sleeping. That was good news.

McGee reached forward and hit a switch, and aviation fuel started dumping from the outboard tanks, which were fastened on the craft, one on each side, like two long, metallic hot dogs. As the fuel drained and the load lightened, the whine of the rotors sweetened in pitch. If they were going to make it over the mountain, they had to drop fuel. At 12,000 feet, every ounce on board counted. It was conceivable you could fly up a snow-draped valley in total darkness and reach a point where you had neither the power in the engine nor the lift left in the blades to go farther. You had essentially flown to the dead end of a physics equation. If you were lucky, you had just enough room to hover the craft and turn exactly on its axis and retreat down the invisible stair-step in the air to K2.

The helicopter, as long as a freight car, entered the slot in the mountain and soared through and punched out the back side. It began sliding down the ragged, frozen carapace of the world, toward the Afghan border.

Ten miles out from the border, after about two hours of flying (they'd thus far flown a circuitous route of about 220 miles), lead pilot McGee got ready to do an inflight refuel. He was running low on gas after dumping the tanks on the way up the mountain. McGee had radioed for a fuel tanker, an MC-130, to come on station after they had cleared the Bear. Now the plane came roaring in.

McGee was clattering ahead fast at 5,000 feet when the plane raced up from behind and passed overhead. It cast a shadow—a looming presence—even in the dark. Essex woke up immediately, as if somebody were staring at him. He looked out the window and could see the cold, dimpled side of the plane, its wings silently bouncing out there in the dark as it went by. Essex felt an oily mist on his face, unburned fuel, as the plane's exhaust swirled through the cabin and swabbed the walls. He was already light-headed from the fumes.

The plane slowed and, like a boat, settled down into the air just off the helo's nose. McGee was at the controls, still flying totally blacked out. Because of the optics of his night vision goggles, he had poor depth of field. The MC-130 looked like a silhouette of a plane pasted against the sky. McGee could look up and see his rotors chopping at the air close behind the plane. He looked down at his speed—he was traveling at 160 mph. One error and they'd crash. McGee was trailed by two Black Hawk helicopters, his security escorts, and he heard an increasing nervousness in the pilots' voices. He had to get the refueling done before they ran into the sand- and snowstorm called the Black Stratus.

For reasons the weatherman at K2 still couldn't explain, the Stratus had been loitering around the border, above the sandflats that ran from the Amu Darya River (which also marked the Afghan/ Uzbekistan border) for thirty miles all the way to Mazar-i-Sharif. This was a bone-strewn no-man's-land, sand and more sand. Through it traveled the occasional tribe of nomads marching in frayed capes and drawing with their passing a dogged calligraphy in the dust. Trailing them were legions of exhausted livestock, the wall-eyed goats, the ornery, terrified mules, the animals' brass bells tolling a sullen note that turned the light of noon into ancient dusk. This was the ground they flew over, not through time, but backward in time. Essex looked down and saw the ground passing beneath him in blurred panes.

The smaller, faster Black Hawks following them did not have the sophisticated radar—the MMR—that the Chinook possessed. This meant that ever since leaving K2, the Black Hawks had been navigating through the mountains by following the glow of the Chinook's engines ahead of them. The engines blew two rings of fire out of their exhausts, side by side—like two flaming wagon wheels—and wherever the rings went, up and down, left and right, the Black Hawks followed. It was boggling to Essex that this was how the pilots had to navigate.

Looking out the windscreen, he could see a long black hose coming out of the left wing of the tanker plane—seeming to grow, actually, like a black tendril. It stiffened out straight in the breeze. At its end was a gray, rubber blossom, about as big as a bicycle tire, called the drogue. It was meant to keep the hose from flapping up and down in the slipstream. It now hung open and empty, waiting for the helo's metal fuel probe, which was about forty feet long and protruded from the front of the craft.

Essex felt the helo drift beneath him and lurch as McGee nudged the stick and the helo jumped and the probe and the orifice of the hose mated with a gentle sucking noise that was lost in the roar.

Fuel began pouring down the hose, through the probe, and into the helo's tanks bolted on its sides. When the helo was full, McGee disengaged the probe and the drogue fell away and was reeled back to the plane. Now it was time to get the hell out of the sky—a fast-moving target flying close to the ground was harder to hit than one flying high. Al Mack pointed the helicopter at the desert floor in a dive.

They were nearly vertical as they plummeted. Nelson grabbed onto his seat with the increasing whine in the engines. They kept dropping. Nelson's stomach jumped in his throat. He could smell the warmer air with its vaguely mineral scent and the ice on the guns started to melt and drip. The wind threw the puddles in silver skeins across the deck. Nelson looked down and they were 300 feet off the ground. It was like looking out a car window at the pavement while racing down the interstate. They were flying like this when they punched into the Black Stratus. It had unexpectedly moved to a new location, lower in altitude.

It swallowed the helo whole. "Can't see, can't see!" radioed the Black Hawk pilots behind them. "We may have to return to base, repeat, may RTB!"

McGee radioed back that they should remain with him if they could. "Can you see my cones?" he asked, meaning could they see the rings of exhaust atop the bird.

"Roger, but, repeat, it is getting very difficult."

McGee switched over to a maneuver called "terrain following," which meant that by using the cues on his screen he was flying close along the grain of the earth, rising and falling over dune, rockpile, and ridge. He looked out and saw the fuel probe starting to spark in the storm. There was so much grit in the air—so much particulate friction—that it began to glow. Soon the probe lit up like a rotisserie rod heated by a torch. McGee could see the tips of the rotors sparking too, making two dazzling gold halos atop the craft. The whole ship was literally glowing in the dark. The Black Hawk pilots radioed that now it was pretty goddamn hard to see and finally in exasperation they broke in: "Silver Team," McGee's call sign, "we are RTB," and they peeled off and turned back through the Stratus and disappeared. McGee, Nelson, Essex, Spencer, Diller, and the rest of the men were alone.

There was no way to tell up from down, left from right. Your inner ear would communicate with your brain that you were turning, but your eyes saw nothing, except more whiteness out front—the whiteness literally flowed at you over the windscreen. You had the sense you were moving and not moving at all. It was like being weightless in a globe filled with odorless smoke.

McGee finally broke out of the clouds at 100 feet. He roared past rock pillars and over cold ground beaten hard as a drumhead by ten thousand years of scorching afternoons. The helicopter was zipping just feet above the earth, lifting a rope of dust that snapped taut in its wake and then dropped to the desert floor. McGee came around a bend in the mountains, straightened the craft, and raced for the landing zone.

Mike Spann heard the clatter of the approaching craft. He studied the infrared strobe placed on the edge of the landing zone. Through his night vision goggles he could see the brilliant pulse of its light; with the naked eye, the tiny, rectangle-shaped beacon was invisible. The helicopter pilot would see the strobe and know that the zone was safe to land on.

At thirty-two, Mike Spann was an intense young man at the pinnacle of his life's goal: deployment in a war. He wanted the Taliban's collective head on a stick. Along with Dave Olson and J.J., he was a member of the CIA's Special Activities Division (SAD), which was a covert unit within its National Clandestine Service. The CIA called men like Mike, Dave, and J.J. "paramilitary officers." They were the heirs of the former Special Operations Group (SOG), an unconventional warfare element of the Military Assistance Command Vietnam (MACV). Mike's work was a national secret, strictly classified. Americans back home did not even know his job existed. His neighbors in his suburban neighborhood in Manassas, Virginia, would occasionally see a banner above Mike's front door reading, WELCOME HOME, DADDY! and think he'd been away on a regular business trip. Because they operated covertly, SAD officers dressed like civilians, preferring jeans and flannel shirts and tennis shoes or hiking boots. They carried non-American weapons, Russian-made AK-47s and Browning 9mm automatic pistols, and toted briefcase-sized satellite phones, a GPS, and a compass. They filed hundreds of intelligence reports from the field, typing with their laptops propped open on their knees in godforsaken caves, cafés, or hotels around the world. Mike was trained to kill or capture terrorists based on information gleaned by the Agency. If he was captured or found dead, nothing about his dress or person would immediately lead back to his being part of a United States force, and certainly not to the CIA.

For his trouble, Mike was paid about $50,000 a year. Back at Langley, there were seventy-eight stars on the CIA's memorial wall at its headquarters—half of these were in honor of paramilitary officers killed in the line of duty.

The CIA's paramilitary officers had been involved in the overthrow of Iran in 1953, placing the Shah on the throne. In 1954, a CIA coup overthrew the government of Guatemala. The 1,400 Cuban exiles who took part in the 1961 Bay of Pigs invasion of Cuba had been trained by the CIA. In 1981, SOG had given support to the Nicaraguan contras fighting the Sandinista government. Since the 1970s, they'd worked in Lebanon, Iran, Syria, Libya, Latin America, the Balkans, and Somalia. They were capable of good or evil. They went where the politicians aimed them.

After well-publicized congressional hearings in 1975, held largely in response to human rights violations by SOG officers against Vietnamese citizens, President Gerald Ford created new oversight of these kinds of CIA operations, and the reach of its militarized spies was put on a short leash. The CIA began focusing more on a white-collar brand of espionage, relying mainly on case officers, posing as diplomats, aid workers, and government factotums, to gather communications intercepts, called "signal intelligence."

Their job would be to convince citizens of foreign countries to spy for the United States. The heyday of the covert gun-slinging spy was over.

That is, until 9/11. CIA director George Tenet had actually begun beefing up the paramilitary division in 1997, as its officers tracked and hunted (to no avail) the Bosnian Serb leader and war criminal Radovan Karadzic. Mike was part of a group of secret warriors who now numbered several hundred men, made up of former Navy SEALs, Army Rangers, Marines, and Army Special Forces soldiers. (By 2002, the CIA's counterterrorism unit would grow substantially, to approximately 900 officers. "We are doing things I never believed we would do—and I mean killing people," one intelligence officer would remark.)

Mike had a new mission in a new CIA.

Since the age of sixteen, all he'd ever wanted to do was jump out of airplanes and chase bad guys (he earned his pilot's license when he was seventeen). Growing up, there wasn't much to do in Winfield, Alabama, on a Saturday night, except hang out at BJ's, a video-game parlor on Main Street, playing Donkey Kong, or you could drive out to the edge of town and swat baseballs at the batting cage, the field lights swarming with moths while bats swooped through the humid air. The towns on the nearby horizon had names like Pull Tight, Rock City, Yampertown, and Gu-Win. Three generations of Mike's family had lived in Winfield (Mike's grandfather had worked in the nearby cotton mills), and about the biggest local attraction was Winfield's annual Mule Days Festival, featuring a cake auction, turkey shoot, and mule-judging contest ("all shapes and sizes of mules are welcome"). One day Mike was hanging out with the football team watching *Top Gun* and announced to everybody in the room, "I'm going to be doing that someday."

He graduated from Winfield High School in 1987, studied law enforcement and earned a criminal justice degree at Auburn. He then joined the Marines. His fellow classmates found him highly disciplined, a man apart. "I don't know if I ever saw him drink," said one. "I always thought he was raised by a preacher." He was part nerd, part jock (he was a ferocious reader of encyclopedias and had been a star running back his senior year in high school), and he hoped he'd get to see battle. After eight years in the Marines and reaching the rank of captain, he was still disappointed. He was thirty years old, married to a hometown girl named Katherine Webb, his high school sweetheart, and the father of two daughters. He felt he was coming up on the short string in his life. He decided he wanted to be part of the CIA.

On his application essay, Mike poured his heart out for several thousand words about the reasons he should be let into this secret society of warriors. "I describe myself as an ordinary person," he wrote, "with a few God-given talents and ample self-confidence. I am a dreamer with lofty goals. I am an action person who feels personally responsible for making any changes in this world that are in my power. Because if I don't, no one else will."

Upon acceptance, he moved Katherine and the girls to Virginia and started training at Camp Peary, aka "the Farm." It was the summer of 1999.

The CIA's premier training facility, the Farm, is a wooded compound outside Williamsburg, Virginia, 9,000 sprawling acres ringed by electrified barbed wire. The training was grueling, eighteen months of work in demolition, shooting, driving, and street-fighting techniques. He was learning to be a case officer, whose job it would be to convince citizens of foreign countries to spy for the United States. After graduation, Mike would be accepted for service in the Special Activities Division. But within six months after starting the training, just as his career was taking off, Mike's home life began to unravel.

Now, as he waited for the helicopter to arrive at the landing zone, he knew, like any obedient officer, that he had to put such distracting thoughts out of his mind.

Alex McGee looked up from the controls and saw dead ahead the helicopter landing zone (HLZ), a bare patch of earth showing up in the night vision goggles as a gray circle of dirt. Thirty minutes had passed since they'd broken out of the fog and vertigo of the Black Stratus, and they were skimming just fifty feet above the river, banking along the serpentine course with valley walls rising abruptly on either side. The *chop-chop-chop* of the helo's double rotors, bounding off the rock, made a terrific racket inside the cabin. Awakened by the noise, Nelson roused himself from a freezing nap and sat up.

Up ahead, he could see through the night vision goggles that the HLZ was about as big as a baseball field. Every several seconds, at its far edge, a bright light went off, a stabbing ray in the pitch dark. This was the infrared beacon that Mike Spann had set out earlier in the dirt.

McGee aimed for the light.

"One minute out," he announced.

Nelson, who had been listening in on the headphones, turned and relayed this down the line, holding up one finger. And then something caught his eye: the oats and corn in the opened tops of the horse feed bags, positioned in the middle of the deck, seemed to be moving. Nelson yelled over to Spencer. "Hey, take a look at this!" Spencer reached in and scooped up a handful.

The oats and corn were crawling with worms. "That's great!" yelled Spencer above the helo's roar. "Just what every warlord needs! Fucking worms!"

They looked at each other, laughed, and hoped that the reportedly violent Afghan general had a sense of humor.

They stood and shouldered their rucks, each load weighing over 100 pounds. They pulled their black watch caps down over their ears and checked and rechecked their weapons. And then they stood stonily in the dark as the airframe bucked beneath them.

"Thirty seconds," said the pilot.

Up ahead, McGee could see that the HLZ was hemmed on three sides by tall cliffs.

There was one approach—straight on. They were going in.

Out of sight, shivering, famished, and eagerly awaiting the arrival of the helicopter, General Mohaqeq and about a dozen Northern Alliance soldiers were hiding behind a small berm at the landing zone.

Some of the men among them had been unable to sleep, so anxious were they for the Americans to arrive. Hiding near General Mohaqeq were the Americans he called Baba David, Baba Mike, and Baba J.J.

Baba J.J., with his gray beard and fierce gaze, was the CIA's team leader. With the Americans' help, Mohaqeq had been eagerly preparing the safe house for the soldiers' arrival.

The Afghans worried that Taliban scouts could be positioned in the surrounding hills, ready to call in tank and artillery fire. It was impossible to tell. In fact, they wouldn't know until the helicopter approached the landing zone and started to land.

Whatever the case, it would be very bad indeed should the Americans be killed as they landed. America would forget Afghanistan and its fight against the Taliban. They would leave and never come back.

Mohaqeq couldn't bear the thought of their leaving after such a long wait for their arrival.

As far back as 1979, when the Soviet Union invaded Afghanistan, bent on owning the mercantile gateway between Europe and the East, Afghans like Mohaqeq had been forced to face the prospect of their own annihilation.

The Soviet invaders wanted to crush Mohaqeq's Hazaras, Massoud's Tajiks, Dostum's Uzbeks, Hekmatyar's Pashtuns (Hekmatyar being another warlord who had variously fought with and against Dostum). The Soviets had wanted to own the country. But the Afghan, it turned out, would not be beaten. He could live in the hills and strike in lightning raids. He was a spook, a ghost who could step into a beam of sunlight and out of the field of fire. The Soviets attacked with an army of a half-million men and lost 50,000 soldiers in fighting that lasted ten years. A million Afghan citizens were dead, and 5 million more had fled in exile to Iran, Pakistan, and Russia. When the fighting was over, the Hazaras, the Tajiks, the Uzbeks, and the Pashtuns started attacking each other.

Massoud's Tajiks rocketed Kabul alongside Hekmatyar's Pashtuns and killed thousands of men, women, and children. Dostum's Uzbeks reportedly raped, tortured, and robbed as they battled. The Hazaras, considered of the lowest standing, fought with them all. Allegiances swirled like smoke. By 1994, the central fact of daily life was primordial terror. The bubbling fountains and fecund gardens of Kandahar, the ancient almond groves of Kabul, the air-conditioned movie palaces on Chicken Street, the late-night restaurants sparked by the laughter of boisterous travelers heading back to Paris at dawn—all this lay in ruins, wind whistling through the ragged filligree of a million bullet holes in a thousand different empty rooms across the land.

Bandits roamed the bomb-cratered highways, hijackers, child

molesters, psychopaths, violent men for whom, after seventeen years of war, the idea of country, state, home had been cauterized from the very center of their being. In its place, the dead emptiness of outer space, quiet as winter. A truck carrying carpets from Kandahar would leave for Pakistan on the national road and within hours every carpet would be stolen, a makeshift "toll booth" sprouting each few miles in the blazing, cracked asphalt. At each checkpoint, grinning armed men demanded goods, money, and sometimes more, as on one day in 1994, when a roving band of militia kidnapped two thirteen-year-old girls and raped them. They left them to die by the side of the road. It was from this chaos, this skyline of smashed buildings and broken bones, electrified by gunfire and dagger slash, that the Taliban had stepped into the international limelight. They had managed to still the entropic universe. *Order, peace, silence. Inshallah—God willing.*

When the residents of a nearby village, outside Kandahar, heard of the girls' rape, they resolved to fight back. They wanted someone to address this injustice. They wanted the highways open again so men could make a living.

They approached a reclusive one-eyed mullah in Kandahar named Mohammed Omar, who had banded together his own fledgling army of former anti-Soviet fighters. These men were Pashtuns, the putative rulers of Afghanistan for the past several hundred years—stern men who, after the Soviets' defeat, dreamed of making Afghanistan a purely Islamic state guided by the laws of the Koran. For them, Mohaqeq, Dostum, Massoud, and Atta Mohammed Noor were secular wretches, infidels deserving either conversion or death. Dostum especially galled them with his large infantry and a reported wagon train of whores who traveled from town to town servicing the militia.

Omar dispatched thirty of his fighters, and they tracked down the bandits who had raped the girls on the highway. When they caught the two men, justice was swift: they were hung from the barrel of an old Soviet tank. Their bodies were later cut down and left as food for marauding dogs.

Over that summer, Omar's stature grew, as did the perceived benevolence of the Taliban. Omar announced that he had posted 1,000 of his men on the Pakistan road, and that they were keeping peace. The workaday Afghan citizens, exhausted and terrified after five years of civil war, couldn't have been happier to be under the watchful eye of the Taliban. They welcomed them with open arms. By the following spring, Omar's army had grown to 25,000 men.

The fighters had poured in from madrassahs in Pakistan, a form of Islamic schooling that dated back to the Middle Ages. In stifling one-room schools, teenage boys and young men whose fathers had been blown up, maimed, and killed by the Soviets, parentless children further galvanized by seemingly endless civil war, were chained to crude desks as they memorized all 6,666 sentences of the true word of God, the Koran. After three years of this, the young men could become mullahs back in their villages, due now all the respect of a religious constable, adjudicating civil disputes and performing marriages and funerals. They were paid in cash or bartered their services for gifts in kind—cattle, sheep, food. They began calling themselves *Taliban*—"seekers of knowledge."

Theirs was the perfect world. Since taking control of Kabul in 1996, the Taliban had banned music, kite-flying, photography, movies, and even perfume. Husbands were ordered to paint their house windows black so no one could see the women inside, and the women themselves were forbidden to leave their homes unaccompanied by a male relative. Women were to be as pliant as cattle, silent as stone. As many as 100,000 girls were ordered not to attend school. Literacy rates among the total population dropped as low as 5 percent. Denied proper obstetrical care, one out of three mothers died in childbirth. The life expectancy for men dropped to forty-two years. Suicide rates among women soared as they were driven mad by privation.

For not wearing a burkha in public, they could be lashed with a rubber hose. For committing adultery, they could be stoned. The Taliban carted women jouncing like livestock in the backs of pickups into crowded soccer stadiums and, before the games began, made them kneel, gasping within the oven of their blue burkhas as heavy footsteps approached behind them.

The women knelt there, shaking, and then the man attached to the footsteps raised his rifle and fired point-blank at the clothed domes of quaking heads.

As a finale, they cut off the hands of thieves, tied the pale index fingers with a ribbon of scrap cloth, and hoisted the hands aloft, the way you would lift for inspection a pair of enormous candles still twinned by their wicks. The hands were pale; the stumps were red.

After this, the soccer games began. Those had been terrible times. Terrible. Mohaqeq remembered them all.

· Now, as he and his men huddled at the berm, they looked skyward, and then at last they heard it—the *thwock-thwock* of the helicopter.

Mohaqeq prayed the Americans would land safely.

McGee slowed the helo and brought it to a hover atop the landing zone. The rotors' blast scoured the ground and sent up boiling clouds of dust that engulfed the craft. Nelson could barely see his hand in front of his face. It was hard to breathe. Up in the cockpit, McGee couldn't see the ground. He had been told in a preflight brief that the HLZ was made of small stones and that visibility should be clear. He was twenty feet aboveground in a brownout and the radar was useless.

From their position on the ground, Mohaqeq and his men could barely see the helo through the dust storm.

On the helo's rear ramp, a machine gunner stood fanning the gun barrel back and forth, ready to fire at the first sight of any muzzle flashes coming from the edge of the landing zone. Two more gunners stood in the side doors, also ready to fire.

Mohaqeq watched as the helo hovered overhead for what seemed long seconds. Then, without warning, it rose straight in the sky as if drawn on a counterweight. And then it was gone.

The fierce Afghan fighter Ali Sarwar, one of Mohaqeq's commanders, also crouched at the edge of the berm, heard the ferocious machine roar up the river. His heart sank.

After flying about a quarter mile, McGee turned the helicopter for another approach. The first one had been a feint meant to draw enemy fire.

McGee's training had taught him that the worst thing that could happen was getting shot at on the ground while soldiers were running off carrying hundreds of pounds of gear. It was far better to be attacked while in a hover, with the engines powered up, when escape would be easiest.

As he hovered again over the HLZ, McGee listened as a door

gunner leaned out and talked him down over the intercom. The gunner could barely see through the dust. If the helo landed cock-eyed, if one or two of the four balloon tires touched down before the others, the craft could roll.

"Ten feet, five, here comes the ground," said the gunner.

And then the helo hit with a hard bounce.

"*Go!*" yelled Spencer. Sam Diller went first, kicking his duffel off ramp. Then came Milo, followed by the others. On their way out, they each grabbed something—sacks of grain, duffels, assorted bags of gear, and carted these off the craft.

As they left, the ramp gunner pulled each man close and yelled in his ear, "Good luck!"

Spencer and medic Scott Black searched the helicopter one last time for anything left behind, then they jumped off.

Spencer thought the drop would be farther and he landed with a *whumpf*, the ground coming up too quickly. It had only been two feet.

He stumbled and went down in a crouch, instinctively pointing his rifle at the dark ahead of him.

Helicopter crew member Will Ferguson stood on the ramp and watched as the soldiers formed a defensive perimeter, the twelve guys lining up in a crescent shape with Captain Nelson at the apex.

The helo lifted, and Ferguson stepped back from the ramp, and the men slid from his sight. He'd been on the ground less than a minute. The helo turned, heading back to K2.

On the ground, as the dust cleared, the silence was eerie. Spencer could hear his own breathing and smell the dry, cold night.

The enormity of where he was—in the desert, several hours before dawn, with eleven other guys upon whom his life now depended—hit him hard. Spencer imagined that at the end of his gun barrel, days and weeks ahead, lay his home. And now all he had to do was get up and start fighting.

And that's when he saw a strange shape . . . then two of them, now three . . . trundling toward him across the landing zone. And these shapes were now standing up and elongating into men, dressed in what looked like long cloaks with gun barrels sticking out of the flowing cloth.

They turned on flashlights and started making what sounded like burbling noises—gibberish to Spencer's ears. He and his team froze.

Spencer found the gibberish otherworldly. Then it dawned on him that it sounded like the language spoken by the "Sand People" in *Star Wars*.

Spencer was fluent in Arabic and some Russian. He knew that this must be Dari, a dialect of Persian, Afghanistan's native tongue. No one back at Fort Campbell had ever thrown a book of Dari verbs and nouns at the team and said, "Here, you might need these someday."

This lacuna reminded Spencer that he had prepared for this mission by acknowledging that, by and large, he was unprepared. If he was going to survive, it would be by his wits.

"I'm in the land of the Sand People," he joked.

And then he felt a tap on his shoulder.

He wheeled around, wide-eyed. Standing before him was one of the Northern Alliance soldiers. He couldn't believe he'd let the guy sneak up on him.

The Sand Man was looking down at him, pointing excitedly at the rucksack on Spencer's back.

I'll carry that for you, he seemed to be saying, and he reached and tried lifting the ruck. And Spencer had to say, "No, that's okay. I got it."

But the man wouldn't take no for an answer and he kept tugging until Spencer stood and said, forcefully, "I can carry it."

The guy shrugged and walked away.

Spencer looked around, relaxed a little. The rest of the team stood up. Back at K2, they'd learned that a CIA welcoming party would greet them. So where were they?

And then out of the dark stepped several larger figures, and as they got closer, Spencer recognized one of the guys as someone he'd met back at K2.

The guy held out his hand.

"I'm J.J.," he said. "Welcome to Afghanistan."

J.J. introduced Spencer and the rest to Dave Olson and Mike Spann.

"Damn glad to be here," said Spencer, relieved.

• • •

For their part, the CIA officers were equally glad to see American faces on Afghan soil.

Like his colleagues, Mike Spann was anxious to start fighting the Taliban. He was worried about whether or not he would survive the combat, but he'd made peace with the possibility that he might not. He was lately spending his rare free time at the Alamo by writing to his new wife of four months, in Virginia, missing her terribly.

Their marriage had capped a passionate romance, during which Spann's old life had come apart and a new one had replaced it.

A year and a half earlier, in April 2000, as Spann was undergoing his CIA training at the Farm, he and his first wife, Katherine, had separated. And several months later, at a Fourth of July picnic held for CIA employees, he fell in love with a fellow student.

Her name was Shannon, and she was someone who, like Mike, thought she'd get a fresh start in life by training to be a spy. A few months earlier she'd been dean of students at a Catholic university in California, riding out the shock waves of a failed marriage, alternately mystified and embarrassed by its collapse. Her marriage's end went against every grain of her born-again Christian sensibility. And then one day she picked up *The Economist* and saw an ad for employment in the CIA. *Do you have what it takes?* the ad asked. *Of course I have what it takes,* Shannon thought. Why not? And now here she was standing next to Mike at the barbecue. He was obviously nervous, but quite a treat on the eyes. His nickname at the Farm was "Silent Mike," and boy, she thought, he was living up to the name. It seemed to take him forever to even say hello.

And then, when he did open his mouth to ask her out, scared, she said no. She couldn't believe it. She felt like clamping her hand over her mouth and running away. What had she just done? She hadn't been on a date since her divorce, and that'd been three years ago! In fact, she was feeling lately that she'd be celibate forever. An old maid with a Browning 9mm pistol in her garter, the spy no one loved. *I blew it,* she thought. So when Mike asked her out again, she was relieved and said yes.

In December 2000, several things happened, some of them messy, the kind of loose ends Mike was not used to contending with. Shannon would later say that getting pregnant earlier that fall was not the normal way for a Christian woman to build a family, but she felt that they "did the best they could at the time." That month, she and Mike finished the training program, and his separation from Katherine was made legal. At the same time, Katherine discovered that she had cancer and refused to sign the divorce agreement.

Mike hired a lawyer and sought custody of the two girls. He showed up at the lawyer's office with a bruised face—he'd smacked it on a parachute strap during a training jump—and his lawyer, a man named Walter Von Klemper, could only shake his head. Some men seemed made to live at home, while others made the world their home. But how in hell was he going to convince the court that his rough-and-ready client should have custody of the kids? He sensed that Mike saw the world in absolutes, in black and white— but Mike was also the man you wanted at your back in a fight. He was the kind of father who drove his grade-school daughters to their bus stop, even though the stop was a half block down the street. He was a bit of a worrier, a bit of a softie. The court awarded him legal custody of the girls.

A few months later, in June, Shannon and Mike's son was born. A few days after that, they got married. Finally, the ship seemed to have righted itself amid the whirlwind of divorce and newfound careers.

And then 9/11 happened. A few days after the attack, Mike sat down at his computer at home in Virginia, trying to convey to his mother, Gail, and his father, Johnny Spann Sr., a prosperous realtor back in Winfield, what was on his mind. He couldn't tell them anything of what he was about to do or where he was about to go. He had already learned from CIA headquarters that a team of eight paramilitary officers was heading to Tashkent, Uzbekistan, to organize the U.S. military's entry into Afghanistan. Shannon used the computer to do the household bills, but Mike had customized the desktop screen to display the U.S. Marine Corps motto, *Semper Fi*—Always Faithful. He got right to the point: "What everyone needs to understand," he told his parents, "is that these people hate you. They hate you because you are an American.

"The U.S. lost the war in Vietnam because of lack of support at home. When you fight wars, people get killed. Americans should keep flying their flags, supporting their government, and writing their Congressmen. God Bless America."

On the day he left for Afghanistan, Mike posed with the kids for a picture in the front yard outside their house, a trim, white clapboard colonial with black shutters. Wearing jeans, a black shirt, and tennis shoes, Mike smiled, looking like a man about to clean and organize his garage rather than a covert operator going to war. He held his young son tightly in his arms, the baby gazing at some vanishing point off camera. His older daughter, nine, was dressed in shorts, a striped shirt, and white anklet socks and tennis shoes, while her sister, four, stood with her hands clasped behind her back, smiling. The past few weeks had been incredibly busy, filled with last-minute demands. Mike was worried that he could die on this mission and he'd recently filed a new will with the courthouse in Alexandria. He worried now how Shannon would carry on in the aftermath if the unspeakable happened.

"It seems I can never find the words to fully tell you how much I sincerely love you," he had written. "The simple fact is, there are no words worthy enough for you. You certainly are God's gift to a hardened and broken-down heart when I needed it most."

He confided to a fellow officer that he had a special request about Shannon. "I don't mean to sound dramatic," he explained, "I'm sure this will be a piece of cake. But if anything should happen to me, I want you to be the one to tell Shannon. I don't want her to hear it from someone she doesn't know."

Around the time that he stood now at the landing zone with Nelson and his team, Shannon, back in Virginia, was curled up in a chair, nursing the flu, pen and paper in hand, and writing to him in a journal. "I can't wait until we're all together," she began. "The house is quiet, and so I'm here speaking to my favorite person."

Mike had called her from Afghanistan and announced he himself was keeping a journal. There were things he was thinking about, he told her, that he wanted to be sure to remember when he got home. Shannon had decided to begin her own recollections: "I miss you so much," she went on, "especially in the evenings." She explained that one of the girls "had to do a little homework for school, a picture of a tree, and we were supposed to decorate it as a family tree, and so we made leaves and printed out pictures of all of our faces, and it made me so happy to see all of our family on one page. I can't wait until we're all together.

"Now I shall go to bed a lucky girl, and dwell on these things and the happiness of belonging to you, and in the hope of your safe return to me."

A few minutes after landing, Spencer and the team were walking down a narrow path. Leading the way were the Northern Alliance soldiers. They walked with a steady, assured gait in the dark, even under the burden of the heavy bags of horse feed and numerous extra bags of gear that had come off the helicopter.

Bringing up the rear, Captain Mitch Nelson could hear the vodka bottles clinking. He hoped none of them broke before he could present them to Dostum.

"Where's the General?" he asked a man walking beside him. Nelson spoke in Russian. Nelson guessed that some of the fighters had learned the language during the Soviet occupation.

The man told him that the general was not in camp. He said he was expected soon. In the morning.

Nelson let the subject drop, seeming satisfied. He was disappointed that Dostum had not been there to greet them. It was important to appear upbeat, determined, focused. Everything he did, the way he ate, the way he talked, the questions he asked, would be scrutinized by the locals. He looked at the dozen Afghan men accompanying him and he didn't trust one of them. But he knew his success depended on working with them.

He wondered how General Dostum would receive him. Back at K2, Colonel Mulholland had raised the specter that Dostum might even kidnap the entire team and hold them for ransom to the highest bidder. As team captain, Nelson had the job of getting inside Dostum's head and predicting what he would do even before Dostum himself knew. He had to convince the warlord that the Americans were here to fight and win. The truth was, without Dostum, they'd be annihilated by the larger Taliban army.

With him, they had a chance.

They had walked about 300 yards when Nelson saw a mud fort silhouetted against the sky. He halted. Somehow, he'd been expecting a longer night's march. He was sweaty and tired of lugging his ruck, but it occurred to him that maybe he and his men were walking straight into a trap, an ambush.

It was possible that J.J. had been taken hostage by the Afghans now accompanying them. They could walk inside, where they'd be attacked.

"What's the name of this place?" he asked J.J.

"The Alamo."

Not necessarily a reassuring name for a refuge.

J.J. explained that the fort would serve as the team's safe house. J.J. seemed at ease. Nothing seemed wrong about the situation.

The fort looked like something out of the American Old West. A crescent moon painted its walls dim silver. Each of the walls was made of smooth mud. They measured about 200 feet long and stood 8 feet high. Five wooden doors were spaced along the front. In the middle stood a wooden gate, wide enough for a team of oxen and a wagon—or a tank—to roll through. A hitching post for horses stood outside the gate. The individual wooden doors were short, maybe only five feet high.

The Northern Alliance soldiers stooped and filed inside first.

Nelson and the rest of the team stared at the small entrance, their guns slung across their chests, fingers poised to curl around the triggers.

Nelson went in next. Behind him, the tall guys on the team like Spencer and Milo had to struggle through by inching with their overstuffed rucksacks in the narrow opening.

Once inside, Nelson scanned the 200-foot-wide courtyard. Off to one side was a large, freshly dug hole with dirt piled on one side, about six feet high. A shovel was stuck up in the dirt. Nelson figured this was a water well the Afghans were digging.

Along the right side of the courtyard, several rooms had been built out from the exterior wall. It was impossible to tell how old the place was; fortresses like this one had been constructed in the nineteenth century as British garrisons or as defensive outposts manned by rebelling Afghans. The place smelled old, dry, dusty. Scattered around were plastic five-gallon water jugs and several black nylon bags, filled with gear belonging to the CIA team.

News of the Americans' arrival spread in camp and soon some

fifty Afghans had stepped from the shadows and stood in front of Nelson and the team. Some wore bandoliers of heavy ammo draped over their shoulders. Nelson thought he'd never seen a rougher bunch of fighters. The guys with the bandoliers looked like they'd stepped out of battle alongside Pancho Villa.

Some were barefoot, others wore sandals, men's dress shoes, or tennis shoes with no laces. The youngest fighter was in his late teens; the oldest was a man in his fifties. Their chapped faces were crowned with colorful scarves, brilliant as peonies, wrapped tight on their heads. Their beards were wild and thick.

Nelson said a few words of welcome, speaking in English as one of the CIA guys, Dave Olson, translated into Dari. The language was filled with explosive sounds, lots of hard consonants like "*chuh!*" and "*juh-zay!*" rising and falling in intensity, like someone chipping excitedly at soft stone.

Nelson told them that he and his soldiers had come to attack the Taliban and that they would supply anything needed to do this job. "This is your war," Nelson told them. "We are here to help you fight it."

The Afghans listened, smiling. They seemed neither to believe Nelson nor disbelieve him. It was like talking to the sky: you knew it was there, but not that it was listening. When Nelson was done, the men drifted away to small depressions in the dirt of the courtyard where they slept uncovered in the cold.

J.J. led them across the courtyard and down the long front wall to their rooms. There were two, one for gear storage and another for sleeping.

The rooms had been swept and neatened. The ceilings were low—only about six feet—and the men had to stoop as they walked around inside. The rooms smelled of dung and animal hair. Nelson turned on his headlamp and saw that the sleeping room had been a horse stable. Crisp, woven carpets had been laid on the flat mud floor. Large, pastel-colored seat cushions had been spread around. Nelson was moved by the precise care with which the place had been made ready.

The men dropped their rucks and made a security plan for the night. Ben Milo and Sam Diller would climb up onto the roofs of the rooms in opposing corners of the fort. They would keep in contact by two-way radio and switch out with another team member after two hours.

As Milo walked to his post, he noticed that an Afghan soldier was following him. Milo was actually trying to find a latrine. Milo turned and said in English that he had to take a piss and then he pantomimed the motion. The Afghan shrugged and kept following close behind.

Milo found two holes dug in a corner of the fort, near a main front gate. The holes were separated by a flimsy wood partition. There was no door. He unzipped and the Afghan turned his back and waited.

The man then followed Milo onto the roof. Milo had to tell him forcefully—using what was called in the Army the "pointy-talky" method of communication—that he needed to leave. The team had made the decision that their security would be their own problem. Milo gave him a gentle shove, but the man wouldn't budge.

He stood to Milo's side as Milo settled into guard duty, which involved scanning the horizon for anything that looked like it shouldn't be there: lights, movement, sounds. He could hear footfalls outside the fort's walls as Dostum's men roamed back and forth, also on lookout. Milo wondered what was making his new friend so anxious. He wondered if the soldier was planning to whack him. He settled in for an uneasy night.

Down in the courtyard, J.J. and Nelson sat on a carpet near the half-dug water well as Nelson peppered the CIA officer with questions about Dostum. J.J. told Nelson that he'd been surprised to learn that Dostum moved around the country by horseback. They were practically a nineteenth-century militia fighting a modern war.

Dostum's men were low on bullets and needed blankets and food. The previous two weeks of U.S.-led bombing of Taliban positions had done nothing but lower their morale. Only the United States' propaganda machine seemed more organized: from transport planes circling the country, fitted with banks of electronic listening gear and a broadcast radio station, the Americans had been beaming messages to the Afghan people, messages the Taliban found comical: "Surrender Taliban invader! Give Afghanistan back to her people!"

Nonetheless, Dostum had remained pragmatic and energetic.

This surprised Nelson. He said he thought Dostum was crippled with diabetes and failing eyesight, among other things. He was worried that he'd keel over as soon as the fighting started.

The truth was, Nelson learned, that no one had been paying much attention to Dostum before the attacks on September 11. He was, in fact, healthy as a horse. Until the previous April, he'd been comfortably living in exile in Turkey with his family (he was married with two young sons). He had returned in the spring when Massoud was losing ground in the Panjshir Valley and as Mohaqeq struggled in the south (in the area where Nelson and J.J. were now sitting, the Darya Suf Valley). Dostum had arrived to try to save the day.

It was meant to be a triumphant return. His exile from the country four years earlier, in 1997, had been ignominious. He'd been driven from the city of Mazar-i-Sharif, his stronghold, as the result of a double-cross by one of his own commanders, a man named Ahmed Malik. Malik had suspected that Dostum had been responsible for his brother's death (an accusation Malik could never prove), and in revenge Malik made a deal with the Taliban to let them enter Mazar-i-Sharif and capture the town, as long as he could share power with them in the aftermath. And as long as they kicked Dostum out of Afghanistan.

As the Taliban poured into the city, Dostum had to flee for his life. He sped in a small convoy of remaining supporters to the border town of Heryaton, forty-five miles north of Mazar-i-Sharif. From this city, a bridge led across the Oxus River to safety in neighboring Uzbekistan.

Upon reaching the bridge, Dostum was forced out of his vehicle at gunpoint. Manning the checkpoint were Malik's men, who relished this insulting turn of events for the previously omnipotent warlord, a man whose own home in Mazar was palatial by comparative terms, with flocks of peacocks imported from France wandering in the gardens and American liquor stocked in crystal decanters on mahogany bars.

Dostum and his men had to drop their weapons, empty their pockets of money and valuables, and walk across the bridge in shame. The men guarding the bridge taunted them as they left.

With Dostum out of the picture, Malik, however, had experienced a sudden change of heart and decided he didn't want to share power with the Taliban after all. After the Taliban moved into the city as victors, Malik launched an attack.

He was aided by 5,000 Hazara troops eager to avenge centuries-old animosity between themselves and the traditional ruling class of Afghanistan, the Pashtuns, who made up the Taliban ranks.

The Hazaras were led by a fierce warlord named Ismail Khan (Mohaqeq's predecessor), and alongside Malik's men, this combined force killed between 6,000 and 8,000 Taliban in a matter of several weeks.

A year later, remustered and reloaded, the Taliban returned with a vengeance.

At 9 a.m. on August 8, 1998, they entered Mazar-i-Sharif firing their weapons wildly in the air. The locals in the middle of the city thought the gunfire was the sound of yet another sectarian gun battle. When they looked up from their work, they saw hundreds of trucks racing through the dirt streets bearing Taliban fighters, all of them dressed in their usual black turbans. This was their first signal that something was terribly wrong.

Some of the trucks had machine guns bolted in the back and men leaped up at the mounts and started firing wildly, spraying the street, houses, anyone caught in the open.

"I came out of my shop," recounted one resident, "and I could see people fleeing. People were running and being hit by cars. Market stalls were overturned. I heard one man say, 'It's hailing!' because of the bullets."

By early afternoon, the Taliban started searching house-to-house for Hazaras. They broke down doors, smashed TVs, tore paintings from the walls, and dragged men into the street and shot them. They broke into hospitals and slit the throats of Hazara patients. They raped Hazara women who ate handfuls of rat poison in the aftermath, preferring death to the shame of their violation.

Broadcasting from loudspeakers around the city, they urged residents to convert on the spot from the Shia version of Islam (the Hazaras' own) to the Sunni brand, practiced by the Taliban.

"Last year you rebelled against us and killed us," explained one Taliban announcer. "From all your homes you shot at us. Now we are here to deal with you."

Cars were stopped at checkpoints and anyone possessing

weapons was arrested. "How many Taliban did you kill in Mazar?" they taunted.

If the suspect could prove that he was a Pashtun (or a Tajik, the other main tribe in the area), either by reciting his prayers in the Sunni manner or by producing an ID card listing his religious affiliation, he was let go.

Otherwise, he was executed.

Residents throughout the city heard an ominous warning on their transistor radios: "Wherever you go, we will catch you. If you go up, we will pull you down by your feet. If you hide below, we will pull you up by your hair."

Burial of the dead was forbidden. Bodies piled up in the streets. Residents who tried fleeing the city were cut down on the highway by Taliban jet fighters, holdovers from the Soviet era of occupation. So many people died in the road that cars had to drive over the bodies.

No one imagined the city could fall so quickly, least of all Malik, who was now a hunted man, a man with his own share of blood on his hands.

One of his counterparts, a Hazara named Al Mazari, had invented what came to be known among Afghans as "The Dead Man's Dance." The macabre practice involved beheading a captured Taliban soldier, pouring gasoline down his neck, and then lighting the fuel. The flaming corpse would jerk and writhe, a life-sized puppet crackling in its own juices. Malik managed to escape a similar fate and fled Mazar-i-Sharif. (He took temporary refuge in Iran.) In 1999, Al Mazari was captured by the Taliban, tortured, and thrown from a helicopter.

As for Dostum, he had escaped death thus far by switching sides whenever his own seemed to be losing. He was a volatile sovereign, but he remained a genial opportunist, and it was this trait that made him a pliable ally.

When Nelson looked up from the conversation, it was nearly dawn. He and J.J. had talked for almost three hours. J.J. explained that Dostum would be riding into camp at eight that morning. Nelson looked at his watch. He was anxious about the meeting. There were still several hours to wait.

He climbed up on one of the roofs overlooking the compound to get the lay of the land. The fort was positioned atop a 500-foot hill, on the east side of the valley. It was nestled in close against a copse of poplar trees, their trunks growing tall and straight, towering above the walls. They looked as big around as flagpoles and their branches clicked against each other in the rising morning breeze. A half mile away, a river winked in the morning sun—this was the Amu Darya, the rushing course they had followed south in the helicopter. Swift, vinegar-hued shadows raced across the valley floor to the river, hit the opposite valley wall, and climbed out of sight.

Up the valley, Nelson knew that perhaps as many as 15,000 Taliban troops were dug into countless hidey-holes, hilltops, and fortified bunkers capable of holding hundreds of men.

Behind him lay the unfettered wilds of deeper Afghanistan, stark mountain country, a land accessible only by donkey and horse.

The plain was packed mud, nearly blond in color, striped with swaths of grazing grass along the river. Nelson heard a rooster crow and he thought that this landscape seemed like something in a John Ford movie. He remembered watching those movies as a kid. In college, he'd even kept a poster of John Wayne on his dorm-room wall.

Now he heard horses neigh in a paddock attached to the fort and they stamped the ground, and he wondered if he and his men would be riding soon across the wild country before him.

Mitch Nelson was standing near the fort doorway alongside Pat Essex when he heard Spencer, outside the wall, shout: "Pat, Mitch, you better get out here!"

Spencer had just stepped outside, holding a cup of coffee and wondering when Dostum would show up. He suddenly heard the beat of hooves, felt a thudding vibration in the warm ground beneath his brand-new hiking boots. Then he looked up to the surrounding hills and saw men popping up in the sun, crouching there, in ragged, loose-fitting black pants, scarves wrapped around their heads. These were lookouts, he guessed, as they began signaling down the valley the arrival of something, or someone, by firing their AKs in the air. The sharp reports rang in the still morning and moved down the valley toward where Spencer stood. Then seven riders stormed up and shouted, "He is coming, the General is here!"

and they waved back, down along the white rocks in the riverbed, a thin tongue of green water lolling under the sun, wiggling its way downstream.

Spencer took the last swig of coffee from the metal cup he'd pulled from his rucksack, knocked the cup dry against his pants leg, then called again over his shoulder to Nelson and Essex.

Inside the fort, Essex turned to Milo, Diller, and Black. "We're gonna go outside," he said. "You guys stay here. And you be ready in case something goes wrong. You be ready to take care of business."

Diller and the rest nodded. They took up positions in the four corners of the fort, on the roofs.

Essex and Nelson walked out the door and over to Spencer.

"He's here," said Spencer.

"Where?" said Essex.

"There he is," said Spencer, nodding upriver.

A white horse emerged from a curtain of dust. The man in the saddle, immense, strong-shouldered, holding the reins in one hand while his other rested gently on his leg, looked straight ahead, the horse churning beneath him while the man floated atop.

Riding along the edges of the river, their gear and weapons clattering, were about fifty more horsemen.

They entered the camp and pulled up in front of the Americans. They sat atop their horses and looked down at them.

Oh boy, here we go, thought Essex. *We either just got a whole lot of new friends, or we're surrounded.*

The lead horseman wheeled his horse to a halt and the animal stilled as if turned to stone.

Nobody moved. Nelson, Essex, and Spencer had no idea what was going to happen next.

Two Afghans ran out and grabbed the reins and the tall, dark-haired man swung off the saddle. He was wearing a long, gray, woolen robe slit up the sides, a red smock, turban, and leather riding boots.

"I am General Dostum," he boomed in English. "I am glad you have come!"

Nelson, Essex, and Spencer were standing in a line, their weapons slung across their chests, their hands at their sides, a sign of comfortable submission, trust, and readiness. Out of the corner

113

of their eyes they watched Dostum's men, waiting for them to make any sudden movement with their guns.

The big man walked up and stopped before Nelson. The general was over six feet tall, weighing maybe 240 pounds. He had a short beard peppered with gray and limpid eyes that swirled into dark focus as he gazed at Nelson.

"An honor," said Nelson. He put his right hand over his heart and said, "*Salaam alaikum.* Peace be unto you, brother."

"And peace be upon you, friend," said Dostum. "Now we will go and kill the Taliban."

Nelson breathed a sigh. They hadn't been killed or kidnapped. Yet.

The general burst into the fort in a flurry of shouts and orders, and immediately several attendants spread two carpets on the dirt berm near to the water well.

On closer inspection, in the daylight, Nelson saw that the hole was quite large, about thirty feet across, and that it was filled with chalky green water, probably fed by a spring bubbling at the bottom. Nelson guessed it also contained rainwater, though rainfall in this part of Afghanistan was less than four inches a year. The crude hole was meant to supply drinking water for the camp. Nelson remembered the filters they'd bought by phone from a camping store, all while sitting in the comfort of their team room back at Fort Campbell. Those days seemed a long time ago.

The general's attendants set ornate glass bowls containing pistachios, almonds, apricots, and chocolates on the carpets. They bowed swiftly and left, refusing to meet Nelson's eyes. Nelson watched as an aged Afghan man in an outdoor kitchen in a corner of the courtyard drew a brass kettle from a fire. Nelson was starved and wanted to reach and grab a handful of the food. He resisted the urge, realizing it would be rude. He watched as the old man hobbled from the fire with the kettle, trailing a rich plume of steam, and knelt at the carpets. With great ceremony, he poured pale yellow tea into chipped cups set in saucers.

Nelson realized the ceremony was as old as war itself. Some corner had been turned. He and his men were inside a guerrilla camp; they were guerrilla fighters now. And they were about to meet their warlord.

Dostum sat first, taking a seat at the head of the red and green carpets. The berm's elevation above the floor of the courtyard, about five feet, gave the forty-seven-year-old general a commanding view of the goings-on around him.

Nelson sat next, to his left, a place of honor, while Spencer crouched down beside Nelson, on his knees. J.J., dressed in green camo pants, green shirt, and hiking boots, hovered at Dostum's shoulder. Nelson and Spencer made a display of unslinging their rifles and laying the stout, lethal-looking M-4s, with their muzzle ends pointing away from the center of the group, on the carpets beside them.

They kept their 9mm pistols holstered and strapped to their right legs. They were dressed in U.S. Army tan fatigues with their insignias and names removed, as they had been for the flight in by helicopter. This was a reminder to them that their mission was classified and that only a few people in Afghanistan, including the men around this camp, and only select military planners back in the United States, knew that they were here.

Nelson had tucked a black-and-white cotton scarf inside his shirt, a first attempt at donning the local mufti. He and Spencer had started growing beards a few days earlier at K2, but they both looked pale, sunburned, and bleary-eyed from too little sleep. Nelson wore a loose, wide-brimmed "boonie" hat that flopped down over his eyes, while Spencer's billed patrol cap was pushed back at a jaunty angle. They watched patiently as one of Dostum's advisers, a stocky, studious-looking man in black wire-rimmed glasses and groomed beard, took a seat at Dostum's right. Next to the adviser, across from Spencer, was Dave Olson, sitting with one leg cocked up, his arm resting on his knee, as if at a campfire. He was wearing a traditional black *shalwar kameez* over green camo pants. Olson planned to translate when the adviser couldn't find the words.

After everyone was seated, they were ready to begin.

Nelson cleared his throat and explained to General Dostum that he had a gift for him. The gift was from all the people of the United States, he said, but he was honored to present it personally. He gestured to the bags of horse feed stacked nearby in the courtyard. (Someone on the team had tried sifting through the grain and picking out the worms.)

Dostum looked at the bags and then back at Nelson. Olson translated.

"I don't need food for my horses," Dostum snorted. "My men are hungry. They also need blankets."

Spencer could see that Nelson was taken aback. Spencer and Essex had agreed they would keep an eye on the less experienced Nelson. They'd have to guide him through awkward moments like this when offhand statements might carry larger portents. Was General Dostum unhappy? It was impossible to tell.

"General," Spencer spoke up, "we have brought you another gift." And here he reached into one of the black bags brought off the helicopter and produced a bottle of Russian vodka.

Dostum's eyes widened. Spencer looked over at Nelson as if to say, This is how you do it.

"Very good," said Dostum. "Excellent."

He set the bottle down next to him, and regarded the two Americans carefully.

He held up one finger. "But what I really need," he said, "is bombs."

"Sir," Nelson said, "we have all kinds of bombs. I got more bombs than you could ever want."

"We will need them all," he said. He let the matter hang in the air. He sat back. He had a way of looking straight into Nelson's eyes that was unsettling. "And what will you need to drop these bombs?" he asked.

"Well," said Nelson, "first off, we need to see the enemy. We need to get up close. Real close."

Dostum considered this. "I can take you to my mountain headquarters," he said. He had just returned from there. The windswept outpost was located six miles up the valley. And there was much Taliban activity in the area. He reached beside him and produced a rolled-up tube. He snapped it flat on the carpet.

It was a topographical map—the biggest map of Afghanistan Nelson had ever seen. It measured maybe six feet long, creased and tattooed by pencil stub and slashes of ink, smudged with sweat and the ash of cookfires, and crawling with hundreds of arrows marking Taliban positions across the paper's soiled grain.

Dostum said he had carried this map for the past month during the fighting at Safid Kotah, which was three and half miles to the northeast, over the mountain ridge now brightening in the harsh morning sun.

Nelson's own map was small, practically worthless in comparison. It was at least twenty years old, with tiny blurred print, written in Russian. In fact, it had been the map used by the Russians during their occupation. And it had been the only map available to the Americans back at K2.

Nelson had thought this was not a good omen, to be carrying a map of the defeated.

Dostum scooped up a handful of pistachios and chewed busily, lost in thought, staring at the map. He jabbed at the paper and started speaking rapidly.

The key to controlling the country, he explained, meant taking Kabul. The key to Kabul was taking Mazar.

Nelson and Spencer nodded in agreement.

The key to Mazar, Dostum continued, was taking the Darya Suf River Valley. And if they took Mazar, the north would fall. All six provinces. Without question.

Next would come Kabul. And with the north in control, you could take Kandahar in the south. In this way, you could capture the country.

Dostum sat back and studied Nelson and Spencer. He next laid a broad thumb near the village of Chapchal. It lay eight miles north of their camp, a wind-scoured, sun-struck place of five hundred souls moving through dusty streets and squat mud huts, nestled in a rolling sea of burnt hills.

"The Taliban are now here," he said, circling the village with his thumb. The Taliban had retreated to this place after their defeat at Safid Kotah. A few roads, little better than goat paths, spidered from the village—a red dot on the general's map.

Nelson pointed to another village, Dehi, about two miles north of them, also up the valley. Nelson asked if Dehi was safe.

Dostum smiled. "But of course. I have captured it last week." After a fierce skirmish with the Taliban, he had ridden through the village, stopped outside the houses while still astride his horse, and announced to the men inside that they should fight for him. That he would pay them for this. That they had nothing to fear from him. That they should fear only the Taliban.

"Let me tell you a story," Dostum said to Nelson. "When I heard of the attacks in the United States, my heart broke."

He had been living in an apartment in the village of Cobaki, north of Chapchal, about twelve miles from where he and Nelson now sat. From this makeshift headquarters, he could look out with binoculars and see the Taliban front lines just one and a half miles away. He shook his head. That's how far north his front line had once extended. He had lost all that ground in the fighting since then, in about two months.

He had been in his apartment listening to the radio when news of the attacks in America broke in. He called an old friend and fellow Afghan, a man named Zalmay Khalilzad, who just happened to be the Unites States' ambassador to Afghanistan, living in Washington.

"Zalmay," he said, "tell the American friends that as far as they have hard enemies, they should know that they have hard friends, too. And that they are us."

And then he hung up and went back to fighting the Taliban. But for the next five weeks, the enemy had beaten him back down the valley, mile by mile.

"We are now here," he said. He dragged his thumb south along the Darya Suf River, where the current churned over car-sized boulders and trilled in silver braids, and stopped near Dehi.

"I want to move today to my mountain headquarters overlooking Chapchal. From there, we will bomb the Taliban. Once the bombing starts, they will break." Then he announced quite unexpectedly, "We will leave immediately."

Nelson asked how he was so sure of his plan.

"The Taliban are like slaves," he said. "They are slaves because they are forced to fight. They threaten to kill a soldier's family if he does not fight."

He explained that many Afghans in the Taliban army, which numbered as high as 50,000, were farmers, teachers, shopkeepers. Men conscripted into service. Men who fought because they were scared. The foreign Taliban, on the other hand, the Pakistanis, the Saudis, the Chechens, even the Chinese, they were fierce men.

Ferocious fighters. They had infested the country from radical madrassahs in Pakistan. They were often joined by bin Laden's Al Qaeda army. Bin Laden's men pulled out grenades and blew themselves up rather than be captured. They called themselves the 055 Brigade, a crack squad of 500 to 600 storm troopers. They were not accepted by the Taliban ranks, who considered them "foreigners." In camp, they did not mix; each group kept to itself. They did not fight for Afghanistan. They fought to convert the world to Islam. The Taliban fought to change Afghanistan.

"When we capture the Afghan Taliban," said Dostum, "they switch sides and they start fighting for us. And we let them live.

"As for the Arab Taliban," he went on, "the foreigners, they prefer death. You can't capture them. You must kill them. They never give up."

Recently, one of Dostum's spies living in Mazar-i-Sharif had contacted him by satellite phone, telling him that he could see more Taliban soldiers in the streets gathering in Toyota trucks; they were shouldering battered AK-47s, shotguns, RPGs, PK machine guns, machetes, sticks and swords, obviously getting ready to fight.

Dostum had given away some fifteen satellite phones to spies across the north, from Herat in the west to Konduz in the east. They were handsomely paying off. The spy was speaking to him at great risk. Taliban soldiers had shown up at his house and told him, "You're coming with us." To save his life, he had joined with the religious fighters.

Now, after several months of being forced to battle Dostum, he was getting his revenge. The spy told Dostum that hundreds of men were loading into T-55 tanks and a similar Russian-made fighting vehicle called a BMP ("Bimpy," for short).

He could hear the snort of their diesel engines tearing out of the gates of the city, heading down the rutted road to a place called the Gap, a notch in the mountain wall that separated Mazar from the country's interior.

The Taliban were pouring through the Gap, headed toward Dostum here in this valley. Thousands of fighters. Coming at them like a storm.

"There is a bounty on your heads," he went on. "One hundred thousand dollars for your body. Fifty thousand for your empty uniform." Dostum said this flatly, without emotion, as if testing a response.

Nelson wondered why Dostum was suddenly telling him this ominous news. Was he suggesting he owed Dostum a favor for not ransoming them to the Taliban?

Nelson and Spencer were shocked by the general's bloodlust and eagerness. (In fact, Dostum had warned his men: "You will guard the Americans. If you fail, you will be killed.") It had been Nelson and Spencer's experience training some armies in Middle Eastern countries that the soldiers were loath to start fighting. They had half-expected Dostum to explain a hundred reasons why his men weren't ready. (In this way, he'd prolong America's presence and subsist on foreign aid.)

Spencer leaned over to Nelson. "Jesus," he said, "this guy's ready to roll. He wants to win."

"We will leave now," Dostum insisted. "I will take you to my mountain headquarters."

"What do you think?" said Nelson.

"I think you should go," said Spencer.

"It's risky."

"I know it is."

"We just got here."

"Hell, it isn't anything you can't figure out."

Nelson turned to Dostum. "Can I take my men?"

Dostum nodded.

"All of them, everybody," Nelson insisted, "we're all going with you?"

Dostum shook his head again. "No, no, there are not enough horses."

"Unless we all go," Nelson told Spencer, "we're not leaving."

He turned to Dostum. "How far a ride is it, General?"

Dostum said it would take several hours.

"We can split the team," said Spencer.

"We got radios?"

"Barely enough. But we got radios."

"You think he's up to something?" asked Nelson.

"Hard to say."

"How many men can I take?"

"Six," said Dostum. "You can take six."

"Horses are hard to come by in this country, aren't they?" Nelson said.

Dostum didn't understand.

Nelson looked over at the horses standing patiently at the fort's entrance. They were shaggy and thin-legged, and short. Roan, white, and gray. In this country, Nelson knew, a man measured his wealth in horses. Each one might cost about a year's wages, one hundred dollars. These were horses descended from the beasts Genghis Khan had ridden out of Uzbekistan, and from farther north, Mongolia. Deep-chested, short-legged animals built for mountain walking.

Nelson counted about fifty in the string, roughly one for each man in camp. He guessed that they had ridden them from home to war.

"When would we leave?" Nelson asked.

At this, Dostum stood up quickly. "In fifteen minutes!"

Before Nelson could stop him, the delighted general had barked orders to his men bent at various tasks in the camp. They dropped what they were doing, walked to their horses, and began saddling them.

Within several minutes they were seated atop the mounts and ready to ride out.

Nelson yelled to the team's Alpha cell, composed of Sam Diller, Vern Michaels, Bill Bennett, Sean Coffers, and Patrick Remington: "You better get out here!"

The men stepped from their quarters. They quickly gathered around the berm.

"Pack your shit."

"We're leaving?" asked Diller.

"We're going to the front lines."

The five men hurried to their rucksacks in the gear room and pawed through clothing, food, ammo, and water, winnowing an essential load to stuff in their green daypacks. They came lugging them back to the center of the courtyard.

They stood stone-faced with their rifles slung on their shoulders, pistols strapped to their legs by means of wide, black elastic bands. They looked around the courtyard at the Afghans gazing back at them. They wore black watch caps pulled down over greasy, stringy hair. A few had tucked Afghan scarves around their collars. Nobody wore a helmet. Helmets were bulky and heavy. For this same reason, nobody had on body armor, either—those forty-pound Kevlar vests.

"How are we getting there?" asked Diller.

"Horses."

"*Horses?*"

"Yeah, we're riding there."

As they spoke, Dostum, across the courtyard, swung into his saddle. The men turned to watch him.

The general sat atop his white stallion, a red pom-pom braided in the coarse ivory hair at its forehead. He carried a soft green blanket and a red carpet lashed to the back of the saddle.

Lined up beside him were a dozen of his men. Their horses stamped the ground, raising and dropping their heads like hammers. They were stallions, all of them.

"He's fixing to leave without us," said Nelson. "We better goddamn hurry up."

Dostum touched his stallion's flanks with his boots, rode over to Nelson, and stopped.

He looked off to the distant mountains, then turned back to Nelson. "I cannot guarantee your safety in Dehi," he said. "There are people there who are not happy about your arrival."

So said the voice of God:

In a different time, in the year 1418 in the Muslim calendar, John Walker Lindh pushed through the glass front doors of the Mill Valley mosque into the night. On his right, a parking lot, eucalyptus trees. A few cars parked inside under the sulfurous glow of sodium lights. High chain link topped with concertina wire. Light traffic on a street in a part of town filled with modest homes and discount stores. Beyond that, the wavelike roar of interstate traffic headed south to L.A. or north along the coast. He kicked off on his mountain bike a new man, headed home.

It was at least a hard hour's ride. When he pulled up in front of his parents' house in San Anselmo, he went inside without saying so much as a word. Went directly upstairs to the bathroom. Reached in and turned on the shower. Undressed. It was September 27, 1997, in the year of the Christian Lord, the god his father worshipped. His mother said she was a Buddhist. He and his father were two continents drifting apart. It was a Saturday, a day he would not forget. He was sixteen.

Back at the mosque, he had stood before the meager huddle of supplicants gathered on the prayer rugs. His friend Nana. Brown faces in the crowd. Men, all. Pakistanis, Arabs, mostly Indonesians. Pilgrims in a land of pilgrims. John Walker Lindh among them, lost.

He had impressed Nana with his studiousness. When Nana told John that he believed music "with a beat" was impure, John agreed. When Nana said he didn't use a fork because Muhammad did not use a fork, John ate with his hands.

The study group at the mosque debated the Sunna, which are writings by Muhammad's contemporaries describing what Muhammad said and did. Lindh eagerly adopted the teachings of Abd al-Wahhab, an eighteenth-century Islamic cleric who believed Islam needed to return to a strict interpretation of the Koran. After study, Lindh and five other teenagers played putt-putt golf at a local amusement park. In his white, flowing robe, his pale face topped with a cotton skullcap, he looked lonely. When they played basketball, he stood watching from the sidelines because he didn't think it fitting to remove his robe in public.

More than anything on earth, he wanted to learn Arabic so he could teach English-speakers about Islam. He wanted to translate the Koran from Arabic into English.

At the mosque he had stood and said *Shahada*, the declaration of his conversion: "I declare that there is no god except God, and I declare that Muhammad is the messenger of God."

Very good, Lindh was told. You are Muslim now.

Go home. You must shower. Cleanse yourself. When you step from the water, a new man will be in your footsteps.

So ends the voice of God.

Before Nelson could protest Dostum leaving without them, the warlord spurred his horse. Nelson watched as Dostum and his men clattered through the fort gate. Once they were outside its walls, they let out a whoop and Nelson listened to the hooves pound; and then silence.

An Afghan soldier came walking up the courtyard leading six horses.

Nelson tried to put Dostum's warning from his mind. *Concentrate. He's messing with you.* "Who's ridden before?" Nelson asked the team.

Only Vern Michaels and Bill Bennett raised their hands. "At summer camp," they said. "When we were kids."

The U.S. Army did not offer "Horsemanship 101" as a matter of course. It did run a program called "Dusty Trails" in the Colorado Rockies, but this taught soldiers how to use packmules in mountain environments. No one in Washington, D.C., had imagined that modern American soldiers would be riding horses to war.

Nelson tried picking out the handsomest, tallest one from the string. It had a sugar-white star whorled on its brown forehead. Its rough-haired legs were knobby at the knees and tapered thin as cypress roots at the ankle. The hooves were cracked, unshod, the color of dishwater. In height, it resembled an overlarge pony, the kind children ride at county fairs in a sawdust circle. The man leading the horses said its name was Suman. He gave Nelson the reins.

The horse was so short that Nelson could walk right up and look directly in its eyes. He tried figuring out the pedigree. Maybe Arabian bred with quarter horse and a mystery animal that shrunk it one-third the size of a normal horse, at least normal by any standard Nelson was used to. Two months earlier, back at home on leave, he'd ridden out across his father's property in Kansas on a big, chestnut quarter horse on a warm summer day, miles of tall wheat whispering against the horse's legs as they passed. He could turn in the saddle and look back to his parents' house where his wife, Jean, waited. She was pregnant with their first child.

He worried about Jean having the baby without him while he was gone. Most people didn't think soldiers cared about that kind of thing. But Nelson did. He had sat through endless sonograms and well-baby checkups. Before being deployed, Spencer's wife, Marcha, had volunteered to be Jeans's labor coach and he'd been relieved. Looking in this horse's eyes, which seemed to contain all the misery and the patience for misery in the world, he knew the future for him had narrowed to include just a thin slice of staying alive through the next day. The wives, though, they had the rest of their lives to worry about if anything should happen to the men.

He tried swinging his boot up in the stirrup and found it was too high. The stirrups were hammered iron rings and they hung down from the saddle on short pieces of leather. There was no way to lengthen them.

He remembered that the Afghans mounted by swinging up in the saddle. He put his left hand on the stained blanket that formed the saddle seat. It smelled rancid. The saddle was made of three boards hinged together, covered by goatskin. He reached up with his right hand to the back and grabbed the edge of the stout board underneath the blanket, jabbed his right toe as best he could in the stirrup, and kicked up and swung down. He landed with a groan.

The saddle was tiny, made for a much smaller man. Nelson saw that the average Afghan weighed about 140 pounds. No one on his team weighed less than 200. On the saddle, there was no pommel horn to grab onto for balance. He gripped the horse's mane with one hand and held the reins with the other. He lifted a boot and jammed it into a stirrup, then jammed the other boot in. Just the toes caught the edges. He was sitting with his knees practically bent up to his ears.

He knew he looked funny. He wondered how in God's name he was ever going to ride this horse.

"Listen up," Nelson croaked, "here's how you make this thing go." He heeled the horse in the ribs and it walked a few steps. "And here's how you turn," he said, pulling a rein and drawing the narrow muzzle around. "And here's how you stop." He pulled back on the reins and sat looking at the guys. "Got it?"

The guys on the team just nodded.

"Now, if your horse runs off," he went on, "and your boot gets stuck in these stirrups and you get thrown, you'll get dragged. And you'll die.

"If that happens," he announced, "you gotta shoot the horse. Just reach out and shoot it in the head."

The guys were now watching him like he'd lost his mind.

"I'm not kidding," he said. "You don't want to get dragged over this rough ground."

Several Afghan soldiers walked up to help the men get on their horses. The Afghans held the reins with their left hand and reached over with their right to steady the stirrup so the soldier could slip his boot inside. The horses started moving in counterclockwise circles, forcing the riders to keep up with one boot in the stirrup and the other hopscotching in the dirt. About every third hop, the guy would try to jump up and swing over the saddle. After several minutes, everybody had managed to scramble up.

Spencer walked up to Nelson sitting on his horse. He thought Nelson must be terribly uncomfortable. The young captain looked wedged into the small saddle.

"I'm keeping the Bravo cell here," Spencer explained. He was holding behind Pat Essex, Charles Jones, Scott Black, Ben Milo, and Fred Falls. They would run the logistics end of Nelson's journey.

"I've got an air drop scheduled for tonight," Spencer said. "Medical equipment. I'm requesting blankets for Dostum."

Nelson nodded.

"Man," Spencer said, "you look funny on that horse."

Nelson didn't say anything.

"We'll keep in hourly contact," said Spencer.

"I'll be on the PRC," said Nelson. This was the interteam handheld radio. "Vern will have the commo package running back to K2."

"Mitch?"

"Yeah?"

"Good luck."

"Luck won't have much to do with it."

"I know."

Nelson kicked his horse in the flanks.

"*Cho!*" he said, remembering what the Afghans had shouted when they'd ridden away. The word meant "Giddy-up" in Dari.

"*Cho! Cho!*"

The horse lurched and started toward the front gate. Soon the five riders were lined up behind him. Looking out the gate, Nelson could see Dostum's trail, a narrow, churned-up path heading north.

He shouted over his shoulder, "You'd better keep up!"

They spurred their horses, and the men rode out the gate.

PART THREE

DANGER CLOSE

Najeeb Quarishy cursed the Taliban tanks and trucks he saw racing from Mazar-i-Sharif, headed south to attack Nelson and his men.

For seven years, during the Taliban's rule in the city, he had lived in daily fear of being arrested and beaten.

At age twenty-one, pudgy, witty, and easy to laugh, Najeeb ran a successful language school on the second floor of a dilapidated office building in downtown Mazar. He had two hundred students who trudged up the stairs at all hours to learn English. The Taliban were constantly threatening to shut the place down. They did not want anyone to learn English, the language of infidels.

One day Najeeb got into a fight on a street corner with a Taliban policeman, a member of a unit known as the "Ministry for the Promotion of Virtue and the Prevention of Vice." Najeeb had refused to grow a beard and he didn't keep his hair cut short. The policeman attacked Najeeb with a riding crop, beating him on the head and shoulders, asking him why he was not a good Muslim.

Najeeb snapped. He hit the Taliban policeman and kept hitting, and he found to his surprise that the man was nothing but a coward who ran away from his roundhouse blows.

Afterward, Najeeb stood on the street, shaking, unable to believe what he had just done. He believed he had just signed his own death warrant. He knew that if he was tortured by the Taliban, he would not walk for a year. His neighbor had been arrested and beaten about his genitals. The man was now paralyzed.

The Horse Soldiers' Ride to Mazar-i-Sharif, October 19–November 10, 2001

N

Bombing of Sultan Razia school, Nov. 10

Siege of Qala-i-Janghi Fortress, Nov. 25–Dec. 1

MAZAR-I-SHARIF

Dostum's forces plus
SF teams enter city, Nov. 10

TANGHI GAP

Essex's team arrives at top of Gap, Nov. 9, to plan airstrikes on
Taliban blocking final U.S. movement to Mazar-i-Sharif

Dean's team and Atta's
forces travel through Gap,
predawn hours, Nov. 10

After the successful Nov. 5 attack,
Afghan and U.S. forces pause
on Nov. 8 and plan attack on Gap

Dostum and Atta arrive, Nov. 8

SHULGAREH

Essex, Milo, and Winehouse nearly
overrun by Taliban during Nov. 5 attack

Bowers and Mitchell land at base camp
"Burro," Nov. 2, and meet Dostum

AK KUPRUK

CHAPCHAL

After landing on Oct. 19, Nelson's team
launches airstrikes while Afghan
horsemen charge Taliban tanks
before final assault, Nov. 5

DARYA BALKH RIVER

Dean's team lands by
helo at base camp Nov. 2 to help
Atta attack Taliban in Ak Kupruk

DEHI

Dostum's headquarters
at Cobaki

Nelson's team lands by helo
at Dostum's base camp
"Alamo," Oct. 19

DARYA SUF RIVER

0 miles   25

0 km   25

Najeeb ran to his father's house. The terrified man begged him to travel quickly to Pakistan and hide from the Taliban. Najeeb deeply respected his father, who made a comfortable living for his family by importing radios and kitchen appliances for the few Afghans who could afford these luxuries. He also dabbled in real estate. His pedigree as a former *mujahideen* fighter during the long war with the Soviets also gave Najeeb's family cachet among Mazar's citizens. None of that mattered, though, under the Taliban. So Najeeb left the city. He stayed away for several months and then crept back, unnoticed. Since that time, four years ago, he'd been living a life under siege, but still refused to close his language school.

The past month, hearing the news of the attacks in America, he felt a strange kind of joy and sadness. He knew that the Americans would be coming for the Taliban. That night, he crept to his rooftop and erected the makeshift satellite dish he'd bought on the black market. It was made of flattened Pepsi cans that had been ducttaped together to form an elephantine dish and fitted with parts cannibalized from other defunct rigs.

Najeeb sat up through the night watching the reports of events unfolding in America. Each morning he had to disassemble the dish and hide it from the Taliban under a tarp. He was filled with a bitterness that he did not like.

Suffering had forced Najeeb to become philosophical at a young age. He knew that if he held terrible thoughts in his heart, he would not have a bright future. If you try to avenge something, he reasoned, it cannot end well.

The funny thing, he thought, was no one ever asked the Taliban why they were so cruel. Why were they hurting people and killing them? They thought the only way to live was by their own beliefs. They did not respect village elders, who traditionally had held leadership positions among the people, fixing land disputes and arguments between neighbors. He couldn't help himself—he wanted to kill them all.

Nelson hadn't ridden far when he saw the dust cloud of Dostum's posse up ahead. Maybe a half mile away. Dostum was keeping ahead of him, out of range. He didn't know why. He worried about losing the trail and pushed hard to catch Dostum. As they rode, they hadn't seen a soul, so far.

They passed empty settlements that had been decimated by the Taliban. Whole families wiped out, the men and boys dragged away, into the army. The water wells poisoned. The shells of houses standing amid mounds of broken mud wherever the Taliban had driven their tanks up and inserted the cannons through the windows and fired.

Soon they came to a crossroads. Nelson lifted his hand and brought the men to a halt. He turned. Riding up behind them were ten Afghan soldiers, Dostum's men, who had left the fort after them. Their security detail. Trailing them were two mules straining under the load of the team's rucksacks. Two more mules carried bottles filled with water. The bottles hung from wood frames around the withers of the mules. Each mule carried several dozen. The bottles made a dull music against the damp hide of the animals as they trudged.

The Afghans passed him on the trail and cut back in and kept riding without stopping. They took the west road to the left and raised storms of dust that Nelson and his men rode through. Soon the Americans had lifted their scarves around their mouths and sat breathing through the machine-oil smell of the tightly loomed fabric. Some of the Afghan riders tucked a corner of a scarf in their mouths and sucked on them, and as they rode, the scarves darkened downward from their chins. After several more miles they approached the village of Dehi. Nelson halted the team.

"Everybody look sharp. Dostum says we may not be welcome here."

The men were nervous and wanted to know what they would do once they rode into town.

"Just be ready," said Nelson. "Everybody lock and load. And if you have to shoot, make sure it's for a very good reason. And for God's sake, don't shoot women or children."

They saw the first gaggle of people standing at the edge of the village, watching them come on. Men with leathery faces hunched in brown blankets. Others in dark suit coats standing with hands clasped behind their backs, as if waiting for the doors to open to some invisible building in front of them.

The main street was dark as ash. Along its edge, rocks the size of fists. A low hump of gray mountains stood in the distance.

Storefronts lined the streets but Nelson thought, *God knows what they are selling.* Hitching posts out front but no horses. The store roofs were low-slung and rested on poplar logs stripped of bark and dug by hand into the hard ground. The walkway along the stores was raw planking that rang dully underfoot, and under the roofs swung the skinned sides of beef and sheep twirling slowly like music-box figures. From somewhere, a back alley, a cook's smoky fire.

They kept riding. They were in two columns about ninety feet apart. Nelson laid his M-4 across the saddle with the barrel pointing at the crowd. He kept one hand on it and waved with the other. At their passing, the villagers parted and reached out filthy hands as the legs of the men brushed past and they swarmed to refill the emptiness behind them. There was no sign of Dostum anywhere.

"These people came to see a parade," said Nelson. "So let's give 'em one."

Now there were at least two hundred armed men on each side of the road. Some of the men wore makeshift uniforms, camo pants and shirts of varying pedigrees from who knew how many different countries. They carried battered AK-47 and RPG tubes. They were an army of castoffs, soldiering in hand-me-downs. The army the United States had forgotten after the Soviet pullout in 1989, once they were no longer needed to fight the Cold War by proxy. *We're going to need you now,* thought Nelson.

Some other men in the crowd were working worry beads between calloused forefinger and broad thumb as they recited prayers under their breaths. More men were walking toward them holding hands. They weren't gay, Nelson knew that. In public, men and women were forbidden to touch. But men could easily show this manner of affection. Nelson thought that they really should just keep moving. He figured the villagers had heard the helicopters land in the night and they'd come out to meet the Americans. Either that, or Dostum had ridden through town and announced the Americans were on their way.

Nelson called out to the team to remain alert.

Without warning, the security detail ahead pulled up and stopped.

Everybody on the team halted, tensed. Diller put his hand on his sidearm in the holster on his leg. Others did the same.

"What's going on?" Diller asked.

"I don't know." Nelson said he'd ask. He spoke in Russian and one of the riders in the security detail answered that they needed to stop for supplies.

Supplies? That sounded dubious. The last place Nelson wanted to be was stuck in the street while the Afghans went inside a store, maybe never to return.

The Afghans dismounted and stepped up into the store and went inside. Nelson could hear them talking excitedly to an unseen storekeeper.

The villagers pressed around the legs of the horses, looking up at Nelson and the men.

"Grip and grin," he told them. He had a few phrases he knew from some pages of Xeroxed material he'd studied back at K2.

He put his hand over his heart and said, "*Salaam alaikum!*"

The villagers looked back and smiled and said the same.

"*Chedor hastee?*" said Nelson. How are you?

"*Namse-chase?*" he said. What is your name?

He kept scanning the crowd and the front door of the store. Up ahead, on the left side of the street, there stood an outpost, like a guard shack, built up on one of the roofs. He kept looking at that, too. Still no sign of Dostum.

He'd gone through about all of his phrases, telling the villagers his name (not his real one), and asking if they liked candy and telling them that he was fine, thanks, when the Afghans emerged from the store hauling bags of horse feed over their shoulders and carrying even more water in five-gallon plastic jugs. *We must be fixing to be gone a long time,* thought Nelson, *if we need that much water.*

The Afghans loaded these items onto the mules, who stood with their knees locked under the weight, trembling. One of the mules would not stand still and the Afghans beat it with a leather crop fitted into an ornate brass handle. Nelson—and he could tell the guys on the team felt the same way—wanted to say something about beating the animal, but he didn't. His job was to be there and not be there.

"You ready?" he called to the Afghans in Russian.

They ignored him, but the question seemed to register because they swung onto their horses and reined them into the middle of the street and started riding.

Nelson rode up to them and pointed to a side street between two squat mud homes. Blue and green tarps rested on tree branches at the street's edge. A woman in a blue burkha sat holding a child. Beggars, mother and child. The father probably killed by the Taliban, or hauled off to fight for them. It was like passing by two sundials churning the sun. A silence without end.

Nelson cut ahead and took the alley between the two houses. He wanted to get off the main drag and ride hard through the backstreets and pick up Dostum's trail. They came out at the other end of town; before them lay more open ground, dark rock piled in jumbled moraines along the valley walls.

Nelson pulled up his horse. The other riders stopped. "Where is the General?" Nelson asked.

The Afghans pointed out across the open ground and then up into the mountains to their left, across the river. It was another four-hour ride, they said. Nelson looked up. The time was about noon. They had maybe six hours of daylight.

Nelson realized that he had to admire what Dostum had done: he'd led him and the team through town as a show of force against the Taliban. Nelson guessed the intended message was, *Here are the Americans. And they're not afraid of you. And they are with me.* That they hadn't been killed was just a bonus for Dostum.

They struck out across the open ground. Wearing his bulky daypack, his load-bearing vest stuffed with ammo clips and grenades and bulging over his stomach, Diller found it was hard to move around much. He could already imagine the formation of saddle sores. In usual Diller fashion, he decided he would ignore the discomfort.

They crossed the cold Darya Suf, the horses plunging in and pulling out the other side wet up to their chests and the men's pants dark up to the knees. The Americans rocked and groaned in the saddles. Diller could feel the bleeding start under him. They rode across a ceramic-hard plain and crossed another braid in the river and came out clopping on the other side and started to climb a mountain. Or a part of it, a shoulder, hunched 6,000 feet high on the thin bank of the river.

They cut in on a trail carved in the rock and started up. The horses put their heads down and didn't lift them. They rode for fifteen minutes headed across the mountain on the trail and then at the trail's end they came to a little cul-de-sac where the horse had to hitch up in its step and turn and bear the rider forward as they struck back across the mountain. Climbing and climbing higher. On one side, to Bill Bennett's right, stood sheer mountain wall, cascading plates of rock frozen in place. To his left, a 1,000-foot drop. He reached his left hand out and underneath was the airiness of the valley floor, the river twisting below like green wax string. The trail was two feet wide. *How in hell does this horse know not to step off?* The animal ground out each step, the cannon bones in the legs driving like small fenceposts on the ground. *Ram. Ram. Ram.*

They rode up the last bit of the hill and the horses strained beneath them and stood on flat ground.

Nelson looked out from the saddle and beheld a landscape stretching as far as he could see on either side. Miles of hills, tinted red. The air was filled with an ocher dust swirled up in winds rising miles away. The dust that now settled on him had flown from far away, Nelson guessed, and he looked at the shellacked faces of the men. They sat like monuments in the saddle, stiff and thirsty, barely moving.

Dostum was still nowhere in sight.

From Kabul, John Walker Lindh flew by helicopter to Konduz. Ancient trade post. Squalor. Thin-lipped boys with sores, bearing silver salvers of smoldering incense through the streets. The smoke meant to keep the *djinn* away, evil spirits that steal men's souls. In the parks, the deranged. Men babbling in tongues. Out of their minds and with no hospital to hold them, wandering freely in their robes striped with excrement. Men thought to be speaking truths, the sagacity of shamans. Seers. Standing in a park and shouting under a noon sun as others went about their work around them.

From Konduz, Lindh got on a bus. The dirt road north swinging and swaying out over the plain as if constructed in a delirium. At one point the road had been straight. Then came the war, and the planting of land mines. A car, a truck, a tank grinding down the thoroughfare would veer around the presence of a mine or the threat of its presence and by this way the road veered like a river seeking least resistance. The road led him to the desolate village of Chichkeh.

The village, or what was left of it after the Taliban's siege, lay along the Amu Darya River. He got off the bus dressed in flowing *shalwar kameez* and his cotton skullcap, a dark rind of beard around his doughy face. *What am I to do? You are to walk to that hill and sit. Here is a rifle. A grenade.* Lindh trudged up the hill.

Facing him, the jagged escarpments of the mountains of Tajikistan across the river, rising 10,000 feet. Snow in the valleys. Bright sun. The ground gritty under his soft palms as he sat in his trench. And waited.

Waited for the invasion of the hordes. The infidels. Dostum's men. They were to come from the north. Enemies of Afghanistan. Enemies of Islam. Drinkers, fornicators. There is no God but God and Allah is his name.

It was September 6, 2001.

So spoke the voice of God.

Nelson sat on the horse and saw that down below the hill there were three trenches. In front of the trenches, facing north, earth had been piled to make protective berms. The trenches were about thirty feet long and five feet deep. A man could stand in them up to his waist. He could lean forward over the berm and rest his rifle on the packed earth and fire. Nelson saw that they were soundly made fighting positions.

At the top of the hill he discovered three caves. Dostum emerged from one of them, the largest. He stepped toward Nelson and with a sweep of his arms said, "Welcome to the mountain headquarters!"

Behind the hilltop, the rest of the team dismounted from their horses. They slid down from the saddles and stood in their tracks, unmoving, as if frozen in place. Many of the men found they couldn't walk, and some were bent over as if catching their breath.

All except Nelson. The ride had invigorated him. Diller could feel the blood seeping through the seat of his pants. The saddle had rubbed the skin clean away. Nelson called out a few good-natured comments about who looked funny now and the guys ignored him. They placed their hands at their backs and attempted to stand up straight. As they did, they stifled groans so as not to raise suspicion among the Afghans that they were not good horsemen. The ruse, of course, failed miserably. Diller saw that a few of Dostum's men were sniggering at them. Diller looked over at his horse and promptly named it "Dumb Ass."

"Well, Dumb Ass," he said, "I'm going over to get my ruck off the mules and when I come back, I hope you're dead."

He hobbled over to help the Afghans unburden the exhausted pack animals. Each of them had carried in excess of 300 pounds, maybe more, when the usual load might've been 100. *I would not want to be a mule in Afghanistan*, thought Diller. *I would not want to be an Afghan, either.*

Under the direction of Dostum's men, he and the team hauled their rucks across the hill to a cave next to Dostum's. Like his, the opening was set into the side of the mountain, the rockface rising another several hundred feet into a promontory. Diller thought it was not smart to be setting up quarters below such high ground, until he saw the face of an armed Afghan peeking over the edge and looking down at him. He toted the ruck inside and dropped it on the cave floor. The accommodations were more than he expected. The cave had a funny odor, like stacked fur in a dry cleaner's shop. Warm. Humid. He could stand up in the middle of the cave with about four feet of headroom. From wall to wall it measured about twenty-five feet. He looked at the walls more closely. They were covered in a strange substance. Diller reached out and stroked the rough, furry surface.

The fur was actually horse dung, maybe mule, too, he figured. For insulation. The blanket of shit was dry as toast and flaked only slightly to the touch. How the Afghans had ever managed to affix it up there and make it stick, Diller couldn't guess. But he found it a marvel of engineering. As he walked around unpacking his gear, he tried not to brush up against the walls.

•   •   •

Outside, Dostum and Nelson sat cross-legged on a red blanket over-looking the valley. Next to them stood a tall wooden pole with a radio antenna wire fastened to the top. Dostum sat by the radio itself and a bank of portable solar cells that could be broken down and carted by mule. Nelson thought it was a pretty slick setup. Dostum was running this part of the insurgency on solar power. That had to be a first.

Nelson guessed their elevation was about 8,000 feet and that the far hills across the valley were ten miles away. The hills, like all the hills, looked denuded. The Taliban hadn't been able to take this high ground from Dostum while he'd been battling them in the south, downriver, at Dehi. This had been hard fighting. Because of the area's three years of drought, much of the drinking water for his soldiers had to be trucked in and carried to battle on mules. Dostum's men had even dammed the Darya Suf to make a pool for their horses to drink from.

At night, he would gather his fighters around a lantern and, unrolling his map, plan the next day's attack. He spread the soldiers out over several miles, in groups of twos and threes, to wait in ambush. If they were discovered, their small number would not give away the fact that five hundred more men were camped nearby.

After a day's fighting, he and his men rode their horses back down rocky trails and picked their way along the river and broke out into a clearing that was their base. They were safe here because the road was nonexistent; there was no way for a tank or Bimpy to reach them. The Taliban would have to walk or ride horses, and he knew they didn't have horses. An infantry attack in the valley would have been suicidal. Dostum could set up firing positions in the rocks and pick off the enemy as they marched down the river, their long black robes swaying, their guns at the ready. He would have killed them all. Now Dostum was going to use this high ground to bomb them.

He despised most of all a tall, bearded Taliban commander named Mullah Faisal, the man who now occupied his former head-quarters at the fortress in Mazar. Faisal commanded the Taliban's 18th Corps, some 10,000 soldiers, a feared man with a perpetually pinched gaze and an inscrutable smile.

Next in command was Mullah Razzak, whose 3,000 to 5,000 soldiers of the Taliban Fifth Corps controlled the Darya Suf and the adjacent Balkh Valley, where they were also battling Atta's men.

Razzak was joined in the Balkh by the fearsome, one-legged Mullah Dadullah (he'd lost the leg fighting the Soviets and wore a wooden peg as a prosthesis). These three commanders were supported by several dozen subcommanders spread across the country.

Dostum pointed across the valley and said, "There they are. Taliban soldiers." These were Mullah Razzak's men.

Nelson raised his set of binos and cranked down the focus with his forefinger and steadied himself for a clear picture. The binoculars were heavy and rubber-coated and powerful, but still Nelson couldn't sight any enemy fighters in the distance. He didn't want to say so. It was like hunting deer or catching fish. Part of your standing in camp meant how well you could spot game. You looked for what shouldn't ordinarily be there.

He lowered the glasses. "I'm sorry, General, but I don't see what you're looking at."

Dostum pointed to a far outcropping on one of the distant ridges. *My God, that's miles away,* Nelson thought. *Does he think we can bomb that from here?*

"Okay, I see it," said Nelson, bringing the dark spot into focus.

"That is the Taliban's position," said Dostum. "One of their bunkers. I know, because I have been fighting them." He asked Nelson, "Can you bomb it?"

"Well, sir, like I said, we can bomb it. But I need to get closer. I can't accurately target it from here."

"No," said Dostum. "You cannot get closer. I cannot let you get killed. Five hundred of my men can be killed before even one of you is scratched."

"I know that, I understand that, but you have to understand I can't call in bombs from this distance. I've got to fix the bunker on a map and call that coordinate to the pilot."

Dostum's aide, Chari, was translating. Nelson wondered if he was getting it all correctly.

Nelson didn't like haggling with Dostum about how to do his job. He wanted to ask him why he'd left them in the lurch back at Dehi. But something stopped him. A feeling he had. That he had passed a test by getting to the mountain headquarters and there was no sense in asking for the test's reason. To do so might signal that he was uncomfortable with accomplishing the unknown.

He thought maybe this argument about dropping bombs was another test. Dostum's men bowed and curtsied to the general's every move and it was clear that he was the bull of the woods in these parts. Nelson could see how this arrangement would work: Dostum would be in charge and Nelson would make him believe that this was true.

"All right," he said, "we can drop bombs. I'll set it up."

Dostum smiled. This made him very happy.

In fact, Dostum was relieved. He had his doubts about the Americans' abilities. He had never seen their bombs. They were reported to fly where you told them to go. By themselves. Like iron birds. This technology had not existed in his long war with the Soviets and then with his fellow Afghans—otherwise, of course, he would have used it. He was hoping the Americans would surprise him again, as they had last night when they landed so expertly on the small dirt pad in complete darkness. Dostum had thought it was most incredible to fly helicopters at night with no lights. Unbelievable.

Nelson asked, "And you're sure these are Taliban, right?"

It was Nelson's responsibility not to drop anything on anybody unless the target was clearly defined as an enemy position. Those were Nelson's rules of engagement.

Otherwise, various warlords could use the Americans and their "wonder bombs" to take out rival factions. Hell, at this distance, Nelson couldn't be sure exactly what he was looking at, but he was determined to find out. The fact that the bunker was so far north, in Taliban-held ground, was one indication it wasn't owned by Mohaqeq or General Atta, Dostum's archrival.

But Nelson wanted more. "You're sure they're Taliban?"

Frustrated, Dostum picked up a handheld radio, a Motorola walkie-talkie.

"Come in, come in, come in," he said, speaking rapidly in Dari. "This is General Dostum."

The small speaker popped to life. Dostum had raised the Taliban on the radio.

Nelson heard shouts and chatter, none of it sounding friendly.

"I am here with the Americans," Dostum said, "and they have come to kill you. What do you think of that?"

The radio roared even louder, as if that were possible. Dostum smiled. "See?" he said to Nelson. "They are listening to me.

"Tell me," said Dostum, speaking back into the radio. "What is your position?"

This was unbelievable to Nelson, that Dostum would ask such a question and that he seemed so sure to expect an answer.

Nelson heard the Taliban talk even more rapidly.

Dostum turned to Nelson and explained that yes, in fact, the bunker they were looking at was the Taliban's. No doubt about it.

And then it dawned on Nelson that the Taliban had no idea what was coming. He had a sickening feeling, elation and fear. That he might get ahead of himself if he got too confident in the fight.

The Taliban believed they were invulnerable. This in itself was an incredible discovery. At the moment, he was possibly the only guy on the planet who understood it. Like Dostum, they had never seen what a relatively inexpensive GPS mounted inside a $20,000 bomb could do. The places it could fly.

He realized they had never fought a war like this one.

Dostum signed off by saying, "Thank you, that is all."

He told Nelson that he talked to the Taliban all the time. Some of his men had brothers or cousins in their army, he said, either by choice or conscription. And sometimes these men came to him and asked, "Sir, could we not attack a certain place so strongly today?" "Why?" Dostum would ask. "Because my brother is there. He is a good man. I do not want him killed." And if Dostum could afford it, if the position could be determined to be of some minor importance, Dostum would hold the attack. That was the way war worked, Dostum said. Everything was possible. Forestalling death was a negotiation.

He explained that the Taliban had called him on his Motorola after the attacks in the United States. "The Americans will be coming," they told Dostum. "Who will you fight for?"

Dostum had laughed. To his mind, the Taliban were fools, ninnies—on top of that, they were boring men. They did not drink. They hated women. They were a social nightmare.

He described how a few weeks earlier he'd even met with some of them in person to discuss the future. Was he worried about being killed? Not really. He knew he was worth more to them alive than dead. In him, they had a known quantity with which to negotiate: a man who would make deals.

If they killed him, who knew who might take his place? Perhaps a man like Usted Atta, who would show no mercy at a critical juncture, a man who was as unbending as a teacher's ruler.

"You are a Muslim," the Taliban scolded him. "Don't work with the infidels." They explained that Osama bin Laden himself had proclaimed jihad against the Americans.

Dostum squared himself and looked the Taliban leader in the eye. He told them, "Your jihad is useless. Don't come to me with your talk of jihad." He practically spit out the words.

He grew angrier: "Even the Muslims hate you. You have committed a crime against humanity that is unforgivable."

He asked the Taliban soldiers how many women they had stoned. "Hundreds? Thousands?" he yelled.

He wanted the Taliban to feel some sense of shame. But he could see they did not.

"Here is what I am going to do," he told them. "I am going to do a 'hometown thing.' Pack up your things. Pack your trucks. Leave the north, leave Mazar-i-Sharif. Go back to where you came from. Don't come face-to-face with me again. Don't bother me.

"That," he said, "is what I will do for you."

He went on: "But if you stay and fight, I will kill you. I will hunt you and kill you."

The Taliban had been flummoxed by Dostum's bravado. They didn't know what to think. The man did not seem to fear them at all.

The wind whipped dust devils up and down the hill that went spinning into the valley and vanished with a visual pop in the air. Nelson knew it was a fearsome thing he was about to do. He stood at the edge of the trench with his hands on his hips looking at the valley and the Taliban positions beyond. He jumped down into the trench and landed waist-deep in the narrow slit. The Taliban were on the other side of the river. The river ran north and south, but at this particular place it took a bend to the left, or west, so that the Taliban were actually on its north side. Ahead he could see the village of Beshcam about three miles away, several dozen mud houses, no people. And beyond that another village, a tiny brown flare on the horizon through the optics of the binoculars.

At the sound of the approaching Taliban tanks, the villagers of these various places had scattered into the hills, hiding. Even farther north was Chapchal. If Nelson could push the Taliban north and reach Chapchal, then they could take Baluch. From there, the Taliban would have to fall back up the river, to Shulgareh. In this way, they could drive the Taliban north to Mazar-i-Sharif, forty miles upstream.

They would start with Beshcam.

One of Dostum's men stepped forward and spread a blanket on the trench berm and Nelson turned and said *"Tashakur"* (thank you) and leaned forward with his elbows on the blanket, still looking at the country through the binoculars.

He let out a breath and waited for the image to settle and focus before his eyes. A collection of eight Taliban pickups loomed up in the water-clear depths of the binoculars, looking as if they were only a quarter mile away. Close enough to see the battered black doors. The blink of the windshields like dusty mirrors. Toyota Hiluxes. In the cramped beds of each, several dozen Taliban soldiers were sitting along the rails with their robed knees touching in the middle. Rifles over their shoulders. His first clear sight of them. He fiddled with the focus wheel to sharpen the image. Their turbans black as crows' wings.

"We'll have to get closer," he said, hoping the older man would relent.

Dostum cut him off. "We will bomb from here."

Nelson shrugged. So be it. He had to give it a try.

He reached down his shirtfront, pulled up the GPS hanging by a lanyard on his neck, and read the pixelated numbers in the gray window of the device. These were the latitude and longitude coordinates that marked his position. He had to shield the small window with his hand to read them in the sun. He read them twice to make sure he was not making a mistake.

He wrote the numbers down in a green, hardback notebook he kept in an oversized pocket on his shirtsleeve and circled them to keep them separate from the rest of the numbers he would be writing down. He did not want to mistakenly give them to the pilot, who might confuse his position with the enemy's.

Nelson asked Dostum to unroll the huge map, and the anxious warlord did so.

"We are here," Nelson said, pointing at their ridgetop on the paper, one of thousands of elevation lines on the map. He reached into his rucksack, lifted his range finder to his eyes like a mariner's spyglass, and shot the distance to the pickups. He read the numbers in the scope's reticule and marked them down: they were eight kilometers, or about five miles, away.

Nelson looked down at the map and counted the grid squares from his ridgetop until he'd marched out eight squares with his finger (each grid square equaling one kilometer) and found the Taliban's approximate position on the far hill.

He set the range finder in the dirt beside him and looked up at the position across the valley with the naked eye. Yellow sun. The far hills looking sheared as if by tiny, incessant jaws. The Taliban were dug into the hillside and their trucks were parked below the bunker about 100 yards off. A small path led up from the hill and into the mouth of the bunker. The doorway was framed up by thick timbers with a heavy piece of wood overhead as its lintel. Nelson studied the scene. He wanted to absorb the raw image of it. When he felt he had done so, he looked down at the map and translated the image onto the elevation lines fanned on the paper. He did this several times, looking back and forth between map and hill until he felt he'd found the position on the paper that corresponded to the features of rock and slope across the valley. He now had a fix on the Taliban's position.

He wrote the position's grid coordinates in the notebook and scribbled "Enemy CP" (Command Post) beside them. These were the numbers that he would radio up to the B-52 overhead.

He looked up, squinting. There it was, the jet, miles above, barely visible. It looked like the silver nub on the screen of a child's Etch A Sketch scoring a wide oval in the sky, about twenty miles long and ten miles across. The pilots called this the plane's "race track." The plane was waiting up there. All Nelson had to do was pick up the radio and relay the numbers.

He hoped by God that he didn't screw it up.

The satellite radio was heavy and boxy like a home stereo receiver, powered by a green battery about the size of a half-gallon ice cream carton. A cable snaked from the radio across the dirt to a black, spindly contraption that communications sergeant Vern

Michaels had erected as Nelson plotted the Taliban position. This was the radio's satellite antenna. It was shaped like a small, charred Christmas tree. Michaels had set it in the dirt and carefully pointed the crown at the sky in the direction of government satellites circling overhead.

Nelson keyed the mike and identified himself as "Tiger 02," his call sign. The pilot came back acknowledging himself as "Buick 82," and said he was cleared to drop.

Nelson read off the coordinates and the pilot repeated them back. Up in the cockpit, the pilot reached over with a gloved hand to a console and typed in the numbers.

He then pressed "send."

The coordinates circled backward in the plane through the skeleton of circuits and entered the bomb bay, where they dropped into the bomb and came to rest in the GPS, which was about the size of a paperback novel, strapped inside the bomb near the tail.

The bomb awoke. It was now armed.

The pilot announced, "Pickle, pickle, pickle"—the traditional announcement of bombs away—then pushed another button on the console and the bomb fell from the belly of the plane and started flying to the ground.

"Thirty seconds," radioed the pilot.

"Roger. Thirty seconds."

As the bomb fell, the GPS monitored its position and checked this against its destination. As the bomb rocked and hummed in the slipstream, the GPS sent signals to the tail fins, which feathered the breeze and ruddered the projectile on course.

"Twenty seconds," said the pilot.

The bomb was 12 feet long and filled with about 1,200 pounds of explosives. It was tapered at its green nose to a sharp point and could fly fifteen miles from drop point to target. It was called a JDAM, short for joint direct attack munition, but it was known informally among the Americans as a "smart bomb," as opposed to the "dumb bombs" dropped by the millions over Europe and Japan during World War II.

"Ten seconds," said the pilot.

Several weeks earlier, Nelson reflected, he'd been sitting in his car at the Wendy's drive-through at Fort Campbell waiting for his order. *O Lord God, do not let me miss.*

Nothing was moving upon the far hills. None of the men in the trench was moving. They were silent. They stared at the Taliban position.

And then: the mushroom cloud.

They saw the explosion before they heard it, and then came the *boom*.

It came rolling up the hill and over them on the ridgetop like a train, and kept going and receded behind them and faded.

The violence was fearsome. Scary. Nelson had called in air strikes before, but never on people. Always in practice.

He scanned the Taliban position with the binoculars, anxious to see the damage done.

As the smoke cleared, he saw that something had gone wrong. He scanned the hill again. The bomb had missed the bunker. By a lot. Maybe a mile, maybe more. It had landed between them and the target.

He wondered if Dostum realized the mistake. He was about to explain what had happened when he noticed that one of the general's most trusted aides, a man named Fakir, was moving up and down the line of Afghan troops slapping high-fives. The men with him were laughing. Dostum himself was grinning.

For the moment, Nelson decided that he'd keep his disappointment to himself.

As he was thinking this, the second bomb hit.

This explosion was even bigger and Nelson saw that this one, too, had fallen short.

The Taliban were filing out of the bunker and looking around—up at the sky and across the desert—unsure of where the big noises were coming from. After a short while, they went back inside. Dostum's men continued laughing at the sight.

Nelson was sure he'd figured the coordinates right. Maybe the B-52's crew had plugged them in wrong. . . . He got on the radio and told the pilot to correct for elevation.

The pilot dropped again. This third bomb hit closer. Nelson figured it had fallen about 600 feet from the bunker—two football fields away. He was going to have to do a whole lot better than this.

At the sound of this closer explosion, and with the air still thick with smoke, the Taliban poured out of the bunker—maybe a hundred men or more. They came running out in a crouch with their weapons at the ready, as if under infantry attack.

As soon as they saw the smoking crater, they stopped. If they saw the B-52 overhead, they didn't seem to connect its presence with the sudden appearance of the ten-foot-deep crater at their feet. Nelson felt like he'd traveled back in time. Here he was riding a horse loaded with sophisticated electronic gear and ordering bombs to be dropped from planes that were flown from Diego Garcia, 2,800 miles away, in the Indian Ocean. As the men on the team would later say, this was like the Flintstones were meeting the Jetsons.

The Taliban stood puzzled at the crater's edge. And then a few of them started walking around inside the smoking hole, shaking their heads, as if to divine its origin. Nelson was getting madder by the minute. He might as well have been standing on Mars and phoning the war in. They didn't know he was even there.

He decided he would recalibrate the drop. Before he could reach the pilot, however, two more bombs exploded. These were even farther off target and landed two miles or more from the bunker. Nelson barked into the radio that the pilot should check fire.

He turned to Dostum ready to apologize. He wanted to say, *That's not who I am, I am capable of more.* But he knew this would starkly draw their relationship and make Nelson into a man looking for approval. It would shift an unspoken balance in Dostum's favor.

Fakir saw the disappointment in Nelson's face. "Don't worry," he said.

"What do you mean?"

"You have made explosions come from the sky. The Taliban are afraid!"

Dostum was speaking happily into his radio with the enemy: "I warned you that I had the Americans here! How do you like me now?"

Nelson saw his opening. "Well, I can do a whole lot better than that."

Dostum wanted to know how.

"Get me closer to those sonofabitches."

Dostum wondered what choice he had. He knew he himself didn't know anything about dropping bombs. The young man seemed serious. He liked his aggressiveness. He was tireless, like himself.

He announced that he would take him to the Taliban.

• • •

That night, before turning in to bed (they maintained guard duty two hours on, two hours off), Nelson looked down the hill behind him where the Afghans were standing beside their horses and removing the saddles. They reached into bags and lifted out what looked like iron nails, measuring at least two feet in length, and as they lifted them in the dusk they looked like men in miniature handling the tools of giants. They dropped the nails on the ground with a thud and bent and grabbed them with both hands and righted them with the point sticking into the dirt. They hitched their smocks at the knee and raised a foot to stomp on the blunt end. They then tested the nails by trying to jiggle them back and forth.

They unspooled long leather leads and ran one end through the horses' bridles and tied the other ends around the mushroomed heads of iron. The horses began walking in circles around these new centers as the riders rummaged in their meager kits, which looked to Nelson like colorful piles of rags, containing some food—bread, nuts, dried meat—and maybe an extra blouse and a spare shoe. Walking back to the horses, they snapped open feed bags sewn from UN flour sacks and dropped in handfuls of oats and corn, lifted the sacks up around the horses' muzzles, and reached up and tucked the straps over the ears and stood back as the horses chewed hungrily, making a muted, wet music inside the steaming burlap.

The riders walked up to a campfire and sat on their haunches, eating and staring at the flames. The air smelled cold. The stars drifted up from the horizon as if loosed from a zoo and swarmed the dark above them. When they were done, the riders stood and brushed the breadcrumbs from their smocks and walked down to their horses, talking quietly. They threw their blankets over the animals' withers and patted their heads and said good night. They walked up the hill to their cave and went inside, and Nelson could hear them talking low as they lay down shoulder to jowl uncovered in the cold night. When he realized that they had given their blankets to their horses, Nelson believed that if he fought with the same selflessness, none of them could be beaten. And that he would live.

They turned out of the camp at midmorning and rode down the hill with the sun hot on their backs and the horses rolling beneath them along the path. Then the riding got hard, the path rocky.

They climbed several thousand feet across the face of the mountain and descended again. Going where there was little sign of a trail. Where no truck or tank had ever been.

As Dostum rode, Nelson heard him on his satellite phone talking to congressmen back in the United States, politicians in Pakistan and Russia. One of the men he talked to was a hearty character named General Habib Bullah, who at that moment was fighting in the Taliban army. Habib had, in fact, forty men under his command and was dug in in a trench in Chapchal, the village Dostum and Nelson were now trying to take.

A former general in the Afghan army (the anti-Soviet army), Habib had been imprisoned after the Taliban took power in 1996. His family urged his release by promising that Habib would join the Taliban army. Reluctantly, Habib agreed. What else was there to do? Habib had rented his skills as a soldier for his freedom.

He loathed the Taliban. He was one of the select few Dostum had given a phone to in order to communicate secretly about their movements. Mendicants, zealots, misfits, revolutionaries, much of the Taliban army had dropped through so many hoops and wickets to finally find themselves resting at the bottom of the Taliban trench that their provenance was a lesson in Middle East geopolitics.

Their reasons for coming ranged from the absurd to the sublime. One soldier fled his native Iraq to avoid embarrassment after being accused of having a "small penis." He felt fighting jihad would make him seem a man in his neighbors' eyes. He was sorely disappointed in the hard work this entailed. "The Taliban is fucked up, I'm serious," he would say after his capture in late 2001. "They pray like twenty times a day. That's too hard for me."

A soldier from China, a Uighur, part of a politically oppressed minority in that country, had found it easy to travel to Afghanistan. There were practically no immigration rules. He was shocked when the United States started bombing in October. He had come to Afghanistan in order to escape oppression in China—he thought that if the Americans knew he was there, they would stop. Didn't they know that the Chinese government would take a woman pregnant with a girl and open her up and throw the baby in the street? He called himself a "normal" Muslim, a man who wanted to live as one and be free of such homicidal horrors.

Another soldier had come from the United Kingdom to attend the military training camp called Al Farooq. "I am a Muslim and Afghanistan is a Muslim state. I belong there." He had been in trouble with police back home and thought that by moving to Afghanistan, he could start a new life. Near the end of his training, when he learned that the camp was funded in part by Osama bin Laden, he told himself, "I will finish anyway. I am almost graduated." He was afraid that if he didn't, people would say that "I wasn't a man and I couldn't handle the training."

As he rode, Dostum talked to journalists, too, from around the world. They wanted to know if he had American soldiers with him. "No, of course not!" he lied. "I have only some humanitarian aid workers. They are here helping me hand out 'lead' to the Taliban."

Lead?

"Yes, lead! It is in short supply here." The joke either confused or mystified the reporters. Dostum thought it was hilarious.

The outpost at Cobaki sat about two miles from the mountain headquarters, but because of the terrain, they rode at least five miles along a crenellated maze of switchbacks, dead ends, precipices. Nelson trusted that Dostum would lead them safely. He knew he was breaking rule number one, which was: Trust no one. He and Diller, Jones, Bennett, Coffers, and Michaels kept their hands near their weapons.

As they rode, the Americans struggled with their horsemanship. At times they could be downright comical. At one point later in the campaign, Fred Falls's horse, an irascible stallion, leaped off the trail without warning, ignoring the switchbacks, and started running down the mountain face. Falls would later remember leaning back in the saddle because he had seen an actor in the movie *The Man from Snowy River* do the same thing—and survive just such a ride.

Falls's head was bouncing up and down on the horse's butt while his hiking boots were flailing up around its ears. He was yelling at the top of his lungs, "I don't want to die!"

At the bottom of the run, the horse spied an eight-foot-wide ravine; Falls saw it, too, and yanked on the reins. The horse leaped and was airborne, sailing downhill, making a perfect landing, and galloping to the bottom. Falls pulled the reins and the horse began making a circle, as if Falls were on a merry-go-round, the horse going faster and faster then finally slowing, until it stopped and began grazing a few sparse stalks of grass.

Falls sat up, amazed that he had survived. He had covered so much ground so quickly with his shortcut that it took General Dostum and Nelson ten minutes to reach him.

When they did, Dostum rode up, gazing at Falls. He said something quietly in Dari as he passed and rode on without stopping.

"What'd he say?" Falls asked a translator.

"He said, 'Truly, you are the finest horseman he has ever seen.'"

"Tell the General thank you," said Falls.

After a four hours' ride, they arrived at the austere, windswept outpost, altitude 4,800 feet.

In the distance, tucked on top of a hill, stood about forty mud houses. The window openings dark and empty. No people about. No animals. The village looked like something recently excavated from the earth. Through the binoculars, the edges of the buildings were sharp and straight and the walls smooth. Flat roofs. The dust was rising and coloring the air amber as the morning drafts roared up from the valley floor. Nelson set to work preparing the bomb strike on Beshcam.

The Taliban had dug a trench line into a hill near the village, which was about two miles off. Nelson saw a collection of brownish-looking pickups that the Taliban had camouflaged that color, he would later learn, by pouring gasoline over the body and then adding shovelfuls of dirt and mixing the two in a thick paste. Water was in precious supply.

Nelson glassed the country and saw several Taliban tanks, Russian models called T-52s, parked on a hill behind the trenches. The tanks could fire a six-inch shell, at five to seven rounds a minute, just over a mile. Over rough ground, the 45-ton behemoths could travel 35 mph; on roads, they could speed at 50 mph. They would be a formidable opponent for men attacking them on horses.

Nelson also spotted several Bimpies. They were designed to protect infantry during an assault, armed with a 100mm gun, a 30mm cannon, and three machine guns bristling from the blunt metal bow. Like tanks, their reach was far—they could hurl fragmenting shells up to two and half miles.

And there were several ZSU-23s—called "Zeuses" by the Americans—which, with their turrets sprouting four 23mm cannons, were normally used as antiaircraft weapons. The Taliban had learned to back these up a hillside so that the uplifted barrels were tipped horizontal to the ground and could be fired, at a combined demonic rate of 3,400 to 4,000 rounds a minute, creating a furious wall of lead in the air. It was through this wall that men and charging horses would try to ride.

Nelson wanted Dostum's men to attack immediately. Dostum had other ideas. He suggested they wait until later in the afternoon. The sun would set at 6 p.m., he said. They would launch the assault at two.

Nelson wanted to know why.

"Because there will only be four hours of daylight left once we are upon them."

Nelson didn't understand.

"This means they will not have time to regroup and counterattack." Dostum's men would fight them as darkness fell and use the night as cover for their outgunned men.

Nelson called in his first aircraft. Because he was about two miles from the Taliban position (the day before he'd been at least five), he felt he could be more accurate in his plotting.

When the first bomb exploded, he was sure of it.

The hit had been direct. The Taliban pickups disappeared in a cloud of smoke. Twisted steel and body parts littered the blackened ground. He set up the next strike. This time he took out several of the Bimpies. After that, he went after the tanks. He was setting up targets and knocking them down.

"How many men do you have?" he asked Dostum.

Dostum said he had three divisions led by Commanders Ahmed Lal, Kamal, and Ahmed Khan. Fifteen hundred men on horses and an equal number on foot.

"What about the horses?" Nelson asked. "How will they react when bombs start dropping?"

"They will not be nervous," Dostum said.

"Why?"

"Because they will know that these are American bombs."

Nelson thought that over. Dostum appeared serious. And then the warlord smiled wryly and slapped him on the back.

That morning, he had sent the first men from the area—300 of them—up the cliffside trail, to the top of the gorge, onto the north side of the river.

The riders had come to the top, turned right, and ridden through a draw or valley screened from the Taliban by a tall hill. They were able to secretly position themselves near the middle of the plain. The Taliban knew Dostum's men had come out to the battlefield, but they didn't know their location.

Through the binoculars, Nelson looked east across the valley and saw the horsemen all carrying weapons, AKs and RPG tubes, glittering belts of ammo wrapped around their shoulders. They milled about the plain, their horses picketed nearby chewing ravenously at the grass. Nelson thought back to fifteen years earlier when he'd been commissioned as an officer in a ceremony on the Shiloh Civil War battlefield in Tennessee. He'd studied the cavalry tactics of Jeb Stuart and John Mosby, whose "Mosby's Raiders" had ridden circles around Union troops in lightning attacks. Now he sat ringside to the first cavalry charge of the twenty-first century.

Dostum was yelling in Uzbek; the radio buzzed with anxious voices. Half the men had mounted their horses and ridden them in a line a quarter-mile long on the back side of the hill. They were still hidden from the Taliban. It took at least a half hour for all the horsemen to align themselves properly, about 150 of them. They sat on their horses lifting their calloused, scarred hands to their mouths and speaking into their walkie-talkies, with more men joining the line and then being told to ride out and wait behind the screening hill to form the second wave. It was thrilling.

"Charge!" Dostum shouted into his radio.

At this, the men bolted. They shot up in their saddles climbing the back side of the hill and crested it and let out a yell and then dropped down and disappeared, at least from the Taliban soldiers' view. Nelson could watch all of it from the side, in profile. Dostum was beside him nervously muttering and speaking into the radio and ordering further corrections in the flying mass of thunderous horses and shouting men.

There was about a mile of ground between them and the Taliban line, a helluva distance to cover, Nelson thought, without getting shot. What he saw as he looked at the plain was a series of seven bare ridges, between 50 and 100 feet high, with about 200 yards of bare ground separating most of them, calm water between tall waves. As they rode, the horsemen continued rising up and down the hills, appearing and disappearing. When they got about halfway across the field, the Taliban guns opened up.

The noise was deafening, shells and bullets whistling over the ground at head height. Men would be riding in the saddle and then suddenly fly backward as if yanked and tumble to the ground and lie motionless as more horses approached from behind and leaped over them, charging toward the firing line. Sometimes a wounded man got up and limped away or held out a hand and swung up into the saddle of a fighter smashing past at full gallop. The riders leaned out over the stretched necks of the horses, firing as they ran, the long, dark reins clamped in their teeth.

The last four ridges comprised a half-mile stretch of the battle-field, with maybe as much as 1,000 feet of open ground between some of the ridges. The Taliban tanks sat squat and black, belching thick yarns of smoke on the last ridge.

Nelson could see the Taliban raising and lowering the tanks' bar- rels, trying to adjust the fire. It was hard, often impossible, as the horsemen swarmed toward them. They were moving fast, growing larger and larger, from the Taliban point of view, as they approached.

As they reached the second to last ridge, the riders halted and jumped down from their horses. They threw the reins on the ground, stood on them, lifted their rifles, and began firing methodically at the Taliban line. Some of them were frightened and sprayed the line at full auto. Others shouldered RPGs and fired. Smoke trails whizzed over the open ground as the flying grenades exploded among the Taliban.

As they did this, the second wave of horsemen advanced under the covering fire. They were riding hard from behind and overtook the men on the ground and blew past, shouting, galloping straight at the Taliban line. The standing fighters swung back into their sad-dles and beat their horses to catch up, and the two waves joined in one line as they drew near, within shouting distance of the Taliban.

Standing near Dostum, Fakir was listening to his radio, tuned to the Taliban frequency. He could hear them yelling, "We can't resist. We have to move!"

As the horsemen charged, many of the Taliban stood up and looked behind them and then at the horsemen, and they threw down their weapons and started running, their black smocks flapping, the men slipping on rocks in their cracked dress shoes, falling and rising quickly as the horsemen thudded behind them.

The horsemen leaned from the saddle to reach out and club them with their rifles and dismounted to finish them with knives. Or shot them in the back, the Taliban throwing their arms wide, turbans unraveling, as they fell face-first in the hard orange dirt.

The fight raged as the light failed, the moon coming up. Just before dark, Nelson spotted a Bimpy and a tank still untouched. They had crept over the edge of the hill where they'd been hiding in reserve. Now they were raising hell, their turrets sweeping the field, firing in measured booms. Nelson and Dostum had been trying to raise the commanders on the field to tell them to attack the vehicles. Either the commanders couldn't hear the radio transmission amid the din of gunfire or things were too chaotic for them to organize an attack, but they didn't respond.

In the meantime, Nelson managed to talk a pilot overhead onto the two targets.

Watching Nelson at work, Ali Sarwar, one of Dostum's commanders, was fascinated by the bomb strikes. He saw Nelson speak on his radio and then scribble numbers in a notebook. Overhead, high up, there were jets circling. They dropped bombs after he scribbled down the numbers. It also appeared to Ali that the back end of one large plane would open and four smaller jets would come out and then start flying around and bombing the Taliban.

Fakir, too, was confused by the sight. Thinking back to the attacks several weeks earlier in the United States, and the way those planes had been rammed into buildings, he wondered if the big plane was being hijacked.

He asked General Dostum on his radio. "Sir, I see something I do not like. Smaller planes are chasing a big plane. What is happening?" Dostum was not exactly sure how aerial refuelings worked, either. It didn't seem likely there was something wrong, but he asked Nelson, "Fakir sees little planes chasing a big one. Is there a problem?"

"The small planes you see are jets," Nelson said, "and they're being refueled by the larger one, which is a fuel tanker. Everything's fine." He was surprised by the innocence of the question.

Now, as the pilot seemed ready to drop on the tank and the Bimpy, he told Nelson over the radio, "Dude, I'm sorry, but I am bingo."

"Bingo? You're kidding me."

"No, I'm sorry."

The pilot had reached that point in his fuel consumption where he had just enough left in the tank to return to base.

"But we're close. You gotta drop, man, I'm begging you."

"Sir, I am out. Bingo."

And the bomber peeled off its race track and headed back to base. He would be arriving just before sunrise, out in the Indian Ocean, on Diego Garcia.

Nelson was furious. He watched in dismay as Dostum's men wheeled and, looking over their shoulder at the charging Bimpy and tank, hunched down in the saddle and kicked their horses and beat them with whips to make them move faster as they abandoned the enemy line they had swarmed and captured.

They rode back over the plain through dead bodies and parts of bodies, heads setting upright in the dirt as if the men had been buried up to the neck. The faces slack and impassive as the horses' hooves passed, inches away. The dead eyes brimming with the sudden flare of sunset.

But Nelson realized: *We can win.* An idea hit him. *If we can coordinate air support, we can beat these guys and kill the armor they bring up out of reserve once we've kicked their ass.*

He knew that the tank and Bimpy hadn't been in the Taliban line when the charge began. They had driven to the battlefield from the west. That's when he had his second realization: *I've got to split the team again. I've got to send somebody north so they can blow up these tanks before they get to us.* He would have to send a man farther behind Taliban lines, someone who could spot the tanks and who had the skills to precisely call in bombs. That man was Sam Diller.

He got on the radio and called Spencer back at the Alamo. "Come on up," he said. "I'm sending Diller downrange."

Nelson and Dostum rode back to the outpost at Cobaki. Men limped into camp bloody and exhausted. The badly wounded in need of medical care were carried down the cliffside wrapped in blankets as provisional stretchers. There was a steady stream of them moving down the trail in the night, screaming and moaning, accompanied by the rattle of the men's gear and the occasional snort of a horse coming up behind them in the dark, passing down the trail, and moving on.

At the foot of the trail, the stretcher bearers turned left, or south, and picked their way downstream along the river, the sound of the rushing water covering their movement. They had a long way to go, several miles, until they appeared at the gates of the Alamo, where they burst in and laid the stretchers in the courtyard, blankets now slick with blood.

At the Cobaki outpost, Dostum told Nelson that he wanted to send his men across the gorge again the following day. He realized they could use the bombers as a new kind of artillery fire. They would have to time the bomb strikes perfectly so that the horsemen started their charge just after the bombs landed.

Dostum also said he had made a decision. "I want you to come with me," he told Nelson.

"Tomorrow," he said, "we will go to the battlefield together."

As Nelson brought the fight to the Taliban, Pat Essex and Cal Spencer were running the logistics train at the Alamo. It was boring work, though neither man would admit as much, but the log train was no less important for Nelson and Dostum than to be with them on the front line.

Essex pumped water to be sent out into the field, using a hand pump and filling five-gallon water cans. He would pump for thirty minutes, then somebody else would take over.

Scott Black, the medic, was busy treating the maladies of the local villagers in Dehi, while Spencer retrieved the supply drops that came in. Black's trips into Dehi were a simple and perhaps obvious subterfuge. As he examined the teeth of the local children and listened to the heartbeats of their parents through his stethoscope, he would ask questions. "Are there any Al Qaeda in the area? Taliban? Who are the good guys and the bad?"

Black and Spencer both noticed that the locals seemed to genuinely like them. The Americans made educated attempts to speak the local language, which endeared them. The team took pains to look away whenever a woman passed by, even though she might be dressed in a full-length burkha, her face hovering unseen behind a piece of latticed gauze. They were trying to become the Other without going native. The truth was, the people in the village saw the Taliban as the invader, even though many of them were Afghan citizens. Black was part of the insurgency trying to overthrow the Taliban.

Getting supplies air-dropped to Nelson and Dostum was proving more complicated. The first drops had thundered in from planes flying at 20,000 feet, and Spencer screamed and yelled in his e-mails to headquarters that you couldn't drop from that height, but the pilots kept saying, "We hear there are SAMs [surface-to-air missiles] on the ground, so that's our hard deck, twenty thousand feet." The bundles had come whistling down and hit with a god-awful *whoompf*—if anyone was around to hear them: they often missed their mark by a half mile. Spencer worried that one would land on a house. He imagined it would flatten the modest mud structures.

He and Essex approached the drop zones at the scheduled hour, usually to see the bundles miss their mark or explode on impact. It was then a mad race—sometimes in minefields—to collect the scattered water jugs, boxes of MREs, and sundries before the locals arrived. The locals would shoot at each other with their AKs as they stripped the crates bare, making off with rice, MREs, horse feed, and bandages. They even dragged away the aluminum crates and used them as roofs for their houses.

The pilots had finally conceded and agreed to drop the bundles from 800 feet, once it was clear the SAM threat was minimal. Not surprisingly, the drops' accuracy went up as the planes came down. Of course, bundles being kicked out the door at that height took less time to land, making for an even crazier dash for Spencer and Essex to gather everything before the locals rushed in. What they missed they had to buy back on the black market. Sleeping bags went for $10, cooking stoves for $15.

Spencer and Essex could have strangled the Air Force guys back in Turkey who were assembling the bundles. There was no rhyme or reason as to how they packed things. The boxes weren't marked or they were poorly marked. Occasionally, radically different items were mashed into the same boxes—some piece of equipment might be stuffed in a carton filled with MREs. Spencer and Essex nearly gave such a box away to the locals. It contained computer cables.

Once, at a drop near the Amu Darya River, Spencer rushed to where the bundle had fallen and spied what he thought were containers full of gasoline and said to himself, "That would be cool. That's something we could use. And we could give some to the locals. They sure do need gas."

An Afghan man unscrewed one of the containers and sniffed, a dumbfounded look on his face. He smelled again and said, "Water?"

He stared at Spencer and then motioned to the river. "You want water? We have the river!"

Spencer just turned away, thinking, *Nope. I can't explain this one.*

Sometimes, bundles hit the ground and vanished as if by magic. Without a trace. Once they couldn't find a bundle containing mail from home. Spencer and the team knocked on doors and asked if anyone had mail for the Americans. The men held out dollars, American dollars, and kept going from door to door. They came upon an Afghan man driving a donkey cart along a rutted path who was wearing a chemical warfare jumpsuit complete with gas mask and hood. He was also toting a black tool crate with "ODA 595," Spencer's team number, painted on the back. It was packed with twelve sets of camo clothing for the team. They never found the mail.

As Nelson and Dostum maneuvered across the mountains, Essex spent mornings around the cookfire heating water for tea and chewing warm, freshly made nan, one of the team's few comforts as they adapted to the harsh conditions. Like the Northern Alliance soldiers, their beards grew coarse and foul, their hair stringy. They raked it with cheap black pocket combs and brushed their teeth using their fingers. Ben Milo, the B-team weapons specialist, had brought in three bars of soap stuffed in an old gym sock for the entire deployment and he apportioned them sparingly. He took what he called "ho" baths, washing his armpits and privates. All of the men began to reek.

A rustic existence, yes, but they had trained for this. The men used a provisional shitter in the courtyard, although with its door torn off by the wind, they might as well have been shitting in the open air. Not the Afghans. They crapped everywhere, wherever they felt like it. Their knee-length smocks allowed them to squat at will and do their business. Essex didn't comment on the habits of the Afghans. As a guerrilla soldier, he'd been taught to let things be as they were. He wasn't there to change how anyone relieved himself.

Hour by hour, Dostum sent couriers from his position, riders who galloped up in the dirt yard to inform everyone what was going on. Essex was having trouble arranging for supplies to get to Nelson. He needed horses to carry the gear.

Normally, a healthy animal would cost $300, but when Essex attempted any sort of haggling in Dehi, he found the price had gone to $1,000. Still, Essex wondered, what choice did he have? His guys had to get their beans and bullets. He figured he would buy twenty horses for the team to ride and to transport supplies. But soon there weren't any for sale. Whatever animals were left behind had been snapped up by Afghans riding them into battle.

Making matters worse, Essex was apprised of a command decision at K2 to force two Air Force personnel onto the team. Sergeants Mick Winehouse, twenty-eight, and thirty-three-year-old Sonny Tatum had been trained to call in aircraft on bomb targets, a job they did expertly. Essex learned they were being added because Nelson and the team weren't hitting targets, or at least enough of them. Essex was short on horses, and now he would have to find two more for these guys! He was livid.

And insulted. Didn't the higher-ups understand the Afghan way of war? They *were* winning. Nelson didn't have to blow up everything he aimed at. He had to make the Taliban *believe* he could smash any target, in any place. And this he was doing quite well.

Nevertheless, Essex would live with the decision. And he looked on the bright side. Tatum and Winehouse were bringing something called a SOFLAM, short for special operations laser marker. Nelson's team hadn't brought the bulky piece of equipment along because they'd figured to travel light.

Targeting a Taliban tank or truck was easier with the SOFLAM than the method Nelson was using. The device controlled laser-guided bombs—munitions right out of science fiction—and was packed into a green metal box about two feet square and six inches high. Attached to it, by means of a long cable, was a trigger grip. The box rested on a tripod and you stood behind it looking through a scope mounted on its top, inside which were crosshairs. You trained the crosshairs on your target and pulled the trigger.

This shot a laser from a lens, at which time the box began to make a *cheeping* noise, like a bird trapped in a cold oven. The laser was invisible to the naked eye, but it contained a code that corresponded to one programmed in a bomb carried overhead by a waiting plane.

The drop of the bomb was positioned so that it "landed" on top of the laser. The bomb then rode the light beam to its destination. You could move the laser as the bomb dropped and the bomb would follow, redirected in midflight.

This was helpful when "lazing" a truck whose driver had sped away. You could follow the vehicle: the last thing the driver would see was the missile suddenly appearing in the rearview mirror.

At dawn on October 23, Nelson and Dostum started riding from Cobaki to the battlefield across the river, out on the plain.

They traveled down the narrow trail to the valley floor and picked among the river rocks and crossed the river, which split into three separate braids, shallow and fast. Even the river had been mined by the Taliban and they had to be careful where the horses walked. They rode into the shadow of the far valley wall and in the cooler dark along the rocks, with the horses' hooves grating in the sand and ticking dully over the stones. They were heading south, hunting the trailhead to the top.

After an hour, they started to climb. With them were several of Dostum's commanders, among them Ak Yasin, leading sixty men on horses, and Ali Sawar, who had watched Nelson and his team step off the helicopter at Dehi four nights earlier. Upon seeing them, Ali had thought, *We will win this. We will beat the Taliban.* He still felt the same way now. As they rose into the sun, Nelson could feel the heat in the rock wall, and then it started to rain. Light at first and then heavy. The trail turned to mud. The horses started slipping, the pounding of their legs coming up through the saddle as they drove their feet under them. The riding was dangerous as hell. Nelson thought he would fall and roll off the mountain ledge, a drop of several thousand feet.

"Get off your horse and walk," Dostum scolded.

Nelson insisted that he would keep riding. It would seem embarrassing to walk. It would mean Dostum was a better horseman.

But Dostum was genuinely worried. Several weeks earlier, he had been riding up a mountain pass and his horse had slipped and the general had rolled downhill, catching himself just before he went off a ledge. As he lay on his back, he could hear rocks tumbling down the mountainside. If he had fallen, he knew that his men would be picking up pieces of him for weeks.

Now Chari, the translator and one of Dostum's aides, begged the general, too, to please stop riding and walk his horse. "You cannot be hurt, General." Chari thought not even a goat would try climbing this trail. "Who will lead us if you are hurt?"

"My body is not worth more than yours." Dostum was glad for the concern of his men. As a matter of habit he absorbed such adoration gladly and took it as a sign of loyalty.

Chari was a chubby man of thirty-six, with a neatly trimmed mustache. He had worried that the Americans would not be able to ride horses and fretted that some of them hadn't even known how to get on one. He had been fighting alongside Dostum for twenty-three years, and in the three years since the Taliban captured Mazar-i-Sharif, Chari had seen over thirty of his friends lose arms and legs in battle and to land mines. Now his life was in the Americans' hands and he had vowed that he would help them succeed. Captain Nelson, he felt, was like a brother to him.

For his part, Nelson took notice of the fact that Dostum was scolding him like a father and worried for his safety. Ever since their arrival, Nelson had been sleeping with a pistol under his sleeping bag, just in case. *Maybe the old man could be trusted. Maybe.* Nelson didn't ponder this for long. *Trust no one.*

He was riding behind Dostum when his horse slipped on the trail. It reared on its hind legs and paddled the air and tipped and Nelson was thrown. He hit the ground and rolled, taking care that he was clear of the horse if it was tumbling toward him. It wasn't.

He watched it flail and finally right itself. Nelson stood, brushed off his pants, pulled his cap right, and walked back to the horse, determined to get back on.

The horse stood with its chest plate of muscles shaking. Nelson talked slowly and reassuringly. He cocked his head. Above, from higher on the mountain, came the sound of wailing.

Nelson could look up and see a line of men on horses moving up the trail. At the top, on the rim, he saw specks, more men, moving around, milling, preparing for the fight. Nelson remounted and they resumed their climb.

As they rose higher, they heard the steady boom of an artillery gun. Dostum told him it was being fired by the mayor of a nearby village, who was lobbing shells at the Taliban line, randomly, but enough to distract the Taliban from the approach of Dostum's men up the cliff.

"Hurry, we must keep moving," said Dostum. He wanted to get to the top quickly.

He was worried that the Taliban would find them. They would come at them with their old Russian fighter jets and shoot them like flies.

The wailing grew and Nelson and Dostum soon passed men headed down the trail, the wounded and the dazed, vacant-eyed men coming back from the battlefield to the Alamo. They stumbled along. Some passed silent as statues, faces stitched tight in masks. Others whimpered like babies.

Dostum leaned from his saddle and saw a man groaning in the pouch of a coarse, rank blanket that was held tight at its corners by four straining men. The soldier's skull was cracked open. He rolled his head to the side and Dostum could see the brain, gleaming white, and then he rolled back and he looked just fine.

They came out on top of the valley with the plain spread beneath them in all directions. The day before, Dostum had gotten on the radio and put out the word that any man who could fight the Taliban should come to this place and be prepared to die.

Now there were some 600 men on horses and on foot moving out onto the plain, readying for battle. They remained screened from the Taliban's view by the hills. Dostum couldn't believe his luck that they hadn't been discovered. The enemy was just a half mile away. He and Nelson rode several hundred yards to a rock promontory and beheld the spectacle.

Using his radio, the excited general began directing traffic. First, 100 riders lined up behind the first hill. And then a second hundred on foot positioned themselves behind the second hill. Soon they were spread in six lines behind six hills, the horses rearing, the men shouting, cracking whips. The dust they raised drifted above the battlefield, and Nelson wondered if the Taliban could see it.

He glassed their position. Three tanks in waiting. And two ZSU-23s, one on each end of the Taliban line, which stretched about the length of a football field. The Zeuses had given them hell the day before. They had to take those out. Nelson guessed there were about 1,000 Taliban dug into the trenches, armed with RPGs, AK-47s, and mortars.

He realized that the Taliban had reinforced themselves and that this was going to be a bigger fight than yesterday's. Dostum's men would have to swarm the Taliban line; but if they paused, they'd be chopped down by the increased volume of fire. They had to make the Taliban break their position and flee. They had to attack the armor and disable it. They had to do what seemed impossible.

"We are in a good position," Dostum assured him. "Because if we can break them out of here, they will have to run. And they will keep running all the way to Mazar-i-Sharif."

"Also," he added, "after yesterday's bombing, their morale is low."

Fakir, Dostum's senior leader and most trusted confidant, had sent scouts ahead at night to probe the Taliban positions. By intercepting their radio calls, he had learned that they were terrified. They were afraid to sleep at night for fear of attack.

Dostum was delighted by the news. There was an old saying in Afghanistan: Death comes at any time, on the street, in a war zone. You never know. If today was the day, so be it. He was ready.

And with a man like Fakir at his side, how could he go wrong? They had been fighting together for fourteen years. Bearded, with piercing brown eyes and a wry smile, Fakir was from Dostum's hometown, Sheberghan, a dusty city clotted with gas wells and lashed by lonesome winds slicing off the steppe.

It was now midafternoon. In several hours, the plain would be dark. Dostum said they would start the battle soon.

• • •

CIA officers Mike Spann and Dave Olson were standing nearby on a rock outcropping of their own, overlooking the battlefield. Standing down below in the stiff grass was J.J. Mike was holding their three horses.

One of Dostum's men rode up, a thin, excited fellow in riding boots and a green field jacket. He announced that the attack would begin any minute. And that they should be ready. And then he galloped away.

J.J. asked his two friends if they were ready.

They said they were.

The CIA officers had spent the past days sending so many e-mails back to K2 and Langley, and attending so many meetings with warlords, arranging alliances, that one of their Afghan counterparts thought they might be glorified clerks. "I have these Americans with me," he had remarked, "but I don't think they're soldiers. They spend all their time with laptops."

Now they were going to prove otherwise.

To pass the time at the Alamo, Spann had followed a nightly ritual of fifty push-ups and sit-ups before bed, followed by twenty minutes of Bible reading. After that, he wrote in his diary, in which he recorded the comings and goings of a mouse he had befriended, and whose antics he reveled in recording for his children, and wife, Shannon, back home in Virginia.

To Shannon, he had written that he wished he could see her so they could slow-dance to a favorite song.

"One thing has troubled me," he wrote. "I'm not afraid of dying, but I have a terrible fear of not being with you and our son . . . I think about holding you and touching you. I also think about holding that round boy of ours. . . . It would be cool to have a slow dance with you . . ."

Now came the sound of the artillery gun starting the fight.

Nelson had been setting up the satellite antenna while Vern Michaels unpacked the heavy radio when they heard the gun. Nelson would give the order to drop the first bomb. He had already radioed the target coordinates of the Taliban line to a fighter jet overhead. With him was weapons specialist Charles Jones, who had arrived in time for the battle from base camp in the river valley. It was Jones's job to stick with Nelson as they rode into battle while Michaels manned the radios.

Nelson said goodbye to Michaels and turned his horse to follow Dostum. Dressed in khaki pants and black coat, a blue turban wrapped atop his head, Dostum sat upright as he glided smartly in the saddle onto the field.

Nelson felt his heart pound in his chest.

He and Dostum pulled up their horses beside J.J., Spann, and Olson.

J.J. was carrying a Browning 9mm pistol in a holster on his right thigh and an AK-47 on a strap around his neck, hanging down within easy reach. A pouch stuffed with ammunition was slung over his shoulder. He was dressed in jeans and hiking boots from L.L.Bean; a knit cap was pulled down over his ears. Spann was sitting on a white horse that was too small for him. He was dressed in jeans, black T-shirt, and a gray overshirt with a pair of binoculars in the large front pocket.

Dostum explained the battle plan. The horsemen would charge the middle, the infantry would attack the flanks, and machine guns set on adjoining hills would spray covering fire.

"Come, let's follow the attack," Dostum said.

Spann, Olson, and J.J. looked at each other. Was the general serious?

And then Dostum spoke into the radio: "Charge!"

A wave of horsemen climbed the back side of the first hill, crested it, and rode down it, quickly picking up speed.

Before them lay a half mile of ground folded into hills. At the end lay the Taliban guns, eerily silent.

And then they opened up.

Mortars started dropping around the horsemen, sending up fountains of red dirt. Rocket-propelled grenades whizzed upward as the Taliban tried timing their impact with the arrival of the Afghans on the crest of each hill. They were missing, for now.

Dostum kicked his horse and broke into a gallop. Nelson and Jones followed, with the three CIA officers behind.

Nelson didn't know exactly where Dostum was going, but he wanted to follow. He figured the old man would ride several hundred yards, over two or three hilltops, and watch the battle from this better vantage point.

As he rode, Nelson saw men topple in their saddles, punched by rifle fire. He heard the pop and whine of rounds passing by his head. He got on the radio and called back to Michaels.

"Drop the bombs now," he said. He wanted to time the strike so the bombs hit before the horsemen arrived.

Ahead, the horsemen charged the middle of the line, about 600 yards away. The men on foot trotted behind, grimacing, gripping their rifles and RPG tubes, ducking whenever they heard an explosion or the whine of a passing bullet.

Nelson looked up just as the Taliban line exploded. The bombs from the jet overhead smashed near the tanks and also destroyed one of the ZSU-23s. Dostum's men let out a cheer and quickened their pace.

As he rode, J.J. started passing Taliban fighters who had been hiding in the grass. They jumped up shooting, and J.J. spun in his saddle, firing his AK. Spann came upon a Taliban who was running away, back to his line, when suddenly the soldier turned and took aim. Spann shot the man in the head.

Nelson rode past dead and dying men, the air misting with the iron scent of blood, the burnt sting of gunpowder. Smoke hovered above the field. The charging horsemen raised their RPG tubes and fired at the Taliban. The explosions rocked them in their saddles.

Up ahead, Nelson could see the Taliban line breaking in places. Here and there, like a sand wall crumbling. Nelson was amazed when he saw that some of the Taliban were running toward Dostum's men, their hands held high in surrender.

He was equally surprised when they started falling face-forward, dead, in the dirt. He would later learn that they had been shot in the back by their commanders still on the line.

Dostum reined his horse and cut across the field to its right flank, then pulled up and stopped. The general did not like what he was seeing. The Taliban had dialed the range on their remaining ZSU-23. The rapid banging of the antiaircraft gun hurled through the Afghan line. Men blew apart in their saddles and were lifted off the ground as they walked, cut in two.

Of the 600 men who had started the charge, Nelson guessed

that maybe there were 300 still in the fight. The remainder had been wounded, killed, or had scattered. And Dostum's men were close, within striking distance for victory. One last hill separated them from the Taliban, about 100 yards. But Nelson sensed they were losing momentum.

The horsemen halted, uncertain what to do, trapped by gunfire. Some of them jumped from their saddles and crouched at the feet of their nervous horses, trying to make themselves smaller targets.

Dostum was furious. "We are losing!" he said. He yelled into his radio: "Attack! Attack!"

His men did not move. Nelson watched as Dostum leaped from his horse, reached into a saddlebag, and retrieved several magazines of ammunition for his AK-47. And then he started to run.

Straight down the hill toward the Taliban line.

Worried for his safety, one of Dostum's men ordered Spann, Olson, and J.J., as well as about fifteen Afghans, to form a perimeter around the rear of the general's advance.

Nelson watched Dostum as he ran. He was running and firing at the Taliban line. Nelson expected Dostum to drop at any moment, fatally wounded. He watched him halt to change magazines and start running again. He was passing by his own men, who looked up amazed and, finally, embarrassed. They mounted their horses or took off on foot, forming a line with their general, the horsemen firing over the horses' heads and racing on. Nelson could feel the battlefield swell. It had taken on new life.

"What do we do?" asked Jones.

"We gotta go with him," said Nelson. "If he gets killed, we're in a world of hurt."

They rode down the hill but pulled up before they got too close to the Taliban line. Nelson watched as the Afghans coalesced as a smoky swarm, bristling with gun barrels, flashing with explosions. They descended on the Taliban line with a roar. Nelson gazed in awe.

Dostum's men attacked the Zeus and killed its terrified gunners. The remaining Taliban threw down their weapons and ran. They were shot unless they surrendered first.

On the hill, one Afghan soldier reached to the ground with a knife and made a swift sawing motion.

He stood and thrust aloft a head, the head of a Taliban soldier, swinging by a fistful of black hair, a dripping pendulum, as the sun drained from the day.

They had won.

As the battle raged, Dean back at K2 in Uzbekistan was listening to the action on a radio. Sergeant First Class Brian Lyle had set it up on a lark on a plywood table, wondering if he could even tune into anything. He was dialing through the frequencies when he heard the sound of gunfire and excited American voices calling in bomb strikes. They all stepped closer to the table.

There were twelve of them, men in their thirties and forties in rough beards, and since leaving Fort Campbell three weeks earlier, it had been a trying vigil for Dean wondering if he would get into the fight. When he learned that Nelson had gone first, his heart sank, a feeling he hid from his friend. Dean and Nelson had been grooms-men at each other's wedding, and before that they'd suffered through Ranger School together, and had been roommates during various other schools the Army had put them through to make them elite soldiers.

Earlier in the month, back at K2, they'd been forbidden to frater-nize as they awaited their missions, and Dean had been dying to talk to his buddy about his upcoming assignment.

One day, Dean walked up to Nelson as he was standing at one of the "piss tubes," open-air camp urinals made of plastic pipe, one end buried in the ground and the other stuck up at waist height.

Dean said, "Hey, man. How you doing?"

"Fine," was all Nelson would say. It was clear he was taking the ban on fraternization seriously.

Dean didn't blame him. "Okay," he said, "take care."

And that was the last thing he'd said to Nelson.

Now, listening to the battle on the radio, Dean wondered how he'd react when the bullets started flying. He didn't want to kill people, didn't relish that thought. What he loved was politics and watching the way governments turned on the end of an idea, and often at the end of a gun.

Since 1999, when he entered Special Forces, he felt he'd never had a dull moment. He'd survived being exposed to live nerve agents during a mock biological attack. He'd spent weeks in the Nevada desert outrunning and hiding from UAVs, unmanned aerial vehicles. He'd survived the Army's SERE school, aka Survival Escape, Resistance, and Evasion training. The grueling event, held in rural North Carolina, lasted eleven days. On the run behind "enemy" lines, Dean lived on raw pumpkins and barely cooked chickens stolen from local farmers. He ate them hunkered in muddy ditches, clawing at the food with his bare hands before being "captured." Men in SERE training are buried alive, berated, and ground to the barest nub of self. When it was over, Dean emerged from the pinewoods at Camp McCall having seen God, or at least the U.S. Army's version.

Dean had been led into this hell of total privation so that if he was ever captured, he'd have already experienced the inferno. This was Special Forces' rule number one: let men experience failure so they never fail again. And by failing, they will learn how to be successful soldiers. Dean pretty much believed there wasn't something he couldn't improve, himself most of all. Growing up on a small farm in Minnesota, his father had drilled this work ethic into him. And something else, too: that no one accomplished anything without the help of others.

For the past year, Dean had been bugging his superiors at Fort Campbell for a training mission in the Middle East. He campaigned relentlessly. In January 2001, he and his team traveled to Uzbekistan. Dean was fluent in Russian, and the political challenges facing the former Soviet republic were a heady fit with his omnivorous intellect.

He was able to learn from the Uzbekistan army firsthand about the Islamic Militant Union insurgency, and this made him think of terrorism broadly. He started asking his intel sergeant, a tall, taciturn Texan named Darrin Clous, about a group of fundamentalists in Afghanistan called the Taliban, and their enemy, the Northern Alliance.

Dean loved parsing the political matrix. Whenever he and the team returned from a trip overseas, Dean posted his findings—pie charts about terror attacks, informal intel summaries about the Taliban, and general musings about geopolitics—on a bulletin board in a back hallway at Fort Campbell's Fifth Group Headquarters. But not many people paid attention. Afghanistan wasn't on anyone's radar.

And still, sitting in K2, Dean didn't have a mission.

He hadn't given up yet. He was the personification of persistence. Several years earlier he'd been rear-ended by a drunk at a stoplight and had gotten out of his car to have a talk with the driver. As he approached, the driver sped away. Dean gave chase on foot.

He ran alongside the vehicle and managed to wrangle himself up into the passenger window and onto the front seat, whereupon he persuaded the driver to pull over. The driver just looked at him, amazed. Dean ached to do an excellent job as a soldier. *Ached*.

But the higher-ups at K2 and the Pentagon had yet to make a decision. What Dean needed was a warlord eager for American expertise.

A hundred miles to the south from where Dean now sat in his tent watching *Moulin Rouge* on DVD for the tenth time (and hating it), the warlord Usted Atta was meeting with CIA officers and discussing just such a plan.

After Dostum's charge up the hill, Nelson and Jones had to turn from the fight and head back to Cobaki, to plan the next day's movement of the team. Nelson hated to leave.

He stood on the hilltop and watched the Afghans continue fighting, running and riding toward the village of Chapchal, several miles to the north. The Taliban were in pell-mell retreat. Shouts filled the dusk as they ran.

Nelson had to get Sam Diller riding to a village called Oimitan, in order to set up bomb strikes on Taliban tanks that Nelson and Dostum were sure would be coming down the valley. And as Diller would leave, Spencer would come in, riding up from the base camp at Dehi. Nelson could hear the eagerness in his voice whenever they talked on the radio: Spencer wanted to see the battlefield.

Nelson and Jones rode to the rim of the valley at the trailhead and they picked their way in the dark with the horses' hooves clicking on the rocks. At times, they couldn't see anything except a skim of stars overhead, like ice crystals thrown against a black dome.

At the bottom of the canyon, Fakir met them. He told Nelson that he admired his courage. They rode in silence to the trailhead

Flying into remote Afghanistan locations required the 160th SOAR pilots of Chinook helicopters to operate in strange weather conditions that presented enormous dangers. Threading the aircraft through the 14,000-foot-plus mountains, the pilots had never before attempted such a daring mission.

The Darya Suf region was utterly beautiful yet laced with hidden Taliban encampments and land mines. The enemy knew the Americans had landed.

Northern Alliance soldiers at rest. At the time of the arrival of U.S. Special Forces, they had almost no ammunition left with which to fight the Taliban. The Horse Soldiers, because of their unique training, were able to unite the different ethnic groups against the Taliban.

Mules loaded with U.S. soldiers' gear. The sure-footed animals could carry heavy loads and were essential to the Americans' historic efforts to repel the Taliban. The Americans, armed with lasers and GPS, rode to war on horses.

The Special Forces were able to airdrop two Gator vehicles into the fighting zone. They were used to ferry people, food, weapons, ammunition, and medical supplies. These soldiers are headed to the calamitous assault at the Gap.

Armed with automatic rifles and rocket-powered grenades, some of the troops loyal to Northern Alliance commander Dostum travel to war. Many of these men had been at war for twenty years.

Lieutenant Colonel Bowers and Commander Dostum had to learn to understand and trust each other to fight the Taliban successfully. War planners would later call the Special Forces' culturally nuanced and Afghan-centered campaign a groundbreaking template for resolving future global conflicts.

Commander Dostum and U.S. soldiers on horseback in the Darya Suf Valley. They would soon encounter stiff Taliban resistance as Northern Alliance horsemen charged the enemy's trenches.

Beautiful, desolate, and extremely difficult to control, the valley of the Darya Suf and its small villages had to be captured before the Americans pressed on. Many of the villages had been decimated by earlier Taliban attacks.

Using a Dshka ("Dishka") gun and SOFLAM, American soldiers were able to rain explosives down on distant Taliban forces that had never before seen such high-tech firepower.

American soldiers in a high trench scanning the horizon for Taliban targets. Using laser-guided weapons proved to be both science and art. Northern Alliance and Taliban soldiers bantered with one another using walkie-talkies to critique the accuracy of the strikes.

The narrow and dangerous Tiangi Gap was the most direct route to Mazar-i-Sharif from the Darya Balkh Valley, and was the scene of a bloody Taliban counterattack. Thousands of friendly Afghan soldiers, along with Special Forces, poured through this valley in pursuit of the Taliban.

Staff Sergeant Brett Walden, thirty-six, from Florida, as he rode into Mazar-i-Sharif. Walden, as did all of the Special Forces, took great care to recognize the customs and culture of local Afghan citizens. He would survive the conflict in Aghanistan and was later tragically killed in Iraq.

The massive Qala-i-Janghi Fortress, shown in an aerial reconnaissance shot. More than 600 yards long, the fort is bisected by a large interior wall. Made of mud, straw, and timber, the fort was completed in 1889, after twelve years of labor by eighteen thousand workers.

The main gate of Qala-i-Janghi, through which captured Taliban and Al Qaeda soldiers, including John Walker Lindh, were taken.

The southeast corner of Qala-i-Janghi, showing horse stables at left and the middle wall at right. The Pink House and the site of the siege's beginning are center-left in the photograph.

Northern Alliance soldiers climbing up the outside of the fortress while the battle raged on the other side of the thick mud walls. (© Getty Images/Oleg Nikishin)

Northern Alliance fighters battle pro-Taliban forces. Surprisingly, a number of the fighters arrived at the battle by taxi from nearby Mazar-i-Sharif, weapons at the ready. (© Getty Images/Oleg Nikishin)

Northern Alliance general Ali Sarwar on the parapet overlooking the Pink House in the southern courtyard, aiming a rifle at pro-Taliban forces, November 27, 2001. (© Getty Images/Oleg Nikishin)

American and British Special Operations soldiers observe fighting between Northern Alliance troops and pro-Taliban forces on November 27, 2001. Major Mark Mitchell is third from the left. (© Getty Images/Oleg Nikishin)

Staff Sergeant Brett Walden, thirty-six, from Florida, as he rode into Mazar-i-Sharif. Walden, as did all of the Special Forces, took great care to recognize the customs and culture of local Afghan citizens. He would survive the conflict in Aghanistan and was later tragically killed in Iraq.

The massive Qala-i-Janghi Fortress, shown in an aerial reconnaissance shot. More than 600 yards long, the fort is bisected by a large interior wall. Made of mud, straw, and timber, the fort was completed in 1889, after twelve years of labor by eighteen thousand workers.

The main gate of Qala-i-Janghi, through which captured Taliban and Al Qaeda soldiers, including John Walker Lindh, were taken.

The southeast corner of Qala-i-Janghi, showing horse stables at left and the middle wall at right. The Pink House and the site of the siege's beginning are center-left in the photograph.

Northern Alliance soldiers climbing up the outside of the fortress while the battle raged on the other side of the thick mud walls. (© Getty Images/Oleg Nikishin)

Northern Alliance fighters battle pro-Taliban forces. Surprisingly, a number of the fighters arrived at the battle by taxi from nearby Mazar-i-Sharif, weapons at the ready. (© Getty Images/Oleg Nikishin)

Northern Alliance general Ali Sarwar on the parapet overlooking the Pink House in the southern courtyard, aiming a rifle at pro-Taliban forces, November 27, 2001. (© Getty Images/Oleg Nikishin)

American and British Special Operations soldiers observe fighting between Northern Alliance troops and pro-Taliban forces on November 27, 2001. Major Mark Mitchell is third from the left. (© Getty Images/Oleg Nikishin)

In the struggle to suppress the Taliban revolt, U.S. forces dropped an enormous bomb on the northeast corner of the fort, nearly killing five Americans. The explosion was earth-shattering and the damage to the main wall enormous, but the Taliban fought on. The injured were rescued, in part, by the U.S. Army's Tenth Mountain Division, which had to dodge furious gunfire.

The Pink House in the fortress stood atop the underground room where the Taliban and Al Qaeda prisoners were housed. During the battle, they shot out of the small holes in the base of the walls.

The remains of one of the massive weapons caches seized by the Taliban prisoners from the containers inside the fort and used in the uprising.

John Walker Lindh shortly after his capture by Northern Alliance soldiers at the Qala-i-Janghi Fortress on December 1, 2001. Lindh had traveled from California to unexpectedly find himself in the bloody battle, and in the global media spotlight.(© The Sun/Terry Richards)

Johnny "Mike" Spann, the first American to be killed in post-9/11 battle, in a family photograph. A former Marine officer, Spann had recently joined the CIA as a paramilitary officer. (© Corbis)

leading up the valley wall, Nelson bobbing in the saddle, exhausted. After four hours, they rode back into the outpost at Cobaki, the wind rising off the river with its chalk smell and the current's hush brushing the canyon's walls.

Back in the States, news had finally hit the papers that U.S. Special Forces were in Afghanistan. As Nelson was riding back into camp, Americans logging onto their computers could read a dismal assessment issued by the Center for Defense Information, an independent think tank made up of academics and retired military, about the military's effort: ". . . The Northern Alliance advance has been described as 'stalled,' largely because the opposition force is still grossly outnumbered and their transport is unreliable and slow.

"Afghan opposition troops in the area around Mazar are reportedly almost out of ammunition, food, and medical supplies, which would hamper any attempts to convert U.S. strikes into permanent military gains."

Secretary of Defense Rumsfeld called Colonel Mulholland at K2, blistering him with questions. What the hell was going on? Where was the progress? Why weren't more Taliban positions being destroyed? Mulholland didn't have an immediate answer. But he'd get one.

When Nelson threw down his rucksack, ready for sleep, one of his teammates nearby stirred and sat up. He said he had a message for Nelson from an intel specialist at K2 who worked as their interface with Colonel Mulholland. The message wanted to know, as Nelson would later say, "When are you guys gonna get off your ass and do something?"

Nelson couldn't believe it. How dare the old man question what they were accomplishing? He didn't know where to begin.

He walked around the camp, getting angrier by the minute. He thought of going to bed. He needed sleep badly. He decided that he wouldn't respond until morning, with a cooler head. He climbed into his sleeping bag. His mind wandered to his wife, Jean, who was pregnant and working mightily to keep the house going. He'd give anything to tell her about this. This goddamned message. He couldn't get it out of his mind. What did it mean? When are we gonna start doing something?

Annoyed by his fidgeting, his teammate asked, "Are you gonna answer it tonight, or go to sleep?"

Nelson decided to get up. He drank some water and ate part of an MRE, thinking about all that he had seen these last few days. Goddamnit all. He'd been fighting his heart out, and meeting the Afghans had nearly broken it. Seeing them—the lame, the scarred, the broken—had touched something in him, deep. Men riding horses into sheets of gunfire. These were men who had nothing, yet they offered him everything: their lives. They would die for him. He wanted to tell the Pentagon that he was doing the best he knew how to do. He set his Panasonic Toughbook on his knees and opened it, the screen lighting up his face. He began typing faster and faster. It read in part:

I am advising a man on how to best employ light infantry and horse cavalry in the attack against Taliban T-55s [tanks], mortars, artillery, personnel carriers, and machine guns—a tactic which I think became outdated with the invention of the Gatling gun. [The *mujahideen*] have done this every day we have been on the ground. They have attacked with 10 rounds of ammunition per man, with snipers having less than 100 rounds—little water and less food. I have observed a PK gunner who walked 10-plus miles to get to the fight, who was proud to show me his artificial right leg from the knee down.

We have witnessed the horse cavalry bounding overwatch from spur to spur to attack Taliban strongpoints—the last several kilometers under mortar, artillery, and sniper fire. There is little medical care if injured, only a donkey ride to the aid station, which is a dirt hut. I think [the *mujahideen*] are doing very well with what they have.

We could not do what we are doing without the close air support—everywhere I go the civilians and *mujahideen* soldiers are always telling me they are glad the U.S.A. has come. They all speak of their hopes for a better Afghanistan once the Taliban are gone.

Finished, he closed the laptop, feeling better.

The e-mail would become, in the words of Major General Geoffrey Lambert back at Fort Bragg, the most famous intel report of the war. Several days later, standing before reporters and television cameras at a Pentagon press conference, Secretary of Defense Rumsfeld would hold it up and read out portions to the nation.

The crowd was moved by the poignancy of this young, anonymous soldier. Nelson's words flew around the newswires. Nightly anchors quoted from it. Nelson had managed to sum up the frustration, fear, and hope among his own men and among the Afghans. That he had written in a fit of pique, when he was tired, hungry, and pissed off, would remain his secret.

Later that night, Dostum, having also returned to Cobaki, told Nelson that at the very end, the fighting had been at close quarters, the opposing sides as near as sixty-five feet from each other. "I have never seen the Taliban fight so hard," he said. He shook his head in amazement. He believed the Taliban knew they were losing the war. His men had killed 123 Pakistanis and captured 2. Dostum had lost several of his men. One had taken a grenade, the last one remaining among the entire force, and run into the Taliban lines to blow them up, killing himself.

When he heard this, Nelson was upset. He and Jones had carried eight grenades and returned to Cobaki with them. Nelson was worried that the general had put himself in harm's way. He regretted not staying with him. He realized how protective he had become of the warlord.

Back at the Alamo in Dehi, as more wounded Afghan soldiers trailed into camp, medic Scott Black found himself knee-deep in gore. For the past several days, he had been working frantically to care for both Afghan and Taliban fighters. Black didn't think he'd ever seen a tougher people in his life. Their eyes were full of terror but they rarely made a sound. These guys, it seemed, could withstand any kind of pain.

He was asleep well after midnight when he was shaken awake. Black woke with a start, looking around the dim room in the fort. He saw before him a middle-aged Afghan man, one of Dostum's soldiers, holding a lantern and looking concerned.

"Commander Scott, we need you now. There is emergency."

He scrambled out of his sleeping bag and walked in flip-flops, camo pants, and brown T-shirt across the dusty courtyard and out the front gate. Before him sat a parked Nissan pickup with all its doors flung open. The truck's pale dome light was on. It cast a feeble light on the ghastly scene inside.

On the rear seat in the cab of the truck lay a young Afghan boy, maybe only fourteen, fifteen years old, it was hard to tell. He'd been shot in the stomach and was rolling his head back and forth on the vinyl seat, which was slick with blood. He was opening his mouth to moan. But no sound was coming out. Black saw that he had to act fast.

He shot the kid up with morphine while another boy, maybe eighteen, who turned out to be his cousin, held his hand. Black stood at the kid's feet. It was as good a place as any to treat him. He probed the wound—it was deep. He felt around inside trying to find the bleeder, the vein that was gushing.

He couldn't feel it with his slippery hands. There were no guts hanging out, so he couldn't examine them for wounds. He guessed an AK-47 round had penetrated the kid's peritoneum, but he couldn't find the hole. The bullet had likely nicked an artery. It was dark in the cab of the pickup, except for the glow of Black's halogen headlamp—and it would be a tough place to do surgery. But Black was pretty sure he was going to have to cut the kid open if he couldn't stop the bleeding.

He first tried packing the wound with an absorbent material called Curlex. The idea was to pack it against the wound with your index finger, as if you were stuffing Kleenex down a Coke bottle. Black packed about two rolls inside the kid but the bleeding didn't stop. He'd lost an immense amount of blood and fresh plasma was something Black didn't have on hand. The kid was quickly slipping out of his control. If he died during the surgery, the Afghans would blame Black for his death. Black knew that and worried about it. This worry had little to do with medicine and everything to do with fighting the kind of war he was in, where his relationship with the locals could mean life or death—his own. Black decided there was nothing he could do, short of transferring him back to a hospital in K2. Unless he could stop the bleeding with the Curlex.

He worked for two more hours, packing and repacking the wound, until he reluctantly accepted that the bleeding just wouldn't stop. Blood had splashed all over his feet in his flip-flops. He'd never lost a patient before in the field. In fact, this was his first battlefield casualty in his six years in Special Forces. In the tiny compartment of the pickup, he could smell the blood and hear the kid's troubled breathing, and he felt a wave of despair sweep over him.

The cousin was still holding the young boy's hand as Black shot him with even more painkillers and through an interpreter told the cousin that he should take him home to his family. Better to let him die in peace surrounded by the people he loved. The young man just nodded. Black helped him lift the wounded kid's legs up in a bent position on the seat so they could shut the truck doors. Black stood in the dirt and watched the taillights of the Nissan disappear over the ruts in the road.

He was running low on med supplies again and had ordered an air drop. This was supposed to come any night now. With the fighting picking up, he realized he had to brace himself for dozens, maybe hundreds, more casualties like the one he'd just treated. It was a daunting thought.

He had believed himself to be a pretty hardened customer, but the kid's impending death ate at him. Just a teenager. . . . He walked back to his sleeping bag in the horse stable and tried to sleep but couldn't.

As Black treated the wounded, Nightstalker pilot Greg Gibson and mission commander John Garfield had taken off from K2 on the resupply mission bringing in the equipment that Black and the team had requested—ammo, blankets, winter coats, sterile water, IV sets, rubbing alcohol, latex gloves, nylon sutures, bandages. The glue of war, the tools for healing men.

The flight in was terrible, the visibility zero. An hour after taking off, Gibson had flown straight into the gullet of the Black Stratus, the wide slough of fog, snow, and dust whose existence had mystified weather forecasters at K2. Gibson was flying strictly by cues presented to him on the tiny screens in the dim cockpit. The world outside the windscreen was a depthless white.

After three hours at the controls, Gibson's nerves were fried, but he had entered a zone where he remained completely in control of the craft and all of his faculties. He completed the majority of the flight without incident, and as he neared the HLZ at Dehi, Gibson handed the controls to his copilot, Aaron Smith. They would land, offload the gear, and zoom away, turning around to head back through the lousy weather to K2.

As they dropped altitude, Gibson, sitting in the pilot's seat with his arms folded, heard a voice from the back of the ship. One of the crew members, standing in the open ramp, was yelling over the intercom: "Pull up! Pull up! Damnit!"

Gibson waited for Smith to do as he was told. The crew in back provided a second set of eyes and ears for the pilots, who implicitly trusted their directions.

But when Smith hadn't corrected course, for whatever reason, Gibson grabbed the stick between his knees and put the aircraft into a serious bank. They were practically tipped on their side, with Gibson staring up at the sky over his left shoulder as they turned.

It was then that Gibson saw they had nearly run straight into a mountain.

The crew in back was knocked to the deck. One of them, Tom Dingman, had seen the rockface coming at them and had been the one to yell just in time. They had been seconds from hitting it head-on at 90 mph.

However, they turned so sharply that the helo's tires hit the mountain wall. The large aircraft bounced as if off a trampoline. They hit so hard that the FLIR, or radar, probe on the nose of the craft was knocked loose; it was hanging now by just heavy wires.

As they finished turning, Gibson caught sight of an Afghan man sitting atop a horse on the ridgetop they'd just struck. He was blown backward out of his saddle by turbulence from the rotor blades.

Gibson leveled the helicopter out. It was shaking violently. The twin rotors were out of synchronization after the impact. The cabin was shaking so hard that Gibson found it difficult to read the instruments.

John Garfield, sitting in the jump seat, got on the intercom. The unit's official motto was "Nightstalkers don't quit," but Garfield felt they would be pushing the limits of the craft's endurance if they tried to land. He didn't want their epitaph to read: *They just didn't know when to quit.* Garfield announced that the conditions were just too hairy. Screw the supply drop. They were returning to K2.

Gibson got on the radio and told K2 what had happened.

"What's your status?" came the reply.

"Well, we just bounced off a mountain," said Gibson in his cool, southern pilot's voice. "We don't know how bad it is," said Gibson. "It's looking pretty damn bad. When we get things figured out, we'll call you."

"Uh, Roger that."

As they headed into the night, Gibson felt like they were getting beaten to death by the vibrations. They had at least three hours of flight time ahead of them.

After an hour, base called back, wondering how they were making out.

"Like I told you," Gibson said, his voice rattling from the violent movement of the aircraft. "We ran into a damn mountain. But I think we're going to be fine."

But an hour from the Uzebekistan border, one of the crew looked down through an open door and saw a flash. They were flying about 800 feet above the ground. The flash was followed by a burst of light in the clouds behind the escort helicopter that had been trailing them during the mission.

They were under attack by antiaircraft fire.

Gibson saw another flash on the ground.

Any moment, he expected the helicopter's defense measures to kick in. These consisted of pods of metallic chaff secreted around the outside of the aircraft, set to go off when in proximity to an incoming missile. The release of the chaff was accompanied by the automatic popping of bright flares on the sides of the craft. These looked like exploding fireworks. The idea was that the heat and mass of these objects would draw incoming heat-seeking missiles away from the helicopter. And that was how the procedure generally worked, just fine.

The right gunner reported seeing another flash as a missile climbed from the ground and exploded, but not close enough to set off the flares and chaff on Gibson's helo.

Another pilot, Jerry Edwards, who was flying the trailing helicopter, was not so lucky. His helo popped its flares and chaff, and illuminated the night sky around it, bathing the aircraft with bright light. Both helos were now clearly visible from the ground. They'd been spotted. Almost immediately, more missiles exploded around them. The pilots put both craft into a serious climb. The excitement was so intense that everyone on Gibson's helo forgot that it was supposed to be shaking itself apart. About half an hour later, they landed safely back at K2.

After shutting down the aircraft, Gibson returned to his cot in the abandoned aircraft hangar, eager for sleep. It was nearly dawn. Life as a Nightstalker had assumed the predictable rhythm of intense periods of boredom punctuated by sheer terror. Fly all night, deliver supplies or people to different parts of the country, come back to base, debrief, fall asleep. Wake at one or two in the afternoon, make a big pot of coffee, and wander out back to the "veranda," which the Nightstalkers had constructed out of sandbags and wooden packing crates, and outfitted with a barbecue grill. The sandbags kept the prop and rotor blast from hitting the pilots as they sat waking up in their robes and flip-flops while the rest of camp was already at work. Someone had even planted a small lawn around the decking, which they kept green with the cook's dishwater. They lined the lawn with handpainted rocks and somebody stuck in a sign announcing that it was the recent recipient of the "Uzbekistan Yard of the Month" award.

Jerry Edwards had begun keeping a journal, writing down snippets about life in the air. On one of his recent missions they'd actually "blown a hole" through the fog with the Chinook's massive rotors and created a tunnel through which a C-130 fuel tanker was able to safely pass and land. It was wild stuff like that that you had trouble explaining to folks back home.

Edwards worried that he wasn't good enough yet as a pilot to keep up with the rest of the crews, the kind of obsessive attention to perfection that, in fact, made Edwards an A-1 Nightstalker, one of the top 5 percent of all helo pilots in the U.S. Army.

Sometimes, after coming back from a mission, he couldn't sleep and walked around the camp trying to wind down. One night he found a cobra in one of the Port-A-Potties that dotted the camp. He dashed backward out the door, slamming it behind him. Returning to his cot in the hangar, he opened his sleeping bag to find two pit vipers inside.

He missed his wife. He wondered where in hell all the "goodbye sex" had gone. Each man got ten minutes a week to make what the

Army called a "comfort call." He queued up outside the crude booth with its ancient rotary phone to wait his turn. The line was long but Edwards was determined to make his call.

After an hour of waiting, he was next in line, hunched against the chill in his flight jacket. The guy inside was exceeding his ten-minute allotment and clearly not caring. Edwards knocked on the door and said, "Hey, man, we got people out here who need to make a call!"

The guy, an Air Force officer, ignored him.

Edwards saw that there were only enlisted guys in the line. The Air Force dude was pulling rank. Edwards outranked them all. This dickweed needed to be taught a lesson.

After fifteen minutes when the man still hadn't hung up, Edwards yelled, "Get off the phone for everyone's sake!"

The door flew open and the guy started asking Edwards who did he think he was, and what was so important, and so on, and Edwards just stood there, bone-tired. What mattered now was talking to his wife. He pushed past and said, "You're all wrong, pal."

It was 10 p.m. back in Fort Campbell. His wife answered, and after a couple of long yawns, he figured he was waking her up. He could hear his three-year-old daughter babbling in the background.

After the expected small talk, in which Edwards reminded his wife that he needed toothpaste and Handi Wipes sent in his next care package, his wife said, "Jerry, I'm pregnant."

Edwards was stunned. "Are you sure?"

"I've taken three tests."

And then his daughter got on the phone. "Are you at work, Daddy?"

Edwards, still reeling from his wife's news, sputtered, "Yes, sweetie, I'm at work."

"Can you come home from work?"

"No, not now."

"I'm going to see Santa."

Edwards was suddenly sad he wouldn't be seeing Santa with his daughter.

"I love you, Daddy!" she cried. "I miss you so much!"

Then she handed the phone to her mother. "That's great news," said Edwards. "Really great. I'm going to be a father again. Wow."

He could barely feel his feet touching the ground.

He and Diane talked a few more minutes and, looking at his watch, after exactly ten minutes, Edwards said he had to go. He hung up.

He walked back to the hangar in a daze. *Man. I'm going to have a baby.* It was nearly dawn. The guys inside were watching *American Pie* and laughing their asses off.

We have to make it out of here alive, he thought.

The following morning, October 24, Sam Diller left the Cobaki outpost. What he and Nelson had planned the night before was audacious.

Diller, along with medic Bill Bennett, and weapons sergeant Sean Coffers, planned to ride twenty miles west into the mountains and set themselves on a high vantage point overlooking the Taliban's flank. With them were about thirty Afghan soldiers, Dostum's men. They would commence bombing the tanks and soldiers rushing down the valley to reinforce the army fighting Nelson.

The key to Diller's survival was stealth. And speed. They would be deep in Taliban territory. Diller hoped to be resupplied with food and ammo, but that was doubtful. The Air Force demanded twenty-hour scheduling notice on drops, and he knew he'd be unable to schedule that far in advance. They'd have to live off the land. There'd be no hope of quick rescue.

He swung up into the saddle and looked at the mountains and then at Nelson.

"This has got risk to it," he said.

"I know that."

"If it goes bad," Diller said, "you won't be hearing from me again."

"I'll be seeing you again."

Diller had an eight-hour ride ahead. He turned his horse and his men, trailed by several mules hauling their rucksacks. They began riding across the camp and soon disappeared down the hill. Nelson waved and watched them go.

As Diller rode away, Spencer and Black, at the Alamo, rode to Nelson at Cobaki on borrowed horses. This was a social visit, of sorts. Each man wanted to leave the confines of the base camp and survey the battlefield at large. Black had packed his medical gear and Spencer was carrying a radio, in case they were able to stay at the Cobaki outpost and run the logistics base from there. Borrowing the horses had been difficult. There was none to spare among the Afghans, and only Dostum's intercession on the Americans' behalf had landed them in the saddle of two tired mountain ponies.

The ride was exhilarating, at first. It seemed to Spencer that every few steps his horse would turn and look at him and say, "You're sure a big bastard. Why don't you get off and walk?"

Spencer believed the horse was actually huffing and puffing, as if to be melodramatic. The stirrups were short and Spencer's knees were practically in his chest. The horse ambled along the rocky path. Spencer's back started to ache. He was wearing his load-bearing vest crammed with ammo, grenades, water. It weighed about forty pounds. His M-4, which weighed another seven pounds, was slung across his chest.

By the time he got back to Cobaki, his legs were numb. He couldn't move. He sat on the horse, petrified, afraid to get off. He didn't want anyone kidding him about being an old man. At forty, he *was* one of the oldest guys on the team.

Black asked him what was wrong.

"Look," said Spencer, out of the corner of his mouth, "I can't get off this horse."

Black thought he was joking. Then he saw Spencer wince.

"I can't lift my legs. My back is fried. Can you take my foot out of the stirrup?"

Black pulled Spencer's right hiking boot from the iron ring and lifted his leg over the horse's rump. Spencer stood gingerly on the foot.

He still had his left foot in the other stirrup. He gripped the saddle and pulled the boot away from the ring and set that foot down, too. Black could hear him gasp.

Spencer stood with his arms over the saddle, catching his breath.

"You're going to have to help me walk, too," he said.

"Are you serious?"

Spencer put his right arm over Black's shoulder and they walked a short ways to a corner of the camp. Spencer let out a groan.

Black lowered him to the ground. He put a pillow under his legs and then one under his back. He handed Spencer a pill from his medical kit.

"Take this. You should be asleep in about thirty minutes."

Spencer woke the next morning after ten hours of deep sleep. He felt a little better. In fact, he had a herniated disc. But he'd be damned if he would be sidelined by the injury. He struggled to his feet and got to work in the camp.

Diller discovered the horses they were riding didn't mind bullets whizzing overhead, but they hated the boom of Taliban artillery and the roar of American bombs. They bucked and reared when either happened. He found that riding a horse uphill was less bone-jarring than riding it down. He found that a tired horse will kneel and roll over and refuse to move, with a blank, faraway look in its eye. Diller would stand there, hands on hips, feeling sorry for it. The Afghans did not feel sorry. They walked up and beat the animals with stiff whips until the animals rose silently and trudged on. Diller discovered that he was capable of living on nothing but fear, the taste of it tucked like a thin wire at the back of his throat.

By the third day of riding the battlefield, he was nearly out of food and low on ammo. They were moving and bombing and moving on quickly. He had begun to feel like a ghost in the mountain.

They blew a Taliban ammo dump that caterwauled in the air for twenty minutes after the air strike. They saddled up and kept riding. They were picking their way across the spine of the Alma Tak Mountains, through draws and around hills, always nervous they'd be spotted by the Taliban. On this third day, Diller realized they'd ridden beyond the range of any hope for quick reinforcement or rescue. They were true outliers now, men on their own. Diller hunched in his saddle against the cold and spurred the horse on.

They were covering ground so quickly it was difficult for him to schedule supply drops. Diller didn't know where he might be in the next hour. (He would be supplied only twice during the ten days' fighting in the mountains.) He cut rations among the men to one MRE a day. They were in a race to cover ground before exhaustion overcame them all.

He was carrying 300 rounds of .556 ammo, his M-4 rifle fitted with a nightscope, four grenades, a 9mm pistol, an extra clip for the pistol, a team radio and 5 pounds of extra batteries, a satellite radio, three MREs in their beige vinyl pouches, a pack of six extra MREs stuffed in his pack for emergency, an extra pair of white cotton socks, and a sleeping bag. His mission: destroy Taliban trucks, tanks, and artillery, and, in lieu of actual targets, "bomb dirt"—simply make explosions fall from the sky—for psychological effect. If he couldn't kill them, he would scare them shitless.

He was doing a good job. Word was coming back over the radio net that the Taliban in Mazar believed some kind of monster now lived in the valley. That monster was Diller. He was intercepting frantic radio calls speculating on the whereabouts of soldiers who had left Mazar and not returned. The women left behind, their wives, believed they had been eaten by a giant. In the city, when the cold wind knifed from the south, these women could hear a low, dark rumble come up from the valley, the sound of Diller's bombs.

The rocky mountain paths they rode along were laden with mines. Moving down the trail was painstaking work. Through the years, some of the mines had been laid by the Afghans and others by the Taliban. Diller would watch as his scouts stopped to tilt their heads and bend down on hands and knees and peer sideways at the earth. Then they would poke around in the dirt and lift the mines from their hiding place, disarm them, blow the dust away, and tuck them in their saddlebags. *Goddamn but these people are tough. They give new meaning to the word "recycle."*

By the end of the first week, Diller's clothes were hanging off his frame and he felt like he was sleepwalking in them. He had entered a hyper-aware zone where he felt nothing missed his attention. He was getting by on two hours of sleep a night. He wore his boonie cap pulled down over his hollowed eyes, his face smeared with dirt, a few sparse hairs sprouting above his lip. He couldn't grow a beard to save himself, and the mustache—that was a joke. *Screw it.* He figured he had ways of gaining rapport and the respect of these guys, and that was by being the meanest sonofabitch on the mountain. At night, by the reddish glare of his safety light, he read the tattered copy of Sun Tzu's *Art of War* he kept in his shirt pocket. The Afghans couldn't read and they wondered at him as he sat cross-legged on a rock, looking at a page, rubbing his chin, the stub of a pencil gripped between his scarred fingers.

All warfare is based on deception . . .

When [you are] near, make it appear that you are far away . . .

*Be swift as the wind, majestic as the forest; in raiding and
plundering . . . move like a thunderbolt.*

He would then close the book and order the camp to its positions
for night security. They were living in caves, small dugouts in rock
big enough for several men curled in sleeping bags. In the middle of
the packed dirt floor, a place for a fire. And above the fire ring, a
hole that had been carved out of the rock to ventilate the smoke.
The walls of the caves were charred black by years of fires, and the
air inside was rich as a smokehouse's.

Diller slept by the door. He did this because he feared attack at
night by Taliban, who might come creeping up the mountain. And
he feared that his Afghan soldiers might turn him in for the
$100,000 ransom offered by the Taliban. That was a lot of money,
he knew—practically more than double what he made working for
the U.S. Army. Each night before bed he set empty water cans with
spoons in them outside the cave entrance, which rattled when
knocked over, a crude alarm system that he figured might give him
several seconds' warning he was about to be killed. He fully
expected to die. He accepted this with a silent bravado that he
didn't talk about with the rest of the team. They already *knew*.
Each day when he left on patrol, he wondered if the Afghans
around them had betrayed him. Each moment he felt the impending prick of a knife at the base of his neck. He figured he'd never
hear the rifle shot that killed him.

After five days of sleeping in two-hour shifts, it was clear nobody
was getting enough rest. Diller announced everybody would rack
out at night and they'd take their chances with the Taliban, as well
as their own Afghan soldiers.

"Boys," he announced, "I'm going on patrol every day, me and six
Afghans, and they haven't killed me yet. Everybody now gets a good
night of sleep."

Diller also slept near the cave door because he figured the guy
positioned there would be first to get killed in an attack. It was his

idea, so he would take responsibility for it. He lay in his filthy sleeping bag looking up at the vent in the ceiling. He didn't like it. The hole made a perfect place to pitch a grenade into the cave. He looked at the hole, rolled the problem over in his mind, and decided there was nothing to be done. He replayed firefight scenarios in his head if the camp was attacked.

He didn't admire the marksmanship of the Taliban or the Afghans. It seemed they had both attended the same "spray and pray" school of automatic rifle fire, so he figured, given a chance, he'd shoot everybody first and duke it out hand-to-hand. Grappling. As he had trained to do for hours back at Fort Campbell. Diller was a man supremely aware of his limitations, and thereby immensely confident. That, to him, was the key to guerrilla fighting. He was outnumbered and surrounded fifty to one by Taliban soldiers. They had tanks, artillery, and food. He had the element of surprise. And he was traveling light.

They were spidering north along the mountain ridge at 9,000 feet. Mornings were ashen, damp. Deserted colonies of pale clouds raced past their faces and vanished downwind. Diller left on patrol each day after hot tea and a mouthful of flat bread. With his Afghan guides, he probed the trail ahead, crawled onto crumbling outcroppings, and lay in the rising sun, glassing the valley below. He was looking for targets, new ones. That was the drill. Back at camp, Bennett and Coffers would call in the strikes on the targets Diller had scanned the day before. Haji Habib's men would then storm the bunkers afterward and make sure everyone was dead. They left no prisoners. They had neither the men nor the means to corral them. And besides, the enemy did not prefer surrender. He could tell when they'd engaged a militant band of Al Qaeda fighters, fresh graduates, he guessed, from the jihad circuit in Pakistan. He watched from a distance as they blew themselves up with grenades. He would see them evaporate in a fountain of mist and then hear the dull *pop* of the grenades carried up the mountain. He admired them grudgingly.

But Diller felt he was tougher. As long as he and his men could eat. After a week, Diller was famished. He was so hungry he had lost his appetite. And then he heard a bell. A slow tolling in the fog one morning, and in the gray mist he saw an Afghan man herding his sheep. The ragged animals were wearing bells so the herder would not lose them.

"Get down there," he ordered Haji Habib, "and get us a sheep."

Haji Habib came back minutes later with bad news. They were not for sale.

"Whaddya mean? Who else is he going to sell them to?"

"He eats them. They are for his family."

Diller walked down the mountain, about a quarter mile, to deal with the man.

He offered $50 for three sheep.

The herder wanted $500 an animal.

Diller was livid. "Who else you going to sell them to," he asked again, "at that price?"

The herder smiled. He had a face like a wrinkled pear. "No one," he said.

Diller dug in his bag for money: thousands of dollars in marked bills. So the CIA could follow the trail. Diller had thought it would be months before this money was spent. The nearest, even meager, store was three days' ride in any direction.

Haji Habib and his men slit the throats of the sheep and lashed them on the rumps of their horses and spurred up the mountain to the camp. They skinned the animals and laid the hides on the cold ground and scraped the creamy fat into pans that they set sizzling on the fire. The meat was sliced and laid in the grease and fried.

They ate an entire sheep in one sitting. The rest was packed in rolled cloth with the hope it would keep in the temperate air. They ate the sheep for the next two days.

Feeling well fed now, Diller set his mind to the final push. Word crackled over the radio that Dostum was planning a battle for November 5.

In preparation, Dean, back at K2, had finally gotten word he was going in. Usted Atta had agreed to retreat to his stronghold south of Mazar, overlooking the village of Ak Kupruk. The village was about ten miles to the west of the Alamo, overlooking a river called the Darya Balkh. The Taliban had captured Ak Kupruk, and Atta had to retake the town before any Afghan forces could travel north.

The two rivers, the Darya Balkh and the Darya Suf, converged near a village called Pol-i-Barak.

Dostum with his 2,500 horsemen, and Atta with his 1,000 fighters, would each fight along a wishbone of the two rivers. Dostum and Nelson would move north along the east side of the wishbone, while Dean and Atta moved north along the west side. From Pol-i-Barak, they'd move north along the Darya Balkh to take the much larger city of Shulgareh.

At the same time, five hundred fighters commanded by the Hazara warlord Mohaqeq would protect Dostum's flank in the east. Dostum believed that the town of Shulgareh, in the central Balkh Valley, was key to gaining control of northern Afghanistan. Once Shulgareh fell, Dostum predicted, Mazar would fall, and so would the six northern provinces. All that his forces required to achieve this were sufficient arms, ammunition, and air support.

From Shulgareh, they'd continue north up the Darya Balkh River Valley to the Tiangi Gap, where the Taliban would likely put up an intense, final fight to stop the attack on the ultimate prize, Mazar-i-Sharif, twenty miles to the north.

In the midnight hours of October 28, Air Force staff sergeants Sonny Tatum and Mick Winehouse, whose arrival Pat Essex had earlier dreaded, lifted off from K2 with Nightstalker Greg Gibson piloting the helicopter.

After a three-hour flight, Gibson spotted the infrared strobe that he supposed marked the landing zone. He set the helo down in an awful dust storm. Swirling out of the dark were suddenly a dozen or so armed men, looking angry, and motioning that they wanted on the bird.

This was odd, thought Gibson.

Tatum and Winehouse had already pitched their gear off and jumped down. Gibson watched them scramble into the fray.

Tatum felt like he had just jumped into a snowdrift; the finely grained sand at his feet was nearly knee-deep. Carrying their 100-pound rucksacks, he and Winehouse had to struggle with each step. The night was bitter cold. The dust burned at the back of his throat.

Tatum could see only about ten feet ahead in the gloom, even through his night vision goggles. He saw what he thought were Afghan soldiers. They were eagerly waving him in their direction.

They were dressed in thin jackets and sandals. Tatum wondered how in hell they were keeping warm. He couldn't understand the language they were speaking—a jumble of tongues. All of the men had guns. Tatum looked around but he didn't see any horses. He knew that the Americans he was supposed to meet rode horses. But where were they?

He looked over at Winehouse and they both froze. Something was not right. *We're in the wrong place,* Tatum thought. *We're not supposed to be here.*

Back on the Chinook, Gibson looked out and saw an Afghan waving his arms and hurriedly directing his comrades to board the helo. At this same time, Gibson noticed that what he had thought was the infrared strobe was actually a campfire burning at a far edge of the landing zone. It also dawned on Gibson that they were in trouble.

The soldiers had surrounded the aircraft. They weren't shouldering their weapons, but Gibson knew that was only a matter of time, maybe in the next few seconds. Clearly, they expected to get on his aircraft. He was damn sure that wasn't going to happen.

On the ground, Tatum was looking around frantically, scanning through his goggles, for a white guy. "These guys don't look like gringos!" he shouted at Winehouse.

"I know!" answered Winehouse.

About 300 yards away, behind a screen of trees, Nelson and several members of his team stood watching Tatum and Winehouse wander around in the dark. What the hell was going on? Why had they landed there? Didn't they know that place was one of Dostum's prisons, filled with Taliban soldiers?

Nelson thought the Air Force guys would get cut down any minute. Behind him sat the infrared strobe marking the correct landing zone. The pilot must've missed it; must have seen the flickering light of the campfire instead. Now Nelson had a problem on his hands. He and his men had shouldered their weapons and were about to start shooting.

Back on the helicopter, Gibson yelled to Tatum and Winehouse that they needed to get the hell on board. The two men couldn't hear him above the roar of the rotors but they'd already reversed tracks and were churning back through the sand to the helo.

Nelson got on the radio and called Gibson. "Whaddya doing?"

"Yeah, we're on the ground," Gibson said, trying to stay cool. "Uh, we'll be there in a minute." Nelson and his men stood down.

Tatum and Winehouse piled on board, yelling, "There's no white guys here!" As some of the prisoners tried to clamber up the ramp, the Chinook lifted off. Tatum looked down and saw dozens of angry, forlorn faces. Some of the men were shaking their fists at the rising bird.

These were Taliban prisoners, local citizens, who were under Dostum's control. He had captured them earlier in the month before the Americans' arrival. The prisoners had agreed to quit fighting for the Taliban and in return Dostum had allowed them to keep their weapons. Some of them even had day jobs as laborers and returned each night to the mud-walled jail, where they sat around in the freezing dark, bemoaning their fate, but nonetheless alive. They were out of water, food, and medical supplies.

When the American helicopter appeared in the night sky, they thought their prayers had been answered. They rushed it because they wanted to be captured by the Americans, whom they were sure would take better care of them.

A few minutes later, Tatum and Winehouse landed at Nelson's position. Nelson looked over their gear, shaking his head. They had been instructed to come in with a bare minimum, but each man was toting an enormous rucksack and black nylon kit bags containing miscellaneous items. It was too much gear for them to carry themselves and there weren't enough horses or mules to do the job.

"You'll have to ditch some of that stuff," Nelson told them. He said they'd have to fit everything in one rucksack apiece.

"Did you bring any food or water?" he asked.

They hadn't.

Later, when one of them asked Essex where they might find some water, Essex angrily pointed at a mud puddle and said, "There's what you're drinking, just like everybody else."

Over the next two days, Winehouse and Tatum set about unpacking (and pumping their own water), and settling into the team. On October 30, a tall, heavy truck with wooden sides rumbled into camp. In the aftermath of Nelson's bombing, Dostum's men had taken the village of Chapchal. Next, they would attack Baluch, seven miles farther north. Nelson and his men were moving their landing zone and base upriver, to be closer to Dostum's advancing troops.

They loaded their gear into the truck and they climbed a dirt road out of the valley and moved north along the rim. After ten miles, they stopped—below them, about 1,000 feet, lay the site of the new base camp. They loaded the gear onto mules and began the climb down along switchbacks into the valley.

The river plain was wide and baked hard by the sun. Copses of acacia and poplar trees grew in dark green stripes against the mountain wall. They set up camp near the trees. Dostum's tent was a large white canvas affair, stiff as sailcloth, staked out on the ground with manila rope. A cookfire smoldered nearby, upon which sat tea water murmuring in a kettle. The Americans named the place "Helicopter Landing Zone Burrow."

Essex and Nelson and the rest of the team lived in caves overlooking the river plain a half mile distant. From here, the team would divide and ride out to the battle atop the rim of the valley, spreading in a ten-mile-wide crescent.

Essex, Milo, and Winehouse would ride east and north, to prevent the Taliban from getting around Dostum's east flank. Dostum and Nelson would form the bottom of the crescent, in the center of the battlefield, with Diller, Bennett, and Coffers stationed on the western point.

Spencer and his men would switch in and out of the different cells, as the force attacked the approximately 10,000 Taliban troops encamped around Baluch.

Nelson and Spencer stood at the cave's entrance and marveled at their situation, which seemed fantastical. They were standing there trying to remove, of all things, their contact lenses from their tired, red eyes. It seemed so odd to Spencer to be doing this while standing in a cave.

The deployment to Afghanistan had happened so fast that neither of them had gotten the long-delayed laser surgery they wanted to correct their sight. And because they sometimes wore night vision goggles, they couldn't wear glasses. They were worried about how long they could make the bottles of contact lens solution last.

If they ran out, they wouldn't be able to wear the lenses, which meant they wouldn't be able to see. And not being able to see equaled getting shot.

Spencer called home from the cave using a satellite phone. He hadn't spoken to Marcha in three weeks. Spencer was going to try to keep it as simple as possible. There wasn't enough time to convey the experience of what he'd lived through. Or the absurdity of changing your contact lenses in a cave while fighting a war.

Back at Fort Campbell, Marcha and their three sons had been living with the assumption that no news from Spencer was good news.

She was just glad that a white government van hadn't pulled up in front of the house with news that Cal was dead. Whenever she heard a car go by the house, she froze, and peeked out the window, relieved when it was a neighbor passing by slowly. She dreaded the daily possibility of seeing the van.

She remembered walking into the high school cafeteria and seeing Cal sitting at the Formica table, dressed in an Army surplus coat, with blond, stringy hair. He looked up and grinned. They didn't know each other. Marcha said to herself, "I'm going to marry that guy."

The bell rang, and Cal got up and went to class. They didn't speak to each other until two years later, at a Christmas party, during Marcha's senior year. They fell madly in love.

Marcha's parents were nagging her about college. Brunswick, Georgia, was about as boring a place as anyone could imagine. One day, she was driving past the Army recruiter's office.

She walked inside and the gentleman behind the desk asked if he could help her. On impulse, she enlisted.

When she got home, she called Cal. "You'll never guess what I did."

"I dunno, what'd you do?"

"I joined the Army."

"You did what?"

"I joined the Army."

"I didn't even know you liked the Army."

"Well, I guess I do."

"What do we do now?" Cal asked.

"I dunno."

And then she hung up.

Cal rode with her parents to the bus station. She was headed to some Army base in Texas. Watching her leave was the loneliest he'd ever felt in his life.

After about three weeks, he called her. "Marcha," he said, "I want you to marry me."

"Yes!" she said. "Yes!"

And then: "But why didn't you ask me *before* I joined the Army?"

They lived in a tiny apartment at Fort Bliss in El Paso while Marcha worked in missile defense systems. After four years, when Marcha's enlistment was up, she decided to stay home with their children (they had two by now). Cal enlisted as a private, never thinking he'd make a career of it. At the ten-year mark, he was ready to get out. But then he decided to try out for something called Special Forces. A couple of guys in his unit were talking about it. Cal discovered that he loved it. Which was how, fifteen years later, he came to be standing in a cave in Afghanistan, calling Marcha back in the States.

It was three o'clock in the morning when the phone rang on her bedside table.

"Hey, whatcha doing?"

Marcha sat up. "Cal?"

"Love you."

"Oh, honey. I miss you."

"So, what are you doing?"

"Sleeping. You always ask that."

The question was a joke between them. Spencer would be deployed in a secret place around the world, and he'd call out of the blue as if he were away on a weekend business trip.

"Where are you, Cal?"

"You know I can't—"

"What are you doing?"

"Working, honey."

"So, where are you calling from?"

"Every time, you know I can't answer that."

"I know."

It was hard to understand Spencer because of the delay in the satellite link. Marcha kept interrupting him. He sounded like he was talking from underwater, his voice wobbly.

"Well, are you okay?"

"Me and the guys are fine. Just fine."

"Can I send you anything?"

"How about some beef jerky?"

"Beef jerky?"

As he talked, Spencer was looking at a cliff rising straight up across the river valley. He thought it was a grand sight. It was late evening, dusk, and the river valley was quiet. The grass was brown. There was barely any wind.

"How are things going with the boys?" Marcha asked.

"Fine, fine. We're all fine."

"'Bye then. I love you."

"Love you, too."

And Spencer was gone. The connection seemed to melt and the phone went dead in Marcha's hands.

Marcha immediately dialed Sam Diller's wife, Lisa.

"Cal just called. I can't figure out where he is. Did Sam call you?"

Lisa said he hadn't. They agreed to keep in touch and share any information. They operated a phone tree, with each wife or girlfriend calling the next name on a list and spreading any news. Word traveled fast at Fort Campbell. Since saying goodbye to the men at the church parking lot, the families had heard nothing.

By the time Dean and his team were ready to board the helicopters on November 2 for infil, the relentless rain at K2 had turned the camp into a virtual swamp of anxious, bickering men. Dean was nervous. He worried about the lack of solid intel on Atta, the shadowy warlord he was about to "befriend." The intel was so weak that when he asked for information on warlord Atta Mohammed Noor, he received a very grainy fax of Mohammed Atta, a dead man who had flown into the World Trade Center on September 11.

Personalities and politics were the two key aspects of unconventional warfare. If Dean couldn't get inside the shoes of Atta and his men, where could he lead them? Dean had ended up going to the CIA Web page and downloading information about Afghanistan from the *World Fact Book*.

Did Atta have kids? Yes. Where was his wife? In Iran with the kids. Where was his money coming from? Iran and Pakistan, funneled through the Northern Alliance. What was the weather? Drought. How many troops did Atta have? About 5,000 men, some of whom he had "loaned" to Dostum to fill out the Uzbek's thinner ranks. They were armed with AKs, RPGs, mortars, and a few tanks. Atta was thirty-eight years old.

The more Dean read, the more the country's key warlords seemed to resemble the Mafia. Their people had many family ties and some were employed in both legitimate and criminal businesses. They fought one another within their families, but the families stuck together against other families. Dostum's associates were reportedly involved in drug trafficking and poppy production. Atta seemed the more peaceful of the two, described as a schoolteacher and de facto governor of Mazar-i-Sharif.

The team had been briefed about the perils of being captured by the Taliban. "They are going to hang you from the town square upside down and beat you until you die. Then, they're going to scalp you."

Everybody had thought, *Okay, bad guys. Bad idea to meet them,* and then put the thought of capture out of their minds.

The team's intelligence specialist, Sergeant First Class Darrin Clous, set to memorizing all he could about where they were going. He knew they'd be trudging through snow in the mountains, that they might be in-country a year. He memorized Afghanistan's ethnic tribes and terrain, and briefed the team on everything he knew.

Before boarding, Mitchell and Dean's team gathered and saluted Colonel Mulholland, who gave the teams one last pep talk. He told them that although they could get into the country tonight, he might not be able to get them out.

Nearing touchdown, Dean heard "One minute out" over his headset. And then he heard they were taking small-arms fire, which alarmed him. But no rounds hit the Chinook. He didn't know what to make of this information. He later learned it was the Afghans' way of communicating with each other: "The Americans have arrived." The Chinook settled down on its wheels and Dean and the team prepared to exit.

They were greeted by a couple dozen Afghans. They helped lift their heavy rucks off the ramp, and trudged with them through the snow a safe distance away. Dean and his team began handing out green Army blankets to the grateful men. So far, so good.

Dean himself was shivering from the ride in. The other bird came down fifteen seconds later, blasting them with snow and wind. Sergeant Brian Lyle's feet were numb from the cold. He saw four or five of the locals crouched by a tiny fire they fed with small, scrubby bushes that looked like the kind of stuff you'd use in building an electric train diorama. It seemed that some of them were actually sticking their feet into the fire to keep them warm.

Dean couldn't believe what the Afghan fighters were wearing: long gowns and plastic shower shoes. Dean, red-haired, tall, with pinkish skin, realized that he had to look strange to them, too.

The Afghans began packing 300 to 400 pounds of gear on a string of a dozen burros. What shocked weapons sergeant Brett Walden was that after the Afghans packed the gear, they scrambled on top of the heap, planning to ride the straining animal. Walden felt equally sorry for the horses loaded down with oversized Americans and their ammunition and rucksacks.

The Americans struggled in the thin mountain air to keep up with the thinly dressed locals who—no longer astride the burros— were walking swiftly. *They're smoking us out here,* thought medic Jerry Booker. They slipped on ice and laughed nervously as they approached sheer drop-offs, their backs pressed against the rock wall as they inched ahead. Even the burros were slipping with a horrible racket, their hooves pounding the slushy snow. Booker slipped and fell and split his lip on the butt of his M-4 rifle. After six hours of marching up and down ridgelines, Dean and his team entered the base camp of the warlord Atta Mohammed, near the desolate village of Ak Kupruk.

It was about 4 a.m. Atta emerged from a mud house, looking regal, wearing a knee-length blouse, matching cotton pants, a scarf, black leather boots, and a beige Pakol hat.

Atta was stroking his beard, looking overjoyed to see his new American friends. Dean immediately wanted a Pakol hat.

He shook the warlord's hand, which was strong and papery, darkened by years of sun and war. For a thirty-eight-year-old, he looked like an older man. Dean thought he had warm eyes.

"I've heard a lot about you," Dean said, lying through his teeth.

What he knew about the man would've fit on an index card. But he wanted Atta to know that he had the United States' entire attention.

Atta smiled and, through an interpreter, welcomed him and explained that he had been up all night planning his army's movements.

He surprised Dean by announcing that he was going to take a nap.

Dean felt the air go out of the moment. A nap? He wanted to sit down and talk about the war.

The warlord bowed slightly and retreated inside his house. Dean shrugged. He ordered everybody to unload their gear and find escape routes in the camp in case of an attack.

After dropping off Dean's team, the helicopter lifted into the air and pushed about ten miles east to deliver Major Mark Mitchell and Lieutenant Colonel Bowers, and the rest of the command and control element, to Dostum's base camp.

About 100 yards from the landing zone, Nelson and his communications sergeant Vern Michaels crouched behind some rocks, waited for the bird to land. They feared being mistaken for unfriendly forces lurking around the landing zone, and Nelson knew that the guys in the helicopters had a habit of shooting at anything that moved. So they stayed hidden as the aircraft lifted away. He could hear the rumble of a small, motorcycle-like engine starting. Nelson wondered what it was.

Through his night vision goggles, Nelson watched as the guys on Mitchell's team loaded two six-wheeled motorized "buggies" with rucksacks and assorted bags of gear. The vehicles measured about the length of a midsized car. Upon inspection, they looked like big golf carts on steroids.

The green-and-yellow-painted buggies were called Gators, manufactured by the John Deere farm implement company; with their knobby six wheels and rugged frame, they could navigate trails that a truck couldn't travel along. Max Bowers had thought to bring along the vehicles after reading the reports that Nelson and his men had trouble finding enough horses.

Mitchell was surprised to see that they had essentially landed in

Dostum's base camp, where Nelson and his team had been living. Mitchell could make out the gray smudge of Dostum's canvas tent, staked down in the hardpan clay. Horses neighed and shifted nervously somewhere nearby.

Bordering Dostum's tent was a crude mud wall, chest-high; Homer, Mitchell, Bowers, and the rest of the newly arrived team unrolled their sleeping bags at its base.

Mitchell figured it was an hour before sunrise and sleep seemed beside the point. He dozed off, then woke with a start. He'd been asleep several hours. He scrambled out of his sleeping bag. At that moment, Dostum strode out of his tent.

Mitchell marveled at the general's confident stride. Mitchell had seen only a head shot of the man in a uniform. He tried to match the unflattering descriptions he'd read about Dostum with the smiling picture of hospitality standing before him.

General Dostum reached out and shook Mitchell's hand and then introduced himself to Bowers.

With his arrival, Bowers was now Dostum's liaison, taking over for Nelson. About this new arrangement, Nelson was disappointed. He also knew there was nothing that could change it.

Bowers, smiling and sporting a newly grown gray beard, sized up Dostum.

He had prepared for this moment. He'd even brought along a piece of the World Trade Center, a thin candy-bar-sized hunk of metal, which he intended to give to Dostum and Atta, in a bid to bind them against their common enemy, Taliban and Al Qaeda soldiers.

"An honor," said Bowers, firmly shaking Dostum's hand.

Dostum realized he would miss Nelson's company. He and the young American had formed a familial bond. He returned Bowers's compliment.

"What do you want, General? What can we do for you?" Bowers asked.

"I want to get out of this valley. I want to take Mazar."

"We can do that," said Bowers. "We can fight."

Dostum studied the American. He believed him. Many men would die, but they would take the city.

• • •

Back at Atta's base camp in the neighboring Darya Balkh River Valley, west of Nelson's position, Dean and his team had been led to a low-roofed mud house inside a walled compound. The walls stood eight feet high and measured fifty yards on each side; in one corner stood a latrine whose door was hung with a filthy curtain. Dean watched as some of the soldiers, dressed in thin cotton pants and knee-length smocks, stood in the courtyard center warming themselves around a pale fire. They looked up at him, rubbing their hands, and smiled. Dean waved back, *Hey*.

Standing there, looking out at the mountains, which he found beautiful, and back at the fire, smelling the woodsmoke and listening to the horses stamp and whinny in the nearby paddock, Dean felt he had been born so that he might be alive at this moment, *in* this moment.

"Where is Atta?" he asked his chief warrant officer Stu Mansfield.

"Still asleep."

"I'm going outside." He was anxious to get the lay of the land. Dean pushed through the low doorway in the front wall of the compound and walked out onto the rocky ridgetop. The strip of ground was smaller than he imagined walking up to it in the dark, measuring about 200 yards long and 50 yards wide. The compound's back wall stood against a rockface rising several hundred feet over the entire structure. The remaining three edges dropped steeply away from the tabletop for 1,000 feet or more, it was hard to tell.

To the west lay the village of Ak Kupruk. Dean walked to the edge of the ridge, lifted his binoculars, and tried getting a clear shot of the houses tucked along the blue Darya Balkh River. The river looked cold in the morning light. The village lay quiet. No woodsmoke, no one moving about. He guessed it was about two miles away. He panned the binos up and over to the right about 1,000 yards, and spied a Taliban bunker in the mountainside overlooking the river and the village. That was the kind of target he wanted to hit.

He swung his gaze back to the houses and tried imagining the men and women and children cowered in the cold rooms of their homes while the Taliban lay in wait in hideouts around the village. He hoped that the Taliban had not learned of his arrival.

Atta awoke after an hour's nap and summoned Dean, Stu Mansfield, and engineer sergeant Brad Highland to a meeting. They walked with the warlord across the courtyard. Atta's men, some 300 of them garrisoned in the compound, stood up and watched them pass. They turned and started talking excitedly among themselves. They seemed intrigued by the weapons Dean and his men carried, pointing at the long guns, the M-4s; the black, heavy pistols in their nylon holsters; the grenades bulging in their load-bearing vests.

Dean was struck by something akin to deep affection for these men, who were fighting and enduring more than he had ever been asked to endure. He had the urge to walk around and hug all of them, and say, "Don't worry, we will get you shit, beans, bullets, and bombs, and we will make life better." He knew this wasn't your classic tough-guy response to being in combat, but he couldn't help it.

Atta stopped in front of a mud house next door to the Americans' new lodging, and motioned them inside. All of the Afghans stooped and removed their shoes.

Dean hesitated. He looked over at Highland, who cracked a smile. Dean was reluctant to remove his boots as cultural habit dictated. Even as a teenager, he'd been embarrassed by the strong odor of his feet. He'd tried powders, shoe inserts—he'd stopped wearing socks, thinking this might stanch the smell. Nothing had helped. The team kidded him mercilessly. Now he worried that he was about to offend Atta with the smell of his feet.

It seemed odd to be considering the diplomatic repercussions of foot odor, but Dean couldn't help himself. He had lain awake nights in his tent at K2 gaming all the ways he might screw up his command, but the smell of his feet was something he'd forgotten to consider.

Here goes, he thought, and he bent down and slipped off his boots.

The odor filled the doorway immediately. It nearly knocked Highland over. He watched as Atta's men drew back as if struck by a stiff wind. Dean quickly took a seat next to Atta on a woven rug spread smooth on the dirt floor.

He tried tucking his bare feet in the crook of his knees, a painful lotus position that he found nearly impossible to maintain.

Highland watched as Atta sniffed the air and then turned his gaze to Dean, who looked up sheepishly at the warlord.

Atta gave a blank look, and then began the meeting. It was all Highland could do to avoid breaking out in laughter. (In later meetings, Dean noticed that Atta made a point to wear his own boots to the carpet meetings—a hint, Dean figured, that it would be perfectly fine to do the same. Dean did not remove his boots again.)

A soldier appeared bearing a tray of almonds and steaming cups of tea. Dean looked around the room at pictures of Ahmed Shah Massoud, the famous anti-Taliban fighter who had been assassinated on September 9, hanging on the walls. It seemed years ago that Dean, on his honeymoon, had read the headline announcing that Massoud was dead. In many of the pictures, Atta stood smiling next to the great Lion of the Panjshir.

"I have several thousand troops," said Atta, speaking through an interpreter. He explained that these men were farmers by day, guerrilla fighters by night, and lived in the surrounding mountains and villages. He had only to call on his radio and they would appear.

Later, Dean would see Atta's fighters show up carrying AK-47s, and there with them would be their sons, carrying spare magazines. Behind the sons walked even younger sons, carrying nothing. Dean understood that in this kind of fighting, the sons who carried nothing would pick up either a gun or a magazine if the fathers or brothers were killed. The look on the faces of the kids seemed to indicate to Dean that they expected to die.

Atta explained that he had been fighting in this area since he was fourteen years old—first the Soviets, then the Taliban. He said he'd been wounded three times in battle and carried ten pieces of shrapnel in his back. His brothers were all soldiers, still living. His father was a shopkeeper. He said he felt like he'd been telling the world for five years that the Taliban were terrorists, and now that the Americans had arrived, the world was listening. He explained that he had built a network of trusted contacts among the locals, and his soldiers moved themselves and supplies swiftly without much detection by the Taliban. The terrain itself conspired to split and divide and confuse.

Dean explained that he needed his men taken to positions in the mountains where they could watch over enemy bunkers. "I want to bomb the leadership positions," he said.

Dean knew that the Taliban army was increasingly led by fighters from Pakistan, Saudi Arabia, and Chechnya. He figured that if he could get rid of them, the locals would defect from the army, because, as Dean had told his team, "Afghans do not like to be occupied by anybody."

"Do you have a map?" Dean asked.

Atta said he didn't, which shocked Dean. He reached into a bag and pulled one out, written in English. Atta's eyes lit up in delight. He took the map, carefully smoothed it open on the carpet, and began narrating in Dari the places he had fought in, the places he still hoped to win.

"I can bring in bombs," said Dean. "You need to show me the targets—I am not going to just drop bombs anywhere. I don't want to hurt the population. We are here to liberate the Afghan people."

Atta told Dean that a village called Lalami, located about five miles west of Ak Kupruk and about ten miles north of where they now sat, had fallen the day before. His men had also attacked Ak Kupruk, but the battle there was still furious. The Taliban had positioned 5,000 troops around the village. So far, his men had captured 800 enemy soldiers and taken 200 more prisoner. They had watched a line of a dozen or more trucks race from the city under cover of darkness, fleeing north, said Atta, to Mazar-i-Sharif.

Atta explained that the Taliban in Ak Kupruk stood in the way of his army's march to Mazar-i-Sharif.

Dean laid out the battle plan: He and Atta would capture Ak Kupruk and drive the Taliban north up the Darya Balkh River, where they would eventually unite with Dostum's forces. The combined American and Afghan forces would move up this valley to victory.

"I will need to split my team," Dean said.

He suggested that Atta stay behind at base camp with chief warrant officer Stu Mansfield, communications sergeant Brian Lyle, engineer Brad Highland, weapons sergeant Mark House, and medic Jerry Booker, who would run a beans and bullet supply train. From this mountaintop vantage point, Atta could direct Dean as he traveled downrange in combat.

Dean would be accompanied by intel sergeant Darrin Clous, medic James Gold, Air Force combat controller Donny Boyle, weapons sergeant Brett Walden, Staff Sergeant Francis McCourt, and communications sergeant Evan Colt, and a security detail of approximately fifty of Atta's soldiers. Their job: bomb Taliban bunkers surrounding Ak Kupruk while Atta's army attacked the village proper.

Atta smiled. "I will have my men bring up some horses," he said.

While the meeting between Dean and Atta was going on in the mud house, Brian Lyle and Jerry Booker stood outside the door and watched as some of Atta's soldiers rode up the mountain from the battle around Ak Kupruk. The soldiers were astride tired horses, trailing a string of burros, the soldiers looking dirty and exhausted, some of them bleeding, their horses breathing raggedly beneath them. The men dismounted and ate a quick cup of boiled beans and drank some water while other men in the camp loaded handfuls of AK ammo and RPG rounds in the saddlebags lashed to the burros. The resupply took half an hour. Then the riders remounted and turned the animals back down the mountain clinking and clanking, descending into the valley and the struggle for Ak Kupruk.

Lyle and Booker dug through their rucksacks, pulled out several cartons of Marlboro cigarettes, and started handing them around as a gesture of friendship to the men in camp. Neither Lyle nor Booker were smokers, but they lit up, too, to be sociable.

Lyle grew light-headed after about three puffs and dreaded the fact that any marching he'd do tomorrow would feel doubly laborious. But then something happened that made the nausea feel worthwhile. Atta's men started laughing, pleased with the Americans' generosity, and one of them started yelling, "Commando!" as in "Welcome, commando Dr. Booker!" "Welcome, commando radioman Lyle!"

Their usefulness was soon put to the test when a wounded man stretched out on top of a door was carried into the compound.

This man's leg was a mess. Booker drew a breath as he leaned in, examining the wound. The guy had stepped on a mine, the kind known as a "toe-popper." Small, light, and inexpensive, these mines were considered very injurious to an opposing army because they merely wounded the enemy rather than killing them outright. If you blasted a soldier to death, that was that. But a wounded man required several of his comrades to carry and care for him until he reached real medical help. The man before Booker was a perfect example of the strategy—his foot had been peeled back, the jellied flesh dark as a ruby.

Booker was worried about gangrene setting in. He didn't have any bone saws with him, and he definitely was going to have to cut this dude's leg off, or at least part of it. He reached into the sheaf on his web belt and pulled out his Leatherman tool. It did have one good blade, about four inches, serrated. It would have to do.

As Booker was studying the leg, he looked up, startled to see a pretty woman approach, her dark hair uncovered, her face pale. She looked about twenty years old. She stood at the man's head and stroked his hair. She began to cry. She was, he guessed, the man's wife, and in her grief she had neglected to cover herself with a veil in Booker's presence. Realizing this, Booker averted his eyes, not wanting to offend her or the Afghan men.

One of them lifted a blanket between Booker and the woman so that he could still work on the man's leg. He could hear him moaning on the other side of the blanket, the song of pain mixed in with soft reassurances from his wife. Booker knew they were praying, and that they understood that the man might not make it.

The explosion of the mine had taken off most of the flesh on the leg. The blast had started at the base of the foot and the shrapnel had chewed up along the shin and scoured the tibia. Everything below the knee was bare bone, sheathed in a thin coat of blood. Booker took hold of the leg and manipulated it, considering whether or not to disarticulate the knee—in other words, if he should disassemble it, taking it cleanly apart at the joint, like you would pop the leg socket on a chicken bone. But he wanted to leave the guy something, some semblance of a stub, in case he could ever get fitted for a prosthetic. He decided to cut three inches below the knee.

Booker had given the guy Nubane, an opiate-like synthetic morphine, to control the pain. But he was bleeding like a sonofabitch. Booker made up an IV drip, providing as much fluid as he could without thinning the guy's blood too much and blowing a clot, which could result in a hemorrhage. Booker worried that the guy had lost too much blood. Yet he had to do the best he could to save him. He also knew that if the guy were in a trauma center back in the States, he'd make it. The woman on the other side of the blanket was crying and still praying. After a while, it seemed to Booker, they all thought, *What's the use?* and the blanket dropped, and he could see the woman's ashen face. He didn't want to make her feel even more uncomfortable in the midst of this gore and suffering.

Now, as Booker sawed the tibia, the guy lay moaning gently as the knife blade made a dull *whiz-whiz* sound cutting bone. Usually at this point, he'd pack the end of the cut with something called bone wax, a substance meant to cap the cut. But he didn't have any of that either, so he improvised by repacking the marrow in the center of the bone and tried fashioning a plug out of that. It worked. He wrapped the flap of skin over the stub and bandaged it. When he was done, it all looked very neat, but he sensed the guy wasn't going to make it.

Afterward, Booker was surprised to be asked by the man's family if the patient should eat some opium, or inject it. The family was obviously thinking of medicating him, and Booker thought, *God, I don't know*. Finally he said, "I guess I'd tell him to smoke it because it would be a lower dose that way. If he has to have it, smoke it." The relatives nodded their understanding.

As soon as Booker finished with the leg, the Afghans brought in a kid, no more than fourteen, who'd been shot in the shoulder. The bullet had entered the front and exited cleanly out the back, something that was called "a through and through." This kind of wound was a no-brainer, and the kid was bandaged and taken away. More troubling was another young man, who'd been shot in the wrist.

Booker gave him an injection of a local anesthetic, unwrapped a pair of latex gloves, and used the wrapper as a skirt on the kid's lap. Booker was mindful of the need to conserve supplies, and he figured that the sterile inside of the wrapper would suffice as a clean operating surface. He had the kid lay his arm on the white wrapper and then he pulled out his Leatherman knife. The boy's eyes widened as Booker extended the serrated blade and held the knife away from him, over the bare dirt, and poured alcohol on the metal to disinfect it. Booker turned back and saw weariness in his eyes, but not fear. Booker shook his head, admiring the boy's courage.

He pressed down with the knife, careful to cut along the tension line of the skin, on both sides of the wrist, so the arm muscle wouldn't pop open like a jack-in-the-box. He went in with the mayo scissors and cut out the bruised and dead muscle. That's what he knew would kill the kid, the dead tissue causing an anaerobic infection. You might leave an air pocket in the suture and pretty soon the air would escape, leaving the dark, moist place without oxygen. It was those kinds of infections that killed lots of guys in the Civil War and in World War II.

Booker then packed the hole with Curlex to stop the bleeding. The best thing would have been to close the wound with some sutures for three days and then ask the guy to visit him. But Booker had no idea if he'd ever see any of these guys again.

When Booker was done, Atta's men came up and thanked him. There was blood everywhere, on the door, splattered on the ground. Booker noticed a dog circling the courtyard, looking hungry as hell.

When he had finished cutting off the first man's leg, they had set the bone aside, and now Booker spied it lying in the courtyard. He knew it was only a matter of time before the dog made off with it. What was the Muslim custom? he wondered. Did it have to be buried in a special way? And then he started imagining the dog gnawing on the leg bone, and he picked up a rock and threw it at the dog. "Get outta here!"

Finally, an Afghan soldier walked up and carried the bone away tucked under his arm.

Booker had no idea what would become of it, but he was relieved.

As he was considering this macabre turn of events, from across the compound came a disconcerting cry: "The Taliban are coming! The Taliban are coming!"

Booker jumped up, and Dean, Mansfield, and Highland stepped quickly from the house in which they'd been meeting with Atta.

"Are we under attack?" asked Dean.

The shouts were coming from outside the compound wall, but there didn't seem to be a crowd out there.

Holding their weapons at the ready, Dean and the team ran out of the compound, down a small hill, and took up a hidden position in a small depression in the ground. Looking at it, Dean realized they were in a dry moat that had been dug around the compound. He strained to hear the approach of enemy troops.

Peeking above the berm with his binoculars, he saw about twenty men in black turbans walking along a trail on a ridge across the valley, about a half mile off.

They were being guarded by several of Atta's men, who were nudging them along with the barrels of their AKs. The group disappeared behind some rocks, and finally, maybe a minute later, reappeared on the trail, this time closer to the compound on top of the ridge. Dean could see—to his surprise—that the men's hands were bound with scraps of cloth.

Dean and his men climbed out of the moat and hurried into the compound where the prisoners were starting to assemble.

Atta walked up and down their line, speaking slowly and solemnly, as if talking to schoolboys. Dean couldn't figure out what was being said. He talked for several minutes. The POWs stood humbly before Atta, still bound, looking at the ground. And then Atta beckoned one of his lieutenants forward, who approached the POWs.

To Dean's shock, the soldier began untying the prisoners. The team tightened the grip on their weapons.

One by one, each prisoner approached Atta, placed a hand over his heart, and humbly swore allegiance.

When a prisoner was done, he stepped away. Then either he walked out of the compound, headed, Dean figured, for home, or he picked up a weapon and joined Atta's soldiers.

It was clear that many of the Northern Alliance fighters knew some of these Taliban men. They explained to Dean, "No, this man is okay! He is a shopkeeper. He sells oranges. He is my friend." Either the men had joined the Taliban against their will, or they had seen the error of their ways. Maybe they never wanted to be in the Taliban in the first place.

Dean could not wrap his mind around this surrender. He thought he smelled a rat. He and the team were now surrounded by the very soldiers whom, minutes earlier, they had been planning to kill.

Atta took Dean aside and explained that each prisoner needed to be treated well, as defined by Sharia law, which Dean knew from his study of the Koran. "I will treat them as I would wish to be treated, should I ever be captured," said Atta.

Dean nodded.

"Stay alert," he instructed the team. "These prisoners might be fixing to smoke us all."

• • •

A short time later, Dean's team left Atta's compound in rain and fog and rode all afternoon, climbing down the rocky trails into the Balkh River Valley, trails so narrow that a slip meant falling into a chasm hundreds of feet deep. Staff Sergeant Brett Walden's horse reared and the saddle suddenly slid all the way back on the animal, around its ass, and Walden had to grab onto the tail, above the kicking hooves. He let go and rolled off the animal to safety. He remounted the ornery animal and the troops were moving again.

Twelve hours later, they approached the wind-scoured edge of Lalami South, a 5,000-foot mountaintop settlement overlooking Ak Kupruk. Dean judged they'd traveled down the mountain about eight miles since leaving Atta's headquarters.

Ak Kupruk now appeared less than a mile away, darkening in twilight at the bottom of the valley. Their guide turned in the saddle and told them to stay where they were.

He rode ahead with several other soldiers toward a cluster of eight houses about 100 yards down the trail. These crude mud dwellings belonged to Afghans sympathetic to Atta, but the soldiers were double-checking the security of the area.

Dean watched as one of the doors on a house opened just as Atta's men approached. A man stood in the doorway listening as the soldiers spoke, and then he held out his hand as one of the soldiers placed what looked like a wad of money in it. The owner of the house turned, looked in Dean's direction, and waved them on. The Afghans remounted their horses and galloped back.

"It is safe," one of them said. "We can go in."

Dean realized that they had just linked up with what was called the "auxiliary"—a fancy word describing the underground of citizens who were supporting the war against the Taliban. Much like the fighters in France during World War II looking for support among the local Resistance, Dean and the Afghans were at the mercy of the man who owned the house. Dean wondered if they'd been compromised. They planned to spend the night here and bomb the Taliban in the morning.

The owner of the house led them to an empty room. They were completely concealed from anyone passing by outside, and Dean saw that this was a well-coordinated plan. He and the men stacked their rucksacks against the walls and slid down next to the bundles.

Unable to stetch out their legs in the cramped quarters, they slept fitfully for a few hours, until dawn, when they mounted their horses and started riding again.

They rode with the sun rising at their backs. In places it seemed the animals were wider than the trail, forcing them to step carefully, like tightrope walkers. Dean wanted to dismount and lead his horse, but the Afghans had stayed aboard and so he tried swallowing his fear. His animal plodded ahead as if led by a string. The travel was slow going. The pack animals strained under their heavy loads of American gear. He wondered how many times horses had made this journey along this path and he guessed the answer was in the hundreds. Dean looked down from the saddle as they passed firing positions carved in the rock and he thought about the men who had fought and died in them.

The Afghans grew restless and beat the struggling burros. One of them collapsed and rose reluctantly. Dean looked back and saw that one of the horses had broken a leg and was lying on the trail. Later, when they stopped to rest, Dean asked what had happened to the horse. The Afghans professed ignorance. Dean figured he'd be eating it for dinner that night.

They rode onto a ridgetop, crossed its hard expanse, and rode off the edge, heading downhill. Suddenly, one of Atta's men held up his hand.

Dean looked around. They were exposed on a rocky face, visible for miles. *They must know what they're doing,* he thought. *Maybe we're lost. Try not to be the pushy American.*

Dean broke out his phrase book, hoping to talk to them in their native Dari. He looked up from the page to see hundreds of red-painted rocks scattered on the hillside. He calmly replaced the book in his coat pocket and announced to the rest of the team, "We're in deep shit."

They already knew.

They had ridden into a minefield.

Dean looked for white rocks designating areas that had been cleared of the explosives. There weren't any white rocks.

Dean looked to their guide.

"Follow me," the man said. He turned his horse tightly in the confines of its own shadow, and they rode back up the hill and turned and stopped to consider the situation.

Just then, Dean heard the thud of hooves, and riding up came a dark-haired young man galloping confidently, his hat pushed back on his head. He dismounted, walked straight up to Dean, and said in Russian, "I am Mohammed Sihed."

Russian! A language Dean could speak.

"Thank you for being here," said Sihed. "Thank you for helping us with our cause to kill the Al Qaeda."

"So glad, and a privilege to fight with you," said Dean. "I heard there are some bad guys in Lalami."

"There are."

"Well, I want to kill bad guys," said Dean.

"Okay, let's go, me and you."

"I want to bring Boyle," Dean said. "He talks to airplanes."

"No problem."

"There are land mines," warned Dean.

"I know the way. We will walk."

Dean radioed Chief Stu Mansfield that he was about to set up a bomb strike, and then grabbed his CamelBak, the water hydration system worn like a knapsack, from which he could drink through a tube over his shoulder. Boyle shouldered a ruck containing his radios, and they struck across the edge of the hilltop, along a snow-dusted trail.

"If you stay on the trail, no problem," Sihed called over his shoulder.

"Pass it on," joked Dean.

"Will do," said Boyle.

After an hour of walking uphill, they were crouched behind a rock facing a towering mountain wall about 2,000 yards away.

Snapping his binoculars into focus, Dean made out the gray, flinty edges of a Taliban bunker emerging from the rock.

He handed the binos to Boyle, who looked through them and lowered them and turned to Dean.

"Taliban or *dost*?" Dean asked, using the Dari word for "friend." He pronounced *dost* so it rhymed with "toast."

"No *dost*. Taliban." Sihed made a slashing motion across his throat, as if with a knife.

Boyle turned to Dean. "I can't drop bombs on that," he said.

Dean was shocked. "Why not?"

"Because," said Boyle, "how do I know that's an enemy position?"

Dean was incredulous. "What would you like to see to convince you? Listen, man, these people don't wear uniforms. What do you want to see?"

"I can't drop bombs on them, sir," Boyle said. "I am sorry. I just can't."

This was the last thing Dean had expected to hear. He knew Boyle was trying to do the right thing. They dreaded dropping bombs in the wrong place.

Dean pointed at Sihed and asked Boyle, "So how do you know he is not the enemy?"

"Well, I don't."

"Exactly," said Dean. "But he is standing here, and he is telling us that those motherfuckers over there are bad guys."

Dean didn't exactly know what else he should say. "Listen," he went on, "we're never going to know who is a bad guy or a good guy unless they tell us, or someone points a gun at us."

Boyle agreed to drop the bombs. They called in a JDAM, arriving at the grid coordinates by plotting with their range finders. The bomb missed by at least a half mile. Dean and Boyle agreed they'd be more successful using lasers. Dean was embarrassed.

He radioed back to the team at Atta's and asked that someone bring forward the SOFLAM, which was the laser designator.

"And walk, don't ride the horses," said Dean. "There's land mines. Be careful!" It would be dark in several hours. Dean felt the day was slipping out of his control.

Upon arriving, Booker and Highland were anxious to start dropping bombs. But they also worried that the spot where they stood contained land mines. Sihed assured them that it didn't.

To prove his point, he ordered his men to run in circles around the escarpment, scuffing their feet and jumping up and down. "See, all clear!"

Holy Christ. Dean expected any minute that someone would blow up. The afternoon went farther downhill from there.

Using the laser, Boyle painted the bunker they had missed earlier and the bomb inexplicably flew off target and hit another bunker nearby by mistake.

Dean wanted badly to impress Atta and he felt nothing but frustration. At dusk, though, he was able to figure out the source of their targeting problem—he was relieved to learn it was not his fault.

As the drafts from the valley floor circulated during the day, supercharged by the sun's increasing heat, the air had thickened to a brown haze. It was this haze that was actually bending the laser, so that the invisible beam of light was not aimed where Boyle pointed it.

Dean was angry about the impasse. He hadn't eaten in two days. He'd slept little. A cold wind blew up the mountain ridge and flashed around the dark, slashing at them. Looking down from their perch, Dean saw fires flickering in the far distance, around Ak Kupruk.

"What's going on?" he asked Sihed.

Sihed shook his head sadly. The Taliban were coming through town, he said, burning the homes of the men who had joined Atta's forces.

"What about their families?"

"If they're lucky, their families went with them."

Dean realized there was no more to do here tonight except watch Ak Kupruk burn. He couldn't bear the sight. He couldn't bear thinking about the screams and gunfire he imagined arising in the village.

He looked at the fighters standing with him. *These people are starving. Winter is coming. It's cold.* He told Sihed they would return to Atta's compound. They would come back tomorrow and set things right.

They rode back in pitch dark. Getting off his horse, he was shivering. He smelled hot, cooked rice and bread warming on an iron skillet somewhere. His mouth was watering. He practically fell off his horse, dead tired.

He was met at the door by one of Atta's men. There was some bad news. They would not be attacking Ak Kupruk tomorrow. They had to wait. Dean was disappointed and he wanted to know why.

Because Dostum's forces were not yet in place, said the soldier.

Dean thought about the town burning, the children, the women. Maybe they were being killed as he stood there, but there was nothing to be done. He felt both frustration and grief.

Dean and his team entered the door to the eating room and sat down heavily on the carpet, the room lit by guttering candles. An Afghan came by with a towel draped over his arm, bearing a pitcher of water. Each diner held out his hands and washed them, then took the towel from the man's arm, dried off, and replaced it. This was meant to show everyone that each man was clean. They sat cross-legged on a red cotton blanket. Dean tucked his stinky feet under him as best he could, trying to muffle the odor. It didn't work.

God, he was hungry. He smelled something good and greasy: *goat.*

He couldn't remember when he'd been so hungry. He and the team watched as the Afghans they'd been traveling with these past three days reached out with a single hand to cup the rice from a communal bowl, then brought the hand to their mouth in one smooth motion: reach out, stroke to the mouth, suck from the hand, and repeat.

Dean tried doing the same and soon made a mess. The rice grains were falling all over the red carpet beneath him. Nobody seemed to care.

His team had a pile of goatmeat at the end of their blanket, and the Afghans had a pile at their end, and they were all eating like mad. The meat had some kind of marinade on it. He bit into a piece of bread and his teeth crunched on the sand grains in it. Dean thought that he would never taste anything as good again.

He understood that Atta and his fighters were giving him the best they had, and that this food could probably feed their families for a week. He realized that generosity was in proportion to what you had. These were some of the poorest people he had ever met, and yet they would give him—it seemed to him—even their lives if it came to that. In return, he wanted them to win their freedom.

Someone on the team pulled out a Dari phrase book and, their shadows flickering on the walls, the Amercians started saying "hello" in Dari.

"*Namse-chase* Dean," said Dean. My name is Dean.

"*Namse-chase* Darrin," said Sergeant Clous.

"*Namse-chase* Brian," said Sergeant Lyle.

Whenever somebody swore, "Goddamn, this is good food!" or, "Shit, what a fucking mess the bomb strikes were today!" Atta repeated the profanity and laughed.

"How do you say 'mudder-futter'?" he asked.

"Say 'muther-fuck-er'?" repeated Dean.

"Mutha-fooker!" said Atta.

The Afghans and Americans laughed as if they were drunk, but they weren't drunk, except on the wind and the harsh sun and the feeling that at any moment they could all die.

Pretty soon, the goat was gone. They had consumed the entire fifty-pound animal. All that was left was the grease at the bottom of the porcelain dishes, which were decorated with fishes and tiny flowery designs. They swiped pieces of flat bread into the grease and gobbled them down and opened the phrase book to the word "good" and pointed at the empty platters, chewing, and exclaimed, "Good!" Dean had felt so terrible riding back to the safe house—terrible about missing the bomb targets, about the fires burning in Ak Kupruk.

But now they were all laughing.

On November 3, as Dean and his team climbed to their observation post above Ak Kupruk in preparation for bomb strikes, Major Mark Mitchell was busy continuing getting to know General Dostum in his camp in the Darya Suf. Mitchell and Dean were separated by less than twenty miles of roadless mountains, but they had been thrust into vastly different camps.

Dostum turned out to be a gregarious host who laughed easily and turned serious without pause, his eyes narrowing. Atta, thin, shy, introspective, whose ambition to power was no less than Dostum's, smiled slyly when he might've spoken bluntly, more Cheshire cat to Dostum's German shepherd–like visage.

Dean sensed that Atta was truly his own man, regardless of the fact that he seemed loath, at times, to betray exactly what he was feeling. Mitchell grudgingly admired Dostum's uncanny ability to be any man that any situation called for.

Dostum, Mitchell, Bowers, and CIA officer J.J. were sitting on a rug as Bowers explained the coming two days' strategy. On November 5, Dostum's forces would mount an assault on the village of Baluch, about eight miles to the north, and from there link up with Atta's soldiers near Pol-i-Barak. It would be Bowers's and Mitchell's job to coordinate bomb strikes between Dean's team to the west and Nelson's team in the east, as they moved north in parallel lines.

Bowers explained that the danger was that one of the teams could get ahead of the other and risk being bombed by its own friendly forces. Bowers would have oversight of the entire operation.

Bowers hadn't been convinced he needed to explain the battlefield in these simple terms, but he was later glad that he had when he discovered that many of Dostum's men couldn't read a map or discern east from west on a compass. These earnest, illiterate men navigated by orienting themselves by terrain features, such as mountains.

Dostum's men were low on ammunition. They had shot thousands of rounds and launched hundreds of rocket-propelled grenades at Taliban lines. They were so low on bullets that they were picking stray rounds off the ground and pocketing them. If they were dirty, they cleaned them by shaking them in a bucket filled with crushed glass, which polished the shell casing and ensured the round would fire.

"You'll have your ammunition," promised Bowers. Before leaving K2, he had devised a phrase that he thought might be effective in gaining Dostum's trust when they met.

"In this coming battle," he said, "we must be brothers and drink from the same cup."

Dostum considered the sentiment, and then nodded. He offered Bowers eight horses for his men—eight horses that he could barely afford to loan out, given the shortages among his own fighters.

Bowers ordered his team—Major Mitchell, Sergeant Major Martin Homer, combat controllers Malcolm Victors and Burt Docks, Sergeants First Class Chuck Roberts and Pete Bach, and medic Jerome Carl—to saddle up.

Martin Homer, who had grown up in a rural West Texas town, sat comfortably on his horse. His friend Victors struggled atop the saddle and wobbled uneasily in the cramped wooden frame.

The eight Americans, led by Dostum and a dozen of his men, struck along the riverbed, south, toward a trail climbing up from the valley to the battlefield on top, headed for the south rim of the Darya Suf gorge. Their job: survey the battlefield in preparation for the November 5 attack.

Victors yanked back and forth on the horse's reins and confused the animal. It reared and shot down the trail, with a frightened Victors whooping and clutching handfuls of the horse's mane.

Victors raced past Dostum, calling over his shoulder, "I don't know how to ride a horse!" and, galloping farther, finally stopped the animal. Homer rode up and patted Victors on the back and said, "Nice."

They rode across harrowing switchbacks. The rugged terrain astounded Mitchell. Soon the entire U.S. force was spread along the trail, so much so that when Mitchell crested a rise, he had to look ahead and scan the horizon for the bobbing, charging silhouette of Dostum, riding still farther forward.

Mitchell was having slightly better success than Victors on his horse. Mitchell's hiking boots slipped out of the narrow stirrups and he was constantly forced to lean and jam his boot back in, worried he'd lose his balance and fall off the horse, and continue falling until he hit the valley floor far below.

He watched as Burt Docks wended his way along the three-foot-wide mountain trail ahead of him. At a sharp turn, Docks's horse slipped and sent him tumbling from the saddle, and off the edge of the cliff. The horse regained its footing and galloped ahead. Mitchell shook his head, horrified: *We'll never even find his body.*

He peered over the edge, expecting the worst. He saw Docks lying on a narrow ledge a few feet down from the trail. Docks was white as a sheet. Mitchell laughed as they pulled him back up to the trail.

And then the situation worsened. Mitchell discovered that the horses, all stallions, fought constantly. Dostum's horse bucked and shot out a heavy, rear hoof, which struck Bowers, who was following close behind on his horse.

Mitchell couldn't believe the sound of the impact. He thought somebody had snapped a baseball bat in half. Bowers grimaced and doubled over the saddle, grabbing his shin.

"Sir, are you all right?" asked Mitchell.

"Yes," gasped Bowers. "The General's horse"—gasp—"*kicked* me."

Mitchell admired the lieutenant colonel's self-control when he knew he wanted to scream. Any sign of agony might have raised suspicions in Dostum's mind about the Americans' steadfastness. Bowers rode upright and silent in the saddle.

After that, he kept a respectful distance from Dostum's rearing steed. They all did.

On November 5, to kick off the final, coordinated assault on Baluch, Stu Mansfield, positioned with Atta at the warlord's mountaintop compound, ordered the drop of a bomb called a BLU-82, which Mansfield called "the Motherfucker of All Bombs."

A few minutes after dawn, barreling toward earth was the largest non-nuclear explosive device in the United States' arsenal.

Brian Lyle was walking outside Atta's compound scanning the treeless terrain, looking for a place to relieve himself, when he saw a flash in the eastern horizon, followed by a long, sonic bath of crunching thunder. Lyle thought they had come under nuclear attack. A gray, angry mushroom cloud filled the sky.

The bomb, weighing 15,000 pounds and measuring about the length of a VW Beetle, had been rolled out the back end of a C-130 and plummeted thousands of feet before a chute deployed and gently carried it to earth.

Armed with a pressure fuse, the steel, barrel-shaped container exploded aboveground and vaporized any plant or animal within 250 yards of ground zero. The concussion created a flash of over-pressure totaling 2,000 pounds per square inch, the same amount of pressure one would feel standing on the floor of the sea a mile underwater. No one, though, was killed by this explosion.

By design, the bomb had been dropped on an expanse of empty desert solely to terrify the Taliban before the day's battle.

A half hour later, another one of the bombs slid from the back of another C-130, and this explosion also shook the ground beneath Lyle's feet. Lyle looked in the direction of Ak Kupruk and saw several pickup trucks with their headlights on, racing out of the village. It looked like the bastards were on the run. Sixteen days after Nelson's team had stepped off the helo at Dehi and started fighting, all of them—Nelson, Mitchell, and Dean—wondered if victory might be within reach.

The nearly several dozen soldiers who made up the American force were spread over sixty square miles of desert and mountaintop in a U-shaped formation. The center of the "U" was the village of Baluch, the assault's objective.

Diller, on his mountaintop in the west, formed the top end of the "U," and Sergeants Milo and Essex, and combat controller Mick Winehouse, sitting thirty miles across the valley, formed the eastern tip of the U-shaped formation.

Several miles southeast of Diller, Dostum, Mitchell, and Bowers were overwatching the battle from the south rim of the Darya Suf Valley.

Captain Mitch Nelson, demoted from his position of official influence alongside Dostum since Bowers's arrival, was riding with the secondary commander named Ahmed Lal and a hundred of his horsemen, positioned in the center of the U.

Five miles to Nelson's east, combat controller Sonny Tatum, weapons sergeant Patrick Remington, and commo sergeant Fred Falls were riding with another subcommander named Ahmed Khan, leading 150 horsemen of his own.

Situated about a half mile behind Khan, Cal Spencer and Scott Black were running the log train and aid station. In all, this force of 3,000-plus men was facing 20,000 Taliban soldiers arrayed in hundreds of bunkers dug into the surrounding hills. Yelling, screaming their battle cries, weapons raised, the Afghan horsemen spurred their horses and charged the Taliban lines.

By midafternoon of November 5, after several hours of furious gunfighting, Milo, Essex, and Winehouse suddenly found themselves in a fix. They had been anxious to push north of the rest of the fighting force, well past the village of Charsu. With them were some of Dostum's soldiers. Now they were about to be surrounded and overrun.

Milo had looked over the berm and there they were, the Taliban, coming up the hillside, their white robes flapping in the wind as they ran at them. One minute they hadn't been there, the next minute they were firing at Milo and his crew tucked in their trench. Rounds zinged overhead, hitting the dirt around them.

Master Sergeant Pat Essex pressed his cheek to the stock of his rifle, squeezing off shots, yelling at Winehouse to stay on the radio and talk to the pilot, because they needed to drop some bombs, quick.

Milo counted fifty Taliban in all, running up the hill at them, their AKs held at the waist, firing at full auto. Milo wished they had been carrying lightweight mortar tubes. They had left those weapons behind at K2 because they had thought they'd be traveling by foot to Mazar.

The Taliban were about a half mile away, running from the bunker on the left, in the west.

Milo had the SOFLAM perched on the trench's lip and he was pulling the trigger with the laser. The bunker squatted in the dry earth about one mile away. Milo was talking to the laser even though he couldn't actually see it; it was invisible to the naked eye. What he did see were the crosshairs of the viewfinder, which resembled a rifle scope's, and he zeroed those on the wooden crossbeams of the bunker doorway. He'd spent hours going through the designator's manuals making sure the device was correctly adjusted. He was confident it was. "Come on, you sonofabitch, blow the fuck up."

He could feel the steady pump of the Zeus as it threw rounds at them. They exploded around the trench, in front, behind, off to the side. The Taliban had elevated the gun and were lobbing them in Milo's direction, hoping to get lucky.

The pilot came over the radio and said he had dropped the bomb. Milo started talking to the bomb as it fell and latched onto his laser, sailing toward the bunker. "Hit that damn thing, hit that damn thing . . ." he muttered.

When the bomb hit and the Taliban bunker blew up, Milo impulsively jumped up and cheered, and a fusillade of machine-gun fire from the Taliban's position erupted around him.

He could hear the bullets snapping overhead. He stood shouting at the smoking hole. "You pissed my wife off, and you pissed me off!"

And then he gave the smoking hole the finger.

Essex looked up in astonishment.

"Milo, what the hell are you doing up there? Get down!"

He grabbed Milo's arm and yanked him into the trench as the machine-gun fire increased in intensity.

Milo sunk down, and it dawned on him that the situation was getting worse. Essex was popping up and firing at the Taliban coming up the hill, while behind them Winehouse was on the radio, yelling, "Drop again! We are being overrun!"

Rocket-propelled grenades were exploding around the trench.

Essex yelled that he was running out of ammo. Milo started firing, too.

Essex ordered the Afghans with them, terrified men dressed in suit coats, sandals, and turbans, to spread along the trench and return fire. They peeked over the lip and started shooting reluctantly. Essex heard Winehouse on the radio still talking to the plane. Essex was squeezing off rounds methodically, pausing to relay to Winehouse the advance of the Taliban up the hill. Milo had his M-4 in one hand resting on the trench and the trigger grip to the SOFLAM in the other. He was painting a group of men—about 150 Taliban—with the laser and shooting at them at the same time. Essex picked out what he thought was the leader of the Taliban assault, a man in a flowing white robe, who was running along the hill, about halfway up, from left to right. The soldier was trying to flank their position.

Essex started shooting at the guy. His M-4 rifle was accurate at a little over 550 yards. The target was still 800 yards away. Essex elevated the barrel and tried lobbing his shots.

As he was looking through his scope, the Taliban soldier turned and looked directly at Essex. Essex was struck by the confused, nearly comical, look on his face, that said, "I know I am being shot at!" and he dropped suddenly to the ground and started rolling down the hill.

Essex began shooting more rapidly as the man rolled. He knew he could never hit him, yet he kept shooting. And then he lost track of him. The man had rolled out of view.

When he lifted his head from the scope and looked at the hill with the naked eye, he was shocked to see a Taliban fighter crouched down about 100 yards away, ready to fire an RPG. He watched the grenade spiral toward him and explode in the dirt about 50 yards from his position.

Essex squeezed the trigger and the round hit at the guy's feet. The Taliban jumped straight up in the air at least several feet and Essex laughed. More of the Taliban were rushing up the hill.

The pilot's voice cracked over the radio that he had dropped.

The explosion rocked the ground and Milo saw that nearly all the men were dead. He turned around to relay this information to the Afghans. They were gone.

They had fled the hill. They had taken everything—the horses, the rucksacks, all the spare ammo inside them. Milo had just what was in his vest, a couple dozen rounds. Essex and Winehouse yelled out that they were running low, too.

Essex was shocked that, of the Afghans, only their security chief remained. The frightened man was yelling, "Let's go! Let's go!"

Essex wasn't leaving until the aircraft arrived with its load of bombs. The Taliban had to be slowed, so that they would have time to get off the hill.

He looked at the security chief and said, "I ain't leaving!"

The man ignored him and bent to pick up the Americans' rucksacks.

"No!" yelled Essex.

The man next picked up his rifle and threw it on the ground. He picked it up and repeated the frantic gesture. He did this several times, until the gun broke in two pieces. Essex couldn't believe his eyes. The man had lost his grip.

Essex had thought they could hold off the Taliban, until more planes dropped more loads and blew everything up. Now he wasn't sure.

"Just keep shooting," he said. He yelled to Winehouse, "Where are those bombs?"

They needed time to get off the hill. They needed to blow the hill without getting killed themselves.

Winehouse was on the radio arguing with the pilot. He was refusing to drop bombs so close to friendly personnel. They would hit less than 200 feet from Milo, Essex, and Winehouse. Anyone within approximately 900 feet was likely to be killed.

Milo peeked over the ridge. The Taliban were walking steadily up the hill, firing their AKs from the hip.

Milo ducked back down in the trench. Thousands of rounds were hitting the berm. Dirt was flying everywhere.

"Tell him to drop now!" Essex yelled to Winehouse.

"Listen, we're in trouble here," said Winehouse, real calm.

The pilot said he would drop.

"Get ready!" Winehouse screamed, done with the call.

The three men ducked down, hands over their heads. Milo opened his mouth and kept it open to lessen the overpressure that the concussion of the blast would create. Otherwise, his eardrums would burst.

And then the bombs hit, seven of them.

Milo felt the air leave the trench. He couldn't breathe. His head drained of all sound. Silence.

And then they stood up, covered in red dust, and the sound returned as the hard rain of dust and rocks fell around them. Essex madly scooped up the radios, his binos, all the gear, and stuffed it in his bag. Then he started running, with Milo in tow.

Winehouse was still on the radio. And then he closed it, stuffed it in his knapsack, and jumped up and started sprinting down the back side of the hill.

"Come on," he yelled, catching up with the other two guys. "I just called in a helluva strike."

Behind them, the Taliban reached the trench just as the men scooted down the hill behind it.

The Taliban stood at the top firing at Essex and his crew as they ran.

And then the bombs hit around the trench. Six of them.

Essex felt the shock wave pursue them down the hill, then it tumbled them head over heels. They rolled and stood up and kept going. It was hard running through the loose scree. Essex felt like his legs were barely moving.

To make better time, they dropped to their butts and started sliding. Milo sat jouncing over large stones as he slid, sputtering in a cloud of dust. They slid for about twenty seconds down the 200-foot incline and got up and started running again.

They were headed to the ridgetop they had vacated earlier that morning. Essex was calling out, "Clear on the right!"

Milo: "Clear left."

Winehouse was still talking on the radio to the pilot overhead. "Drop, Drop, Drop!" he kept saying. He wanted to bomb the trench some more.

Each man had a sector to scan as they ran. At one point, Milo yelled, "Bad guys out front!" and he dropped and fired. He didn't know if he hit anything. The Taliban had ducked back down behind a hill. They kept running. When they passed the hill, the Taliban were gone.

After ten minutes, they made it back to their previous position on the ridge.

Milo was suddenly weary, more tired than he'd ever remembered being in his life. He looked through the binoculars at the ridgeline they'd just vacated. He watched as a Taliban soldier bent down and picked up an MRE and started eating the Skittles inside. Milo recognized it: that was the MRE he had just opened when they were attacked. *He's eating my fucking lunch*, he thought.

Winehouse announced that they had to replot and blow the hill. About thirty Taliban soldiers now milled around the top.

"We got the coordinates already," said Milo.

"We do?"

"Yeah. On the GPS. I punched them in when we were there."

Milo looked again at the guy munching on his MRE. He was mad. He thought, *Eat up, brother.*

Winehouse picked up the radio and called in the coordinates off the GPS. "Can you set the fuse to proximity?" he asked the pilot.

The pilot said he could. Winehouse, Essex, and Milo sat and waited.

On proximity, a bomb blows up at a preset height aboveground, vaporizing anything within a 500-foot radius.

Milo watched the bomb streak in. Gone was the man eating his lunch.

Meanwhile, ten miles to the west, Lieutenant Colonel Bowers watched through binoculars as Commander Lal and Commander Khan's soldiers massed on the plain for a horse charge. Hundreds of horsemen were gathering in ranks behind the cover of small ridgelines.

The Taliban were well dug in around the village of Baluch. All day the fighting and air campaign had been intense, but with little effect. The Taliban had taken key positions in the avenue of approach to the village, and the Northern Alliance couldn't get past them.

Nelson got on the radio and told two of Dostum's subcommanders to get ready to charge the line. This was meant to be coordinated with a bomb strike. The bombs would hit the fortified Taliban positions, and then the horsemen would swoop through and attack. In fact, as Nelson relayed this plan to Dostum's subcommanders, the pilot was overhead, getting ready to drop.

And then he did.

At the same time, however, Nelson saw that something was going terribly wrong.

The subcommanders had mistaken his order to get ready for the charge as an order to charge immediately.

Nelson couldn't believe it when he saw the first rank of riders shoot ahead. They were heading straight into the strike zone, with the bombs set to fall any second.

There were some 400 riders, in groups of 100 riders each, reins in one hand, rifles in the other, as they charged. Nelson feared they'd all be killed by the bombs when they hit.

Dostum, in the heat of the moment, was on his radio, yelling, "Charge! Charge! Charge!" as his men galloped across some 1,500 yards of grassy ridgelines.

Almost immediately, from the Taliban positions, erupted blasts of tank fire and the rattle of machine guns. Dostum's radio crackled with the excited cries of his commanders pounding over the field.

Nelson saw that the riders had less than a half mile to go before reaching the Taliban lines. He realized again that the bomb might hit the line just as the horsemen approached it.

Seeing hundreds of screaming Afghans galloping at them, some of the Taliban soldiers, in groups of ten and twenty, started jumping up and running away.

The charging horsemen disappeared behind a last hill. Nelson held his breath. And then the bomb landed.

The riders climbed back into view and rode through the debris cloud just seconds after the explosion, leaping over trenches and landing behind the line. Miraculously, they had escaped injury.

Nelson watched as some of the riders disarmed a truck with a machine gun mounted in the back. They circled the vehicle, firing their AKs and killing everyone in the back.

The rest of the Taliban who hadn't been killed by the strike were in full flight. Some of them turned to fire at the riders. The horsemen shot them as they passed. Mitchell thought it was a slaughter. He also understood that this was a turning point in the offensive. His job now was to keep pressing the enemy.

That night, Atta's forces, with the support of Dean's team, would continue their assault on Ak Kupruk and capture it. The demoralized Taliban forces would begin to fall back almost as fast as the Northern Alliance could advance. Despite reinforcement by thousands of volunteers from Pakistani madrassahs and contingents of Al Qaeda forces, the Taliban were soon in headlong retreat for the Tiangi Gap.

The following morning, November 6, Milo, Essex, and Winehouse hit the command bunker again that they'd won and lost the day before. When they went up to inspect it, Milo saw one thing he knew he'd never forget. A Taliban soldier was lying facedown in the dirt, his legs mangled. Milo bent down to study the strange sight of the man's foot.

It had no bones inside it—it was just skin and toenails. No bone, no meat, no ligaments. It lay flopped over like a boot.

From that point on, realized Milo, they'd beaten the Taliban.

The day after Essex and Milo were overrun, Sam Diller was trapped in an ambush at his position twenty miles across the mountains to the west. He and his men were riding across a hillside when a Taliban gun, positioned about 800 yards on their left, on the opposing hillside, fired. Between them lay a valley—or draw—a half mile long, speckled with grass. Diller was pissed. The day before, his Afghan intel officers had informed him that the bunker was empty. Diller had wanted to level the place but he'd left it alone. He had thought he was sneaking successfully through Taliban country. Now he wished he'd bombed the place.

They had only one escape route available to them, and that was up the hill they were crossing. Diller studied the rocky path, about two feet wide. They couldn't charge up the path en masse. Somebody would likely step on a mine.

And they couldn't cut across the hill and ride down it, as had been Diller's plan. He had thought he was sneaking successfully through Taliban country.

Small-arms fire started pecking at the hillside behind them. Light at first but gathering like rain.

"Get down!" Diller shouted.

And then came the spiraling *whoosh* of rocket-propelled grenades. They launched from wooden portals in the bunker and swam speedily across the draw and smashed against the hillside.

The fire was not accurate, but if the Taliban got lucky, it could be deadly. Diller next heard the ugly, jacking sound of a big .50-cal machine gun open up. The big rounds started knocking bucket-sized divots out of the hill.

Diller reasoned they had to climb up the hill to safety. He ordered everyone to take cover behind the scattered, tall boulders along the path. And then in groups of two and three the Afghans and Americans began leapfrogging up the hill, from boulder to boulder. It took ten minutes for everyone to make it to the top. Diller was last. He was carrying the eighty-pound satellite radio in a pack and he couldn't move quickly.

Bennett and Haji Habib ran behind him, pushing him along. Diller expected that at any minute he'd get shot in the back.

"Run, damnit," urged Bennett. Diller realized that his friend was saving his life. He was out of gas when they reached the top. He didn't know how much farther he might've been able to run. He sat on the far side of the hill, out of view of the Taliban bunker, and caught his breath.

He turned to Haji Habib. "All right, get those sonofabitches. Get your men down there and charge that bunker."

Diller, Coffers, and Bennett crawled to the lip of the hill with binoculars and began calculating the bunker's position. Diller was personally going to call this strike in. He was pissed that he'd been ambushed. He had no margin for error. He was low on ammo and food again, and he had at least forty miles to march through hostile mountains to reach Mazar. They were still a long way from home.

Through the binoculars, Diller saw that the Taliban had constructed the bunker poorly. They had limited fields of fire. Because of the size of the firing ports, they could not tilt their gun barrels easily. They could shoot across the draw but not down the hill. Diller realized that if the Taliban had any gumption, they could charge and overwhelm his outnumbered force.

Habib and his men erupted down the path, the Afghans shooting over their horses' heads as they rode. The Taliban soldiers fired their AKs and machine guns but kept missing. Diller figured the Taliban were shocked by the audacity of the maneuver. He waited for Habib's men to slump off their saddles to the ground. None had so far.

By the time they reached the bottom of the draw, they were safer. The Taliban guns couldn't reach them now.

Habib's men spurred their horses and the animals started climbing to the bunker. The machine guns fired impotently out the windows, still hitting the opposite hill. Diller could see Taliban soldiers inside the cramped ports craning to aim their rifles down, but their movement was limited.

Habib's men surrounded the front of the bunker and lobbed grenades through the ports and on top of its dirt roof. Then they shoved their rifles inside and fired on full auto. The firefight lasted a furious ten minutes. And then they mounted their horses and rode down the draw and back up the hill just as the bomb strikes came in. The bunker turned into a tornado of dust and wood. In the swirling cloud, Diller could see that nothing was left but a smoking hole.

Goddamn you, he thought. *I want to get home to my wife. And I will.* They saddled up and kept riding north.

As they made their way along the Darya Balkh River, also headed north, Cal Spencer, Scott Black, Vern Michaels, and their Afghan interpreter, Choffee, were driving one of the John Deere Gators brought in by Lieutenant Colonel Bowers on his helo. Black and Michaels were traveling in one Gator, with Spencer at the wheel of the other, Choffee riding shotgun.

Spencer, nursing the blown disc in his spine and an excruciating backache, had welcomed the chance to get off his horse and drive.

The nearly nine-foot-long vehicle was stacked on the back with cases of MREs and the team's rucksacks. The tower of gear, lashed with rope, rose nearly six feet off the ground. Spencer marveled at how the buggy puttered along, quiet as a garden tractor, over rocks, ditches, and up and down hills.

As they rode, Choffee kept a sharp lookout for the enemy. Tall, thin, easily worried, Choffee was in his late thirties, a former industrial plant manager who had the annoying habit of repeating everything Spencer said.

Spencer would say, "We are going to kick the shit out of the Taliban," and this would confuse Choffee, who did not understand American slang.

"Could you rephrase that?" he'd ask. "What do you mean, 'kick the shit'? Where will the shit go?"

Spencer realized he had finally met someone with as quirky a sense of humor as his own, albeit unintentional.

As they drove, Choffee fretted constantly about being attacked. "There are bad guys all around here," he told Spencer. "We can't see them because we are going up and down hills."

Spencer handed Choffee his M-4 rifle. "You shoot the bad guys because I'm driving."

Choffee looked at Spencer. "Oh, no. I cannot do that." He didn't want to touch the rifle. He was afraid of it.

"Well, why not? You earn your pay."

"Do you really want me to do this?"

"Hell, yeah."

"Okay."

Choffee cradled the rifle gingerly, sitting beside Spencer on the bench seat as they bounced along. Spencer grinned at the sight.

He knew Choffee was thinking, *Am I going to be able to shoot somebody?*

Along the way, they stopped and talked with Nelson, who was traveling north by horse. They met up alongside the river. Nelson wanted to know how they were faring.

"We're doing all right," said Spencer. "Choffee is guarding me."

Choffee broke into a satisfied smile. He slapped the weapon. "Yes, I am doing a good job."

Spencer was glad. He had accomplished his goal: to make Choffee feel good. And to allay the Afghan's fears about being captured and tortured by the Taliban.

Still, Spencer was worried. He knew they could be attacked. They were running low on fuel—maybe a few gallons left in the plastic cans lashed to the buggy. If they did come upon some Taliban, they would be able to outrun them for a while, but they wouldn't be able to scoot up into the hills, into the rough country, where the Taliban's trucks and tanks couldn't follow. They were driving a twenty-first-century gas-powered vehicle, and it had its disadvantages. Part of him longed for his horse.

This was especially true when they crossed the Darya Balkh River after passing through the village of Shulgareh. The water was wide, maybe 200 yards across, rushing in braided streams. Spencer couldn't tell how deep it was.

Black and Michaels drove into the river first and started heading across. Spencer watched as the heavy, bulky vehicle lifted in the current and drifted downstream. Michaels jumped off and grabbed the back end, leaning on it to make the tires grab the river bottom. He howled when he hit the frigid water. Black gunned the engine, and the knobby tires spun and bounced on the gravel. Little by little, Black and Michaels made it across. Black powered the buggy up the riverbank and turned the vehicle, and they sat looking back at Spencer. Michaels stood dripping on the sand, freezing. It was near dusk, orange twilight blooming behind the distant mountains.

Spencer looked at Choffee. "Okay, we're going to do the same thing, Choffee. You're going to drive this thing."

Choffee nodded.

"I'm going to jump off and push. You got me?"

Choffee nodded again.

"Whatever you do, don't stop. You hear me?"

"Yessir!"

"Don't stop on me now, you'll screw me up, and I'll be pissed."

Choffee gunned the engine and in the middle of the current, the buggy began to lift off the river bottom, and Choffee, frightened by the fact that the whole vehicle might tip, decreased the throttle.

"No!" yelled Spencer. "Damnit, keep going!"

Choffee hit the gas and the Gator sputtered.

The rear tailpipe was underwater; when Choffee had slowed, water had flowed into the engine.

Cal jumped off and started pushing. And then the engine quit.

Black and Michaels were laughing at him from the bank, "Hey! What happened?"

"Never mind about that," yelled Spencer. "Get your ass down here and help!"

Spencer was out of breath from pushing the buggy, and he was cold, and he had no idea how they were going to budge the vehicle the rest of the way across the river. He was standing waist-deep in freezing water, hands on hips. He just couldn't believe how cold the river was.

Just then Michaels yelled out, "Behind you!" He jabbed at the air, toward the opposite riverbank.

Behind him, Spencer heard the clopping of hooves on the river stones and he turned to see something that took his breath away.

"My God," he said. "Look at that."

On the far bank, thousands of horses were walking out of the printing press of the night, one after another, looking flat and dimensionless in the dusk. Without stopping, the riders turned the animals and they trotted straight into the river, pushing white collars of foam before them, as Spencer, Choffee, and Black watched them come.

Spencer called out to Michaels, "Are they ours?"

Meaning, were these Dostum's men? Michaels shrugged that he didn't know.

There were so many horses, Spencer had a hard time believing they were fighting in the same war. Men on horseback wrapped in red and blue and green scarves (which he remembered were Dostum's signature colors), propping their rifles on their knees as they rode past where he sat on his dumb machine in the middle of the river. Spencer guessed there were over 4,000 horsemen passing by.

They looked down at him and said nothing as they neared. Some of them laughed at the buggy. The stallions among them reared and paddled at the air and bared their teeth.

Several of the men in the line whistled and threw Spencer ropes, which Spencer fastened to the handlebar of the buggy, then the men on horseback kicked the horses and the horses sunk low on their haunches against the strain and began dragging the buggy through the current to the other side.

Spencer untied the ropes. The men didn't stop, and the ropes zipped over the handlebar and trailed in the dirt in a jerking motion as the riders coiled them sitting in their saddles and kept riding. Soon the horsemen were gone, riding into some shrinking aperture in the night. It was nearly dark.

Michaels and Black towed Spencer's flooded Gator several hundred yards to a two-story mud house along the bank of the river, and they asked the man inside if they could stay while they repaired the Gator. The man silently fed them dinner, boiled goat and rice, and some green onions from his garden, and they ate by the yellow glow of the lamps in the glassless windows of the house. Black had lost twenty pounds in the last two weeks. He hadn't felt this exhausted since training during Ranger School fifteen years before.

After dinner, they walked outdoors and flipped on headlamps, and Michaels and Black began taking apart the first Gator's engine with a screwdriver on the Leatherman tool Michaels wore in a leather holster on his belt. They removed the engine's head and set it aside to dry and covered the open cylinders with a tent made from a tarp they carried, so the pistons would be protected if it rained.

They slept on the packed dirt floor in the house wrapped in their poncho liners and were up at dawn, putting the engines back together. They were out of food and they had little water. But Spencer could feel it: Mazar was within reach.

So said the voice of God:

He was dressed in a white robe and skullcap as he walked the Irish streets, a mendicant. His father, wishing to spend some time with his teenage son, had taken him to Ireland on vacation. It was the summer of 1998; John Walker Lindh was seventeen. On the street, schoolchildren stopped and asked John if he was in some kind of theatrical production. He laughed.

He and his father passed a butcher shop and a sign advertising pork for sale. As a Muslim, John could not eat this. He stood and good-naturedly posed for a picture. His father snapped away with the camera. His father had once said to John, "I don't think you've really converted to Islam as much as you've found it within yourself. You sort of found your inner Muslim." They laughed about the picture.

After the vacation, they returned to California. After that, in July 1998, John departed home for Yemen on the Arabian peninsula. He planned to study at the Yemeni Language Center in the ancient city of Sanaa. He stepped from the plane wearing his robe, his cap, and a fierce fringe of beard.

But in Sanaa, he discovered that some of the students (there were about fifty, from various countries including the United States) were not as serious about a spiritual life as he was. They missed their prayers, they dressed in jeans, not robes. They reportedly imbibed stimulants and worse. The women wore shirts that left their arms exposed. This was annoying and insulting to John. Where were the pure souls to commune with?

He grew unhappy. He told his classmates that he wanted to be called Suleyman al-Faris. Some of them laughed at him. They answered his request by instead referring to him as "Yusef Islam," which was the name adopted by the pop singer Cat Stevens, after his conversion to Islam in 1977. John woke his fellow students at dawn so they would not miss their prayers; he prodded them to worship at midnight. He complained about their indecency: "Dear Inhabitants of This Room," he wrote in a public note, "please abstain from getting naked in front of the window. Our neighbors from the apartment across the street have complained."

The school's director would later call John a "pain in the butt." After five weeks of study, he withdrew from the school.

That summer, August 7, 1998, it also happened that Taliban soldiers were marching into Mazar-i-Sharif in Afghanistan. They slaughtered thousands of citizens, principally Hazaras. The streets of the city piled with bodies. As many as 5,000 people lay dead.

That same day, the U.S. embassies in Nairobi, Kenya, and Dar es Salaam, Tanzania, were bombed by members of Al Qaeda. Four men would be convicted for taking part in the attacks. The Nairobi embassy blew up first, killing 213 people, wounding approximately 4,000. Twelve Americans were killed. In Tanzania, a number of minutes later, the embassy there exploded. Eleven people died.

In Yemen, John Walker Lindh, disquieted by the lack of piety among his fellow students, walked into a different mosque in Sanaa, a place barer, less adorned, more conservative, where he might be taken as seriously as he took himself. He had done some of his study about the Taliban, jihad, and Islam while trolling the Internet at home in California, where ideas about sacrifice and martyrdom were little more than orderly collections of pixels on a computer screen.

For some men, though, these were ideas worth dying for, ideas expressed by bloodshed.

By November 2000, Lindh had escaped the liberal atmosphere of his youth and arrived at one of the strictest maddrassahs he could find. Seated on a stool in his teacher's study, in the dusty village of Bannu, Pakistan, he often refused to discuss his life in California.

The following spring, 2001, he wrote home, "I don't really want to see America again." Some four months after that, including military training at the terrorist camp Al Farooq, he traveled by foot and taxi into Afghanistan.

He would later say about the journey: "I went to Afghanistan with the intention of fighting against terrorism, not to support it." He wanted to help make Islam free from the corruption of warlords like Abdul Rashid Dostum and Atta Mohammed Noor.

By the time he had arrived in Chichkeh in September 2001, he had passed a point of no return. He found himself surrounded by hundreds of men sworn to fight to their deaths in order to defeat infidels. He feared being killed if he showed hesitation in his allegiance to their cause. He feared being accused of being a spy. "There's a kind of paranoia in Afghanistan about spying," he later reported. "If I had spoken up for America, I would have stood out."

He stayed put.

Back in the States, as the men passed through Shulgareh and prepared to attack Mazar-i-Sharif, Karla Milo, Ben Milo's wife, thought she saw her husband on TV. He was riding a horse across an expanse of what looked like prairie in the Dakotas. *Does he look okay? Is he getting enough to eat?*

She got close to the TV to study the picture and she couldn't tell if the man on horseback was Ben. (Milo would later tell her that the person in the picture was not him.)

She wished he would at least call. She wanted to hear his voice. She promised herself that if he did call, she would not bother him with any of her problems because he would feel guilty about not being with her. And if he was in Afghanistan worrying about her, that meant he wasn't worrying enough about himself. And that meant that he could get himself killed.

As a Special Forces soldier, Ben had done only extended training missions overseas, and Karla knew that he'd never fired a bullet in anger or in defense of his own life. She wondered how he would manage. For all of his bluster, Ben was a private man with a loud voice who wouldn't hurt a flea. After spending fourteen years in the military, Karla realized that this would be her husband's first brush with death as a soldier. It scared her.

In the days following the September 11 attacks, the phone had rung constantly, family and friends calling to wonder, "Ben's gone, isn't he?" Curious people, these relatives, and she could tell them nothing. Ben wouldn't even speak about where he was headed.

In the weeks leading up to his departure, he had spread his clothes all over the house, a real mess. He tore the bedroom and basement apart looking for his will. She also overheard him making phone calls to finalize his funeral arrangements. Her job was to make sure that the kids understood that Dad was leaving, but that he would be coming home as soon as he was done with his "work."

That had been almost a month ago. Now Karla had the task of moving the family and all their belongings from their house in Clarksville to less expensive Army housing at Fort Campbell, a split-level town house in a cluttered development called Hammond Heights. The lawns were strewn with tricycles, trampolines, and kids' plastic playhouses.

Karla was on her hands and knees in the kitchen, a scrub brush in hand, dripping over a bucket of Spic and Span, when it finally sank in that Ben was really gone. She was suddenly angry that she had to do all this cleaning by herself.

She rapped the brush on the bucket and scrubbed even harder. "I can't believe you did this to me!" she muttered. And then she felt stupid. Nobody, she knew, had forced her to marry Ben Milo. She knew she would never tell him about these feelings if and when she talked to him on the phone. She'd save this for later, when he returned, after wisdom had replaced anger.

Diller rode off Alma Tak Mountain into the valley, headed for Shulgareh. It had been ten days since he'd said goodbye to Nelson at the Cobaki outpost. He'd lost thirty pounds and, except for the feast of sheep, he hadn't eaten anything except scraps of bread, nuts, raisins, and cheese in three days. He and Bennett and Coffers rode upright in their saddles, solemnly nodding at the passing country-side. The exhausted Afghans sat stonily atop their stumbling horses. It took two days of riding to reach the town. After the first day, the horses had given out.

Diller's horse lay down and rolled over on top of him. He didn't have the heart to beat it back to its feet. He could barely sit any longer in the saddle. He could feel himself bleeding through the seat of his pants. He was glad to walk.

His legs had been contorted for so long in the ill-fitting saddle and stirrups that the latent arthritis in his ankles flared—a painful souvenir of hundreds of parachute jumps. Diller gripped the reins in one hand and walked upright, stiffly.

To lighten the mood, Bennett suddenly belted out: "It just keeps getting better and better!" Here they were, living on fried sheep and filtered ditchwater, talking on expensive radios, and calling in GPS-guided bombs on bunkers built of mud and wood scrap, surrounded by Taliban fighters. "It just gets better and better!" Diller chimed in. He cracked a grin and they plodded ahead.

Several hours later, they stood at the outskirts of Shulgareh beside the tired horses. Diller looked down the crowded main street.

He held up his hand and the men behind him stopped. "Keep your guns up," he said. And then they led the horses through town.

The single main street was lined with thousands of people. They obviously had been awaiting the Americans' arrival. Some of them clapped—all of them stared. They didn't seem to know what to make of these pale, filthy Americans riding out of the desert. Diller smelled piss and shit. They had to walk around the sewage flowing in the dirt street.

They kept moving, looking neither left nor right but scanning both directions with peripheral vision.

Diller decided he could not walk anymore. He got on the radio and called Nelson.

"Dude, where are you?" He gave Nelson his position.

Nelson, traveling north of Diller, was shocked to hear his friend's voice come over the radio.

"We are a mile from you—stay there," said Nelson. "I'm sending somebody down."

Diller was standing by the side of the road when he saw a six-wheeled buggy approaching, one of the Gators he'd heard about.

He smiled when he saw Spencer at the wheel.

They hugged, and Diller and his men slid on board, and Spencer gunned the buggy.

After his days spent on horseback, Diller felt strange sitting in a vehicle, watching the road zip past. He had to fight to stay awake.

In the early predawn hours of November 9, Master Sergeant Pat Essex had bedded down for the night near Shulgareh when the cell phone of one of the Northern Alliance guards rang. Dostum was calling: he needed Americans to help clear the enemy from the Tiangi Gap, five miles up the road, to the north.

Essex, along with Milo and Winehouse, grabbed their gear and crammed into a tiny Russian jeep. They were driven through a cold, steady rain to a staging area at the base of the Gap.

General Dostum was waiting for them with a map spread before him. In their best "pointy talk"—Essex didn't speak Uzbek and Dostum spoke only broken English—Essex figured out what the general wanted. He asked the Americans to climb up the heights to the top of a mountain ridge, about 4,000 feet above them, overlooking the Tiangi Gap and the Darya Balkh River flowing through it. From there, Essex and Milo's job would be to bomb Taliban artillery hiding in wait on the north side of the Gap.

The Tiangi Gap is about a mile-long cut through the mountain ridge that divides the country's wilds from Mazar and its more civilized metropolis, twenty miles to the north.

Over the millennia, the Darya Balkh River had carved its bed through the 6,000-foot wall of rock, which runs for several hundred miles east and west. The cut through the mountain wall was a natural choke point. Whoever controlled the heights around the Gap controlled passage through it.

Essex, Winehouse, and Milo left the warlord and began to make their way up the slick rockface. It was tough going. Their heavy night vision goggles wouldn't stay on during the bruising ride, and Essex clung to the saddle as his horse navigated the steep switchbacks. Finally, the incline got too steep, and the three men got off their horses and walked them to the top. All the while, they could hear bullets zinging by them—Taliban soldiers shooting wildly in the dark. They worried that the Taliban had already taken the ridgetop. They expected a fight at the top.

They finally reached the summit at about 3 a.m. and Essex was shocked to find that the Taliban had made no effort to occupy this rocky outlook. This told him that they were in even greater disarray than he had imagined. One of the first rules of war was: Always control the high ground.

The rain slowed to a drizzle. Ben Milo climbed into his sleeping bag wearing a fleece jacket and long underwear and lay down in one of the rock trenches dug into the mountaintop. Five Afghans, Northern Alliance soldiers who reached the top earlier, had simply wrapped themselves in tattered blankets and trash bags and curled up on the cold ground. Others were making do with cellophane that had been used to wrap the Special Forces' air-dropped supplies. Milo felt pampered in his expensive cold weather gear.

He and Essex were concerned they'd wake up and find the Taliban just several hundred feet down the mountainside, on the slope facing Mazar. They placed Claymore mines around the perimeter and sat listening to the tick of the rain and the wet wind moving through the rocks. Even at first light, around six or seven o'clock, Essex couldn't see anything off the mountain. The mist was too thick. And then around nine the sun appeared and burned away the cover.

Daylight revealed a landscape that Essex thought resembled Colorado. The cliff face, some 2,000 to 3,000 feet high, dropped straight to the river below, which formed the Tiangi Gap. No grass; a few trees sprinkled like matchsticks in the distance.

A bomb-cratered road hugged the green, chalky current of the Darya Balkh, and followed it north. In places, the rock walls closed in and the passageway was only as wide as a three-lane road.

Essex looked down and saw a series of four enemy trenches dug in on the tops of gentle hills. He started calling in the bombs immediately, but it was tough. He would identify a target, a Toyota pickup truck, for example, and the pilot would radio back, "We can't strike that."

"Look, buddy," said Essex. "That's a personnel carrier."

"Personnel carrier? Well, it looks like a . . . *truck*."

Nelson, listening in, got on the line and told the hesitant pilot, "Any military vehicle on the battlefield be at liberty to destroy."

"Wait a minute, come on now," said the pilot.

Nelson explained, "All friendlies are either on foot or they're riding horses."

The pilot asked Nelson to repeat the part about the horses.

"We're riding horses," Nelson explained.

The pilot, obviously new to the conflict, couldn't believe it. "You guys are doing what?"

Essex, Milo, and Winehouse called bombs on anything they could spot, truck-mounted guns, vehicles, and Taliban troops. Milo identified at least twelve targets.

They had blown up about half of them when they suddenly came under counterattack. Essex could hear a faint *whooom, whoooom, whoooom,* and he recognized the sound from his stint in the Gulf War.

Milo and Winehouse looked at him: "What's the matter?"

And then the rockets hit.

Down in the valley, Dostum and Lieutenant Colonel Bowers, along with several hundred of Dostum's men, had begun moving into the pass. Dostum had staggered the departure of his men to keep some fighters in reserve, in case the Taliban attacked.

After the rocket attack, Essex, up in his mountaintop perch, feared that Dostum, Bowers, and all the soldiers were dead.

Nelson, meanwhile, was standing at the south end of the Gap, about a half mile from its entrance, when the Taliban rocket barrage started.

The rockets ricocheted off the canyon's sides, spinning end over end and exploding. Molten skeins of shrapnel clawed at the rock walls. Nelson counted twenty-one rockets in all. He didn't think anyone could survive the barrage.

After the first explosions, Dostum's men had scrambled up the rock and tucked themselves on ledges and hid behind outcroppings. They were unsure of what to do next. Some of them were out of their minds with fear. The rolling explosions had been terrifying.

Nelson knew that these kinds of rockets, called Hails, landed in salvos. They were launched from a wheeled pad capable of holding forty of the nine-foot-long projectiles. Nelson guessed the Taliban wasn't done with them yet. He worried about Dostum and Bowers.

The canyon exploded again, driving even more men up into hiding in the nooks and crannies. Around them, horses lay kicking in the dirt, riddled with shrapnel. Men walked around holding their eyes, blinded. Arms and legs that had been blown high in the air landed with sickening thuds along the riverbank.

Nelson, sitting on his horse, watched smoke drift out the canyon's mouth. He turned in his saddle and surveyed the remaining Afghans with him. They were scared. So was he. He turned back and faced the canyon mouth and tried to think what to do. He'd wait. He figured they were in for more salvos.

After several minutes, none came.

He knew that Diller was somewhere off to his right, in the east. He didn't know if he was dead or alive. Essex was also on his right, up on the ridge.

Nelson tried raising Bowers and Dostum on the radio. Nothing. He then called Essex up on the ridge.

"I don't have comms with Dostum. Can you see him?"

Essex said he couldn't see into the canyon. A ridge was blocking his view.

Nelson worried that everyone was dead.

About one mile west of Nelson in the Gap, Diller was skirting along the crumbling foothills and looking for a trail up and over the mountain, when the rockets starting landing around him. His job, after leaving Shulgareh, was to keep moving east and hit the edge of Mazar not by the main road, as Nelson was doing, but through the backcountry. The rest of his team, as well as Atta and Dean's men, were to travel through the Gap and enter Mazar by the main highway that led into the city.

Along the way, Diller was to hunt for Taliban soldiers in hiding who might circle and attack the approaching main force from its rear.

He heard the rockets launch just as he was getting on his horse. He froze, with his left leg in the stirrup and his right swung over the horse's rump. The rocket made a screeching *chooo* sound and exploded about 100 yards away down a ridge. Diller's horse reared and he was thrown backward and lay a moment looking up at the sky. He scrambled up and started running after the horse. Most of all, he was afraid he'd lose it and that one of the Afghans would claim it for himself. Sure enough, as Diller ran a short distance through the smoke, he saw an Afghan soldier, one of Dostum's men, reaching for the reins.

Diller ran up and snatched them from the man and yelled, "No!" He then tried leading the frightened animal down the hill, but it wouldn't budge. Diller reared back and punched the horse in the jaw. He dragged its head down to the ground just as the second rocket barrage came in.

The explosion threw him on his back in a ditch. The horse was standing over him, its feet straddling the edges of the deep depression in the ground. He looked up at the horse, silhouetted against the sky. The horse was looking down at him as the rockets crashed. Diller thought Bennett and Coffers were probably dead. They'd ridden ahead a few minutes before the rocket attack and he didn't know where they were.

He lay there listening for the telltale screech of a next launch. It was quiet. The barrage had come from the north and west, from the back side of the Gap, about a mile away. He figured the Taliban hadn't targeted him specifically. They were spraying the ground indiscriminately.

He turned at the sound of approaching hoofbeats. Bennett and Coffers were riding up fast. They pulled to a halt and Diller stood up and brushed himself off. They explained that when the barrage started, their horses had taken off and they couldn't stop them running. They had laid out along the necks of the racing animals and held on. The horses ran maybe a half mile, then stopped. Then they'd turned them and raced back to Diller.

Commander Ali Sarwar, who had watched Nelson and his team get off the helicopter three weeks earlier at Dehi, was attacked by BM-21 rockets in the Gap as he moved the twenty-five soldiers under his command. He had been traveling and riding with Commander Kamal, attached to Essex's group. Ali had been fearless throughout the campaign, but the rockets falling around him were terrifying, totally random. The canyon rang with their explosions.

He heard the whooshing sound of their launching, one after another. They started hitting his horsemen. When the smoke cleared, five rockets had hit in the canyon; others had flown over and crashed on hillsides (near Sam Diller and Bill Bennett, traveling north on the east rim of the canyon). Ali could see about sixty-five men lying dead on the canyon floor. Their horses were in pieces around them. The blasts had split one horse in half, lengthwise, so the animal was laid out on a pile of its own guts and its feet were spread in all four directions as if staked to the ground.

After leaving Shulgareh, Ali and his men had moved up the road through the Gap and found it littered with cars that had been bombed by American air strikes. Some of the cars were charred and melted on the road; others were perfectly fine. The Taliban drivers, terrified, had abandoned them with the keys still in the ignition, and ran north, into the hills, to escape the Americans' bombs that seemed to haunt the road.

As Ali rode through the canyon after the rocket attack, he spied something ahead on the road that scared him. His first thought was to flee. He squinted and saw about fifty cars and pickups racing toward him.

The vehicles were still a half mile away, but they were coming, that was for sure. And they were filled with Taliban soldiers.

Ali and his men had fought for three weeks without rest. But here they were going to have another fight, a bad one. Ali looked behind him. He and his men could head back down the river valley. *If we push ahead*, he thought, *we could be killed.*

He inched down in the saddle, jammed the magazine tight in the rifle, making sure it was there.

It's either victory or death, thought Ali. *If God is with me, I will go toward Mazar.*

Ali and his men rode into a furious firefight.

They battled for two hours and Ali lost several men. But they killed many Taliban. He watched as the survivors of his wrath scrambled back the way they had come in their pickups, heading north up the valley, to Mazar.

Nelson, sitting on his horse at the mouth of the Gap, saw that the Taliban rocket attack had quashed the forward momentum of the Northern Alliance. Nelson felt this was a critical moment. Dostum's men had scattered into the hills. He knew he had to do something. He could see men lying up in the rocks, like stunned lizards.

You have to lead these men, he thought. If they stopped here, the Taliban might have time to regroup and attack again.

He swallowed hard and spurred his horse ahead.

Inside the canyon, Taliban vehicles were canted and burning. The drivers had been burned alive and spilled from the doors, dark as wicks. Nelson looked down to the river and saw more men and horses lying in the water. The Taliban had even mined the river.

At the sound of his approach, Nelson saw men stand up on the rock ledges and look at him. They looked startled and watched him pass. And then, slowly, one by one, he heard them scrabbling off the rock. The scrape and trickle of pebbles rolling downhill.

He closed his eyes and thanked God. He was horrified by the sights around him, yet he felt exuberant. It was hard to explain.

He had hoped the Afghans would follow him. If they hadn't, he would've felt that he'd failed as a U.S. Army captain. As a teenager in Kansas, he had always admired a particular painting of a Civil War battle. It depicted a general riding through a battlefield, and following him were his men, hollow-eyed, trusting, hopeful. He felt that whatever happened in his military career, for this moment he was leading these men through a version of hell.

He still had no idea if they'd be attacked again by more rockets. He kept riding.

He looked behind him and saw that about three hundred Afghans were marching with him, some carrying weapons, others walking empty-handed, their rifles having been lost in the explosions.

About halfway through the canyon, he found Lieutenant Colonel Bowers and General Dostum. After walking about a quarter mile, Bowers had been caught in the barrage of rockets, which exploded about twenty-five yards from where he'd been sitting on his horse. All of the animals started to rear, threatening a stampede. Bowers ordered his men to dismount and hide in a nearby hillside, and they crammed into what looked like a crack in the rock, while Bowers stood before them, trying to shield them. One man tried containing all of the horses by grabbing their leads, but most of the animals broke free and started running up and down the canyon.

Nelson hugged Dostum, who asked him where he'd been. The burly warlord had worried that Nelson had been killed in the attacks.

"I got here as soon as I could," said Nelson.

They rode back through the Gap and met Essex, Milo, and Winehouse as they walked down from their observation post atop the mountain. When Major Mark Mitchell arrived in a truck, along with his team, the entire force was ready to move into Mazar.

The Gap was cleared.

Dean and Atta stopped at sunset at the south end of the Gap, for prayer. Atta planned to spend the night here, before heading to Mazar at dawn, with Dostum.

Dostum's men were picketed a half mile farther south, along the same road. Most of them were on horseback, while Atta's men were traveling by vehicle.

Atta had either bought, stolen, or captured all the vehicles he could from the Taliban. Dostum, on the other hand, was overextended. His men and horses needed time to rest. They couldn't keep up with the speed of Atta's mechanized advance.

The two warlords met and agreed that no one would leave for Mazar-i-Sharif until dawn. Victory would be theirs as comrades-in-arms.

Dostum didn't know that Atta had other plans.

Atta already had forces inside Mazar-i-Sharif, holding select pockets, even though the city still crawled with Taliban and Al Qaeda fighters. He was eager to beat Dostum into the prized city.

After the meeting, Dean watched as Atta sat on a carpet on the dirt, surrounded by different satellite phones, all jerry-rigged to run off car batteries, and began talking to different Taliban commanders in Mazar-i-Sharif. He was trying to arrange defections and increase his own troop strength with the newly surrendered soldiers. Dean admired his smooth and efficient diplomacy.

Meanwhile, in Mazar, hundreds of Taliban vehicles—their headlights bobbing in the dark and visible for miles—were fleeing the city for Konduz, eight hours to the east over a bumpy road. The trucks were filled with retreating enemy fighters.

Atta decided he couldn't wait any longer. He had to get to Mazar before any more Taliban fled for Konduz. His prospective power base was slipping away from the city.

When he heard the news that they were leaving early, Dean readily assented. He had immense respect for the bookish, pious warlord. Dean knew it was not his place to get between Atta and his rivalry with Dostum. He climbed into a jeep and took a seat next to the general. Crammed next to Dean was his senior medic James Gold, while other members of his team were following behind on horses and in pickups. Dean's second in command, Warrant Officer Stu Mansfield, was jammed in the dusty cab of a large wooden-sided truck bringing up the rear of a caravan that measured a half mile long. Atta's several thousand men—on foot, on horseback, and behind the wheels of trucks wheezing ahead on the few last precious gallons of gas available to his army—lurched ahead through the dark. His men hadn't eaten in several days. This was an army on its last legs.

Dean was amused thinking that he would beat Mitch Nelson into the city. During the last week, the two young Army captains had grown competitive, each of them hewing closer to the psychological needs of their warlord. Dean looked over and saw that Atta was smiling. The war, it seemed to Dean, was ending. They'd roll into Mazar, clear the streets of any Taliban refusing to surrender, and begin rebuilding the place: water and power plants, schools, a police department, the mechanics of daily life badly needing repair.

They'd scour the city for intel about the whereabouts of Osama bin Laden and other Al Qaeda soldiers. A recent report had placed bin Laden in the town of Balkh, thirty miles west of Mazar. The veracity of the report was doubtful. The teams were getting regular notice of these sightings and most were wishful thinking by Afghan citizens hoping to collect the enormous reward for the terrorist's capture. But still, the news excited the team.

As they proceeded through the canyon, they drove with their headlights off. Dean didn't know if any Taliban had remained behind to snipe at them from the heights of the mountain wall, but they weren't taking chances.

Through his night vision goggles, Sergeant First Class Brian Lyle, sitting in the large truck, could see that the Gap was about a half mile wide and that the road was narrow and gouged by explosions. It was also littered with blown-up Taliban pickups that loomed suddenly out of the night.

The driver kept wrenching the steering wheel to veer around these charred obstacles. The road was cut high into the canyon's side, and down several hundred feet below, lay the Darya Balkh River. Lyle figured that it'd be easy to sail straight off the edge. He tried crouching lower behind the truck's wooden sides, but this was impossible. Lyle's back was still sticking above the flimsy wood. He thought that it would be hell to get shot in the dark, not even knowing where the deadly round had come from. The team's ruck-sacks were piled so high that all of the men had to lie on the equip-ment and none of them could find good cover. Lyle lay with his rifle poking through the slats. He hoped to hell the Taliban hadn't hung around.

About halfway through the canyon, they began to pass the bodies of the Taliban and Afghan soldiers killed in the fighting earlier in the day. Lyle saw severed arms and legs jutting from the ground as if they had been planted there. It was as if they were driving through the unearthly garden of a macabre giant.

Lyle thought it was amazing the way the explosions had severed some of the heads cleanly from the bodies, which often were nowhere to be found. One head watched the men as they approached. Like a painting whose eyes follow you across a room, it tracked them as they passed. Some of the men laughed nervously and snapped pictures with digital cameras. Others turned away in disgust at the sight.

After thirty minutes, they drove out of the cooler canyon into the dry air of the plain. The first light of dawn was seeping above the crumpled edge of the horizon and ahead lay the dark hump of Mazar-i-Sharif. Through his goggles, Lyle could see hundreds more trucks with their headlights on racing from the city, headed to Kon-duz. Lyle had received reports that the Taliban were preparing a final battle there, which was now barricaded by thousands of North-ern Alliance soldiers. All the roads in and out were cut off. It was a place under siege. Hard-core Al Qaeda soldiers from Saudi Arabia and Pakistan had taken to executing the local Taliban in their midst. These men had only wanted to surrender, walk back to their villages, and return to their lives with their families.

Seeing now that Mazar-i-Sharif was so close, the men atop their horses sat up and spurred their reluctant animals and began gallop-ing across the plain. The men in trucks raced their engines and shot off in pursuit. Those on foot began running, swinging up to one of the passing trucks if they could. They were in full flight toward the gates of the city.

Near the outskirts, still several miles from the city center, the first crowds of welcoming Afghans appeared. Hundreds of men, women, and children stood and watched as Atta and Dean passed. They clapped and cheered. They stepped to the trucks with outstretched hands to touch the Americans as they passed. They asked the soldiers to take their pictures. Scott Black had two disposable cameras with him and he did the best he could—the truck was moving briskly as he snapped away. He couldn't figure out why the kids wanted him to take their pictures since he would never see them again. And then he thought that maybe they just wanted someone to remember them.

From his carpet shop, Nadir Shihab spotted the American soldiers as they entered the city.

They came trotting by on horses and zipping past in trucks. They wore scraggly beards and long hair, longer than most American soldiers, and he thought they looked half dead. He noticed, however, that they were gripping their rifles over the tops of their saddles and that they only *looked* asleep. They wore sunglasses and he saw that they constantly scanned the crowd for any sign of trouble. Some of them had wrapped their faces against the dust with long scarves, and there were wet, dark holes where their mouths were.

Nadir was on the verge of tears. He thought, *From now on, Afghanistan will be free—I will be free!*

PART FOUR

GATES OF MAZAR

Mazar-i-Sharif, Afghanistan

November 10, 2001

Thousands of people, many clutching paper flowers and candy as gifts for the Americans, were soon rushing past Nadir Shihab's carpet shop, raising clouds of dust as they ran, shouting, "The Ameriki are here. They have taken the city!"

Shihab watched as one man ran past. Unlike the others, there was a frantic look in his eyes. This man, he guessed, was a Taliban policeman. Or had been, until about five minutes ago, when he'd probably ducked behind a building, removed his black turban, torn off his black smock, and started running, hoping to blend in with the rest of the inhabitants of Mazar-i-Sharif.

And now the people whom he had tormented as a Taliban policeman wanted to kill him. Soon enough, several men rounded a corner, carrying knives, shovels, and clubs. They pressed him against the wall of a fruit shop and beat him. Swiftly. *One, two, three blows*. He slid down the wall, leaving a vertical red stripe as he dropped, then slumped to the ground, dead.

It is over, thought Nadir. *It is finally over*.

News of victory in Mazar raced through the cities of Afghanistan, borne by walkie-talkies and radio news reports. Only two days before, on November 8, Osama bin Laden had been granted Afghan nationality by the struggling Taliban government, led by Mullah Omar.

"Now he is not [just] our guest," announced the Taliban

spokesman. "He is a citizen of Afghanistan, and [we] will not hand him over to the U.S." When people heard this news, they wondered when their country would be free of monsters.

Young men and women in places as far away as Kabul, 200 miles to the southeast, now knelt close to their small TV screens wired to bootleg satellite dishes, and watched as Dostum, Nelson, and Dean entered Mazar amid the cheering crowds. This was thrilling. They had feared the powerful weapons of the United States, but they had also prayed the Americans would bomb their country back to a new beginning, so it could be rebuilt from the ashes.

After more than twenty years of war, this had seemed the only way to start over. A return to zero.

Up the road from Nadir Shihab's carpet shop, still on the outskirts of the city, Atta's pickup truck had broken down. His men pushed it to the roadside. Dean was worried about Atta's safety amid the crowd. Atta assured Dean that his bodyguards could fix the truck and he would soon be driving again. He urged Dean to continue on without him. He would follow shortly.

Atta, meanwhile, was worried about something else—the sudden rumor of several hundred Taliban holdouts barricaded in a school located in a neighborhood of cinder-block apartments. The buildings, two to four stories high, were within sight of the Blue Mosque, a little over a half mile away.

These Taliban, he told Dean, were hard men. Soldiers from Saudi Arabia, Pakistan, Chechnya. They were refusing to surrender. Atta said they would fight to their deaths. He warned Dean to be careful. Trust no one. And make sure none of the fighting harmed the Blue Mosque, the city's most sacred religious site. The Taliban had shut it down when they took control, believing its ornate domes and gilded walls were too flashy and inappropriate as a place of worship. Its beauty was an ugliness to them.

Dean reluctantly drove on with his team. They had to get into town, fast.

About 500 of Atta's men had already entered the city. The streets were empty and strewn with trash, the curbs stacked with dead bodies. The air rang with the cellophane buzz of flies.

Most of the Taliban soldiers who'd occupied Mazar had abandoned it just several hours earlier. In some houses cooling tea still sat in cups, untouched. Articles of clothing—scarves, smocks, sandals—lay scattered. The Taliban had left in a panic.

The city, the men knew, was akin to a giant chessboard, where the opponent was simultaneously losing, resisting, or even advancing, depending upon which square or neighborhood he inhabited. You had to be careful. There were still pockets of frightened fighters hiding in houses, on rooftops, in barricaded storefronts throughout the city. These were men who had been left behind when the majority of the force fled.

Upon entering Mazar, half of Atta's men raced in their Toyota pickups to the airport, on the western edge, and claimed it for their leader. A message was sent out over the radio: *"Atta has captured the airport!"*

This made Atta an immediate player in the city. Anyone wanting access to the airport was now going to have to deal with him.

Dostum was traveling through the Tiangi Gap when he heard the news. He was not happy. He was still several hours from entering the city, leading a tired string of some 1,500 men on horses and on foot.

He decided not to dwell on the loss. He would redouble his efforts to get to the enormous fort known as Qala-i-Janghi. From there, he would rebuild his own power base. At the height of his control in the early 1990s, he had printed his own money, operated his own commercial airline, and even programmed his own TV station. He had funded schools and insisted that women attend. Children had flocked to movie theaters on weekends to watch the latest love story pumped out by India's "Bollywood" movie studios. The city had one of the few sanitary water systems in the country and residents found themselves supplied with electricity twenty-four hours a day. Dostum had supported a burgeoning middle class, liberal by Afghan standards, and the city of 300,000 souls had been a thriving, if backwater, metropolis.

Those were the good old days. Dostum was convinced they would return.

While Dostum schemed and fretted in the Gap, Atta's men continued to probe the heart of the city, street by street. Firefights broke out, carried out around corners, small-arms fire raking the sides of buildings. The families who lived inside hid in closets, under beds, praying the Taliban were finally gone. The gunfights were furious and deadly. Atta lost thirty men over several hours of shooting. But they had the Taliban and Al Qaeda soldiers on the run.

The enemy had retreated to a girls' school, a plain-looking, four-story structure built of gray steel with tall walls of classroom windows. Students hadn't filled its hallways for at least three years. The Taliban didn't believe in educating girls. They had taken knives and scratched out the eyes on all the photographs and pictures hanging in the building. They believed the Koran prohibited the display of such images.

In the late morning, Atta's men snuck across the sparse grass bordering the front. They looked up and down the building, about 100 feet in either direction. They were standing in the middle. They planned to go inside and confront the men, who were suicidally refusing to surrender.

Suddenly, a group of Taliban soldiers rushed out of the doors.

Both sides opened fire.

When they reached Atta's men, the Taliban soldiers pulled the pins from grenades and blew themselves up, taking some of their enemy with them.

More Taliban poured out the doors and Atta's men opened fire again, trying to drop them before they drew close.

Atta's men quickly retreated a safe distance to nearby houses and carried on the firefight from there. The Taliban shattered the windows within the school and started firing down at the street.

The neighborhood popped and rattled with gunfire.

As the shooting raged around the school, Dean's convoy entered the town and made its way through the cheering crowd.

The team was shocked by the hundreds of men, women, and children clapping along the road, the women peeking shyly over the men's shoulders through the eye slits in their blue burkhas. Children ran along the street kicking soccer balls; others stood in nearby fields flying kites.

"What are the people saying?" Dean asked Wasik, one of the translators Atta had provided the team.

"They're shouting blessings," said Wasik. "They like you!"

The carpet shops, the tire repair stations, the innumerable food stalls selling lamb kebobs, these and every other establishment had been shuttered in anticipation of a final battle with the Taliban. Now shop owners were reopening them, sweeping the doorways. Barbers moved their wooden chairs out into the street and men lined up impatiently to have their Taliban-required beards shaved off.

Dean watched as the smiling men stepped fresh from the chair, rubbing their smooth chins in wonder. He could hear Indian pop music being played on a boom box carried along the street. He would later learn that a commercial radio station had begun playing music as early as 3 a.m. that morning. The day before, the purveyors of this entertainment would have been imprisoned.

Dean stopped the convoy, and the team stepped out to shake hands with the people. One of the children ran up and hugged James Gold, who swung the child up in his arms and held him to the sky. Gold was filled with a longing for his own children back in Tennessee as he put the child down, tears in his eyes.

Gunfire suddenly broke out and the team scrambled back into the vehicles. Dean scanned the crowd, looking for the shooter. He saw a man with his AK raised in the air. He had been firing off harmless rounds in celebration. Dean breathed a sigh of relief.

From his second-story office on Mazar's main street, Najeeb Quarishy, twenty-one years old, saw that the rumors of the Taliban retreat were true.

The streets were empty of the usually grim, sanctimonious policemen twirling their leather batons, members of the city's Truth and Prevention of Vice Squad. Certainly, the policemen had been less bold these past several days as the fighting intensified in the mountains south of the city. *Good riddance*, thought Najeeb.

Najeeb had always found sport in making fun of the overly serious Taliban. When he was a few years younger, he'd had the legs to outrun them when they chastised him for cutting his hair in the wildly popular *"Titanic* hair" style, named after the haircut worn by Leonardo DiCaprio in the eponymous movie.

"Why do you have the *Titanic* hair?" the Taliban policemen had pestered Najeeb.

"Because I *like* it, damn you!" And then he'd run away, taunting them over his shoulder.

Najeeb had learned English by listening to the BBC and conversing with any foreign NGO worker who happened to come through Mazar during its war-torn years. This meant that he spoke with a slight British accent, in a reedy voice verging on sarcasm. He'd grown up in Mazar during the long Soviet occupation and the civil war that followed, and he'd witnessed nearly every manner of savage misanthropy, yet he maintained in his jaunty step an irresistible joie de vivre.

He had never been happier in his life than now, when the Taliban were leaving. Earlier that day, his father, the mayor, had told him that their home had been chosen as a safe house for some American soldiers.

The chance of a lifetime! Najeeb dreamed of being a TV news correspondent. He dreamed of visiting America. What better way to practice his English than by making friends with American soldiers!

He jumped on his motor scooter and, borne along the stream of celebrating neighbors, hurried home.

Amid the continuing celebrations, General Dostum and Captain Mitch Nelson rode into Mazar several hours after Atta.

Following close behind were Major Mark Mitchell and his team, led by Lieutenant Colonel Bowers.

Mitchell and the team had turned their horses over to Dostum's men at the south end of the Gap. Burt Docks, Mitchell's combat controller, could barely walk from his horse to the two-ton truck that was to carry them through the Gap. He winced as it jounced over the road through the canyon.

Docks's teammate, fellow combat controller Malcolm Victors, looked around at the dozen Afghan soldiers riding with them, and because they couldn't speak English, he thought they were dumber than him. He knew they weren't, he knew this was an ignorant thing for him to be thinking, but he couldn't help judging them.

One of the Afghans, standing at the front of the truckbed and facing ahead, yelled something Victors couldn't understand.

A translator told him the guy wanted everyone to keep their arms and legs inside the truck.

Yeah, right, thought Victors. *What's this guy talking about?*

A few minutes later, Victors looked up as the truck passed between two rock walls with only inches to spare on either side. *Holy smoke. I've got to pay attention.*

Cal Spencer, wearing his son's student baseball cap, the bill now frayed, embroidered with a maroon "F.C." for Fort Campbell, was overwhelmed by the crowd's reception. He couldn't believe they had captured Mazar so quickly. He was driving one of the Gators loaded with the team's rucksacks. General Dostum approached, having stopped his truck up ahead and walked back to Spencer.

"Take me for a ride," he said.

Spencer understood that Dostum wanted to be seen riding alongside the Americans as they entered the city.

"Hop in, General."

Dostum waved to the crowd as they sped along.

He suggested that they hoist an American flag on a pole attached to the buggy. Spencer and Nelson thought it was a good idea.

Max Bowers disagreed. "This is their victory," he said. He explained that the Afghans shouldn't perceive the Americans as victors. Both Spencer and Nelson realized that Bowers was right.

Dostum, after a short ride, returned to his vehicle. His security detail worried that the crowd had grown too large—nearly 5,000 people. It would be easy for a Taliban soldier still lurking in town to take a shot. Up ahead loomed a strange building, a towering fortress made entirely of mud. It resembled something out of *Arabian Nights*.

"What's that?" Spencer asked, pointing down the road.

"Our new home!" shouted Nelson.

The convoy of victors rumbled through a tall gate and entered the fort.

Spencer asked if the place had a name.

Someone in the group—an Afghan—answered, "Qala-i-Janghi." He explained that in English it meant House of War.

Leaving his carpet shop, Nadir Shihab started running like everyone else to welcome the Americans as they passed through the gates of the city.

Nadir remembered the first time the Taliban appeared at his house in 1998, looking for his father. The elder had been an officer in the Northern Alliance army, and the Taliban suspected that he was still a loyal soldier after his retirement (which he had not been).

Nadir had heard a Toyota pickup pull up in the dusty street and several Taliban policemen got out and walked to his house. Nadir waited for the knock, then swung the door open.

"Is your father in the house?" asked one policeman.

His father had warned him to bite his tongue and not criticize these men. They had killed neighbors who lived nearby and then had the audacity to move into their house. Every day with them was like waking in a land of ghouls. Only you were never sleeping.

"What do you want with my father?"

"We need him to talk."

Nadir called his father on his cell phone. "They are here. They want you to come home."

"We're supposed to search your house," the policeman told Nadir's father when he arrived.

"What are you looking for? Guns? Ammunition? We don't have anything!"

The Taliban soldier raised an ax and started smashing their closets. He threw their clothes in a pile in the street and set them on fire. Then he barged out of the house and returned shortly, carrying a shovel.

He announced that he would find weapons buried in the dirt floor, weapons that Nadir's father was hiding as a favor for the infidel General Dostum.

He drove the shovel into the dirt and started digging. He found nothing in the kitchen. Nothing in the bedrooms. He walked from room to room, leaving piles of dirt everywhere he went.

Finally, he left. Nadir and his father stood in the house and surveyed the damage. Then they retrieved shovels of their own and started filling in the holes. Life was like that, thought Nadir. Somebody keeps digging holes in your life and you keep filling them in.

A week later, the Taliban policeman returned.

He told Nadir's father, "You said you don't have anything, but someone has told me that you do."

He started digging again.

"Oh, brother!" Nadir's father cried. "We do not have anything buried in our house!"

After the man had dug several holes, he walked up to Nadir's father and slapped him across the face.

"We know you have ammunition. We will take your car and your money."

Nadir's father was silent with rage as the Taliban police officer lectured him on being a proper Muslim. "Do you want to scandalize Islam? Do you want to discredit your family?"

The officer said to Nadir, "Brother, tell your sister not to go to school. Otherwise, we will kill her." The man was serious. Recently there had been a chilling message broadcast on the radio in Mazar: "There are only two places for an Afghan woman," the radio announcer had said. "In the husband's home. And in the graveyard."

Nadir had studied English at Najeeb's language school in Mazar-i-Sharif, and he was proud of his education. And he was proud of his sister's.

"We must study," he told the policeman, "because this country needs the educated person!"

"Tell your sister, if she goes to school, we will kill her."

Nadir just shook his head.

One day he was walking down the street when he felt that he was being watched. He spotted a Taliban soldier who was pointing an AK-47 at him, holding it at waist level.

"Boy, come here," the man barked.

"What is the matter, sir?"

"Don't talk! What did you do with your beard? Why have you shaved?"

Nadir considered his answer carefully, but he could not help himself.

"My beard!" he yelled. He pointed at his clean-shaven chin. "This is my beard. It is not your beard!"

The Taliban wrapped Nadir's wrists with a scarf and hauled him off to a jail in Mazar, where the guard beat him with a rubber hose.

Nadir yelled at them, "Islam doesn't depend on only the beard, you know! Islam says, 'Don't kill the person. Don't kill the human. A human is a creature of God.'"

He was in jail for three days. The Taliban called his father and told him, "We have your son." Then they arrested Nadir's father and carted him down to the jail.

His father tried reasoning with the Taliban. "What is wrong with my son?"

"He doesn't have a beard."

"He's young. Besides, it is his choice whether he grows a beard."

The Taliban raised a rubber hose and struck the old man on the shoulder. He held up his arms to ward off the blows. "My son has done nothing wrong!" he yelled.

"Be silent!" They kept hitting him. "He is in violation of God's law."

"He will grow a beard. Will that make you happy?"

The Taliban threw his father into the same jail cell. They beat him in front of Nadir. It was all the young man could do not to cry out. He was amazed and proud of his father's courage.

When the Taliban were done beating him, his father agreed to buy their way out of jail. He gave the Taliban some money, and father and son hobbled out blinking in the sunlight, walking home arm in arm.

Dostum and his security convoy didn't stop at the fort but continued into Mazar-i-Sharif to the Blue Mosque. Dostum had triumphantly arrived to reopen it. The rooms had been shuttered during the Taliban's occupation of the city.

He stepped from his jeep and looked up at the sparkling minarets. They seemed to float overhead in the midday sun. His muddy boots rang across the tiled courtyard, so peerless that it appeared he was walking across the sky itself. Dostum entered the mosque.

He removed his boots and knelt to pray. Beneath the mosque's stone floor, in a sealed crypt, lay the body of Muhammad's son-in-law and cousin, Ali bin Abi Talib, the fourth Caliph of Islam.

After his death, in what is now present-day Iraq, the holy man's followers buried him in Baghdad. But later, fearing desecration of the grave, they dug up and bundled his remains onto a white camel, who was set to wandering for days across miles of desert, before collapsing here, on this spot, in Mazar-i-Sharif.

Dostum had never been a religious man, but now he prayed. He prayed for Captain Nelson, whom he cared for as if he were a son. *I have been at war my entire life and I have killed many men.*

And yet I am alive.

He climbed back into his jeep and motioned for the driver to take him to his headquarters.

By day's end, all of the teams had arrived in Mazar-i-Sharif, except Diller's.

He finally appeared outside Qala-i-Janghi in late afternoon, starved and filthy. In the last twelve hours, he and his team, along with their contingent of Afghans, had marched thirty miles to reach the city. He had been powering up his radio just long enough to raise Nelson and communicate his position before shutting it off. He was low on battery power and conserving every volt. The day before, they'd marched twenty miles.

Now, standing in the road 200 yards from the immense fort before him, his legs were numb. He couldn't feel them. He was bleeding down his pant legs, the result of painful saddle sores.

After his horse had collapsed from exhaustion on the stony trail, Diller had led it the last ten miles.

Diller looked at the fort ahead, and then he looked at the horse.

"Now, I know you don't want to," he told the tired, patient animal, "but, goddamnit, I rode a horse out of Dehi when I started this fight. And I'd like to ride one in."

Diller put his foot in the stirrup and sat gingerly in the saddle. He gave the horse a kick. Every bounce was excruciating. Diller grinned like an idiot to avoid crying out in pain. Soon, he drew up to the four men standing at the fort's entrance, pulled on the reins, and stopped.

Standing there were Nelson, Spencer, Essex, and Milo. Diller sat and just stared at them. It sunk in: he was *here*. In Mazar. He slid from the saddle with a groan.

His teammates rushed forward, shaking Diller's hand and hugging him. It had only been two weeks since the entire twelve-man group had left K2, and eleven days since they'd ridden out of Dostum's camp in the mountains. Not that much time, really. But for most of it, they had expected to die at any minute. Now they were together again.

This feels good, thought Diller, flopping down on a cool strip of grass in the fort's shadow.

Soon he was fast asleep.

• • •

As Nelson and his team settled into the fort, Dean sat down to a feast in Atta's safe house. The men passed around platters of goat, rice, and fruit, washing the mouthfuls down with lots of instant coffee. The house was bordered by gardens and a pond, an oasis within scarred metal security walls. A chubby, dark-haired young man with doleful eyes introduced himself to Dean as the son of the owner. He spoke perfect English and said his name was Najeeb. Najeeb Quarishy.

Dean's interest in meeting an English-speaker and possible interpreter was dampened when Najeeb started asking questions. "Do you have a family? Where do you live in the United States? What kind of military training did you go through?"

Dean wondered (incorrectly) if Najeeb was some kind of spy for the Taliban. If this hadn't been Najeeb's home, he would've kicked him out to the street. Dean resolved that they had to move to a new safe house as soon as possible.

There was, in fact, much more to worry about. Local citizens were walking up to the gate outside with news that suicide bombers had targeted Dean and his team. *How did they find us so quickly?* he wondered.

Dean instructed Brian Lyle to send news of the threat up to K2. They set up a security perimeter around the house and went over an escape route in case of attack. This was called their "go-to-hell plan."

Operations sergeant Brad Highland placed a large orange panel on the roof. Now, if they had to be extracted by helicopter, the pilot could easily find them. They also made up a "go-to-hell" signal, a word that any one of them could utter if they had to leave the house immediately. This word would mean only one thing: *get out now.* The word they chose was "Titans," after the Tennessee Titans NFL team.

As Dean listened to gunfire erupt across the city, he received more bad news. Several of Atta's men had just been gunned down as they walked up to the Sultan Razia girls' school.

Dean told the team to check the situation at the school and radio their report. He would remain at the safe house and direct operations from there.

• • •

The Taliban were firing from the windows when Stu Mansfield and Brian Lyle rolled up in a truck a few minutes later.

Mansfield backed the vehicle behind a corner out of the line of fire. By standing in the bed of the pickup, he could see the entrance to the school, enough of a vantage to get a sense of the situation. The school building was a drab glass and metal complex. Now its rooms were filled with refuse, firewood, weapons, and feces.

Atta had been convinced he could get the enemy fighters inside the school to surrender. They were militant Al Qaeda soldiers, Pakistanis mostly, who'd been abandoned by the Taliban during their retreat.

Atta's thinking was guided by the school's proximity to the Blue Mosque, several blocks away, and to the hundreds of mud and cinder-block houses in the surrounding neighborhoods. He couldn't attack the school without casualties or collateral damage. It was a sticky situation, requiring a light touch—the warlord's genius.

Back at the safe house, in a meeting with Atta, before the team had left, Dean had suggested laser-guided bombs could be dropped on the school. Dangerous, but possible.

Atta had disagreed. He explained that nine more of his soldiers were already in the building trying to negotiate a surrender. This news surprised Dean.

"Well, sir, what do you want to do?" he had asked.

"I want to wait."

They decided they would give diplomacy a second chance, and if it failed, Dean would level the building.

Mansfield now looked at Atta's two envoys lying dead on the grass outside the school's front doors and wondered if diplomacy really was an option here.

The envoys had walked up to the doors carrying the Koran opened before them, the pages facing outward at the fighters inside, a sign of peace. The bullets that killed them had passed through the books and riddled their bodies.

When he heard news of the siege at the school, Dostum, in his office at Qala-i-Janghi, had decided he wanted to bomb the building immediately. Around him his workmen were already hammering, sweeping, and painting, repairing the damage caused during the Taliban occupation. Electrical wire hung from the ceiling, ripped out by the fleeing fighters.

But then Atta had called and informed Dostum that the school had been surrounded by his men, and that the people of Mazar would object to a bomb being dropped in their neighborhood. It would risk lives and jeopardize Atta and Dostum's support among Mazar's citizens. Dostum grunted, reluctantly agreeing.

About this time, Mitch Nelson was standing in the courtyard of the fortress with Lieutenant Colonel Bowers. Neither had been privy to the discussion between Dostum and Atta, nor to Atta's insistence that he was in control of the situation.

Dostum told the Americans that a cleric had been killed in a gunfight at the school, and that the enemy fighters inside had refused to surrender. Dostum explained that he wanted to bomb the place and quickly end the siege.

Nelson and Bowers questioned Dostum about the wisdom of such an attack. Dostum reminded them that he had seen how accurate Nelson and his men could be with bombs.

Bowers, nonetheless, felt Dostum wanted to act too quickly, and both he and Nelson worried that hundreds of civilians could die. However, after talking over the situation, they decided to send a team into the school.

Sonny Tatum, Fred Falls, Patrick Remington, and Charles Jones pulled up a block from the building. With them was CIA officer Mike Spann. While the others stayed at the truck, Tatum and Spann moved forward several hundred yards, dashing from corner to corner, looking for a building from which to watch the school. To blend in, Tatum had changed into jeans and a dark, buttoned shirt.

Tatum picked a nearby building, and he and Spann climbed the back stairs.

At the top, Tatum unpacked his laser gear and radios and set up an approach for two F-18 jets on station overhead. Using the range finder, he measured the distance between himself and the school at 300 yards. Close. Too close.

He and Spann would be in danger of being blown up.

He also didn't have a clear line of sight. Too many of the buildings in the neighborhood were the same height. But to get a straight shot with the laser, he'd have to get even closer. He was screwed.

Tatum aimed the green, box-shaped SOFLAM at the schoolhouse and pressed the trigger.

The device chirped as it shot out the laser, but he couldn't get a good bead on the building—only one corner and part of the roof's southern edge. He needed to place the laser in the middle of the roof, in order to ensure an accurate drop on the target.

This left a final option, most dangerous of all.

Tatum turned to Spann and said, "We need to get to the schoolhouse."

They would need the GPS coordinates of the building's position. Tatum had a plan.

They made their way back down the stairs and dashed from building to building, right up to the school. They stood against it, backs pressed to the east wall, near the corner. Tatum pushed the button on the face of his GPS, logging in the latitude and longitude coordinates of his position.

They next dashed down the wall to the west corner and Tatum pushed the button again. They ran to another corner and repeated the process. You couldn't get more perfect coordinates, except by climbing atop the roof.

Tatum thought that at any minute they'd be spotted by one of the Al Qaeda soldiers inside. Overhead, on the second floor, most of the windows had been busted out. Tatum could hear voices, angry voices. Pushing off the wall, they sprinted a short distance to a nearby building and caught their breath. No gunfire followed them.

Quickly, they ran back up the stairs to the overwatch building and set up the strike. Over his radio, Tatum asked the two pilots, "See that raggedy-ass house with the boarded windows? Now look to your left." And here Tatum was careful to visualize himself inside each pilot's cockpit, orienting himself from above, as the pilot was oriented.

From the air, the school was T-shaped. The lead pilot said he could see the T-shape. Now they had a visual ID. Tatum then relayed the GPS coordinates as well.

Tatum and Spann crouched down on the roof and waited.

Inside the school, the men who had entered as Atta's envoys, after the first group had been attacked, were in the middle of negotiations with a group of enemy soldiers. They had managed to escape being gunned down, so far.

At the same time, Remington, Jones, and Falls were positioned around the block in sight of the school, but out of sight of Mansfield and Lyle.

Each team had no idea that the other was there.

Standing in his pickup, Brian Lyle was trying to reach Nelson's team at Qala-i-Janghi, but for some reason his radio malfunctioned. He couldn't get through. He wanted to let Nelson know that he and Mansfield were at the school.

Lyle had no idea that Nelson was already aware of the siege.

Back at Atta's safe house, Dean was also trying to connect with Dostum and Nelson. It was dawning on Dean that the teams had entered the city so quickly and at such different paces that they hadn't updated each other with their locations. Dean wasn't even sure if Dostum had arrived in Mazar yet.

The situation was fluid. *Too fluid*, Dean thought.

Curious about the unfolding events, he and Brad Highland were ready to drive over to the school when they heard the explosion.

They looked up. Dean was shocked to see how close the mushroom cloud was—just three or four blocks away. *Who dropped that fucking bomb?* Dean wanted to know. And then: *Atta's men just got smoked.*

How was he going to explain to Atta that the United States of America had just killed his men?

Up on the rooftop overlooking the school, Tatum checked back with the pilots, "Olive Thirty-one, this is Tomcat. Request a second drop."

The bomb, guided by the GPS inside it, had hit the center of the school's roof, just as Tatum had planned. The back wall and east side of the building had collapsed. A handful of enemy soldiers had stumbled from the building and disappeared down the street.

The second jet pilot announced he was ready to drop. Tatum ducked just as the projectile streaked in. The earth moved beneath him, beneath the building. Thunder rolled down the city streets.

Two blocks away, the explosion rocked Stu Mansfield and Brian Lyle. Mansfield, the radio still in his hand, was furious and confused. The Afghans with them, Atta's men, glared at Mansfield and Lyle, equally uncertain of what had just happened.

Mansfield didn't know how many of Atta's men were in the building, but he knew they were dead. And that wasn't good news.

He held up the radio. "I didn't do it!" he told Atta's soldiers. "I didn't call in those bombs!"

They didn't seem to be buying the protest.

Mansfield and Lyle agreed they should leave the area immediately. Mansfield felt terrible over what just happened. Dean could've been in that building. What then? They started the truck and sped away, back to the safe house.

From his position, Tatum surveyed the damage of the second air strike. Half the school was a smoking heap. The bomb had entered through the same hole in the roof made by the first bomb, a perfect hit.

More Taliban and Al Qaeda fighters ran from the jagged reefs of concrete, firing their weapons wildly through the smoke at the surrounding buildings. Tatum worried that they would escape into the city. He radioed the pilots overhead.

"Look, we just stirred up a hornet's nest. You guys are going to have to drop another bomb."

"Ready for immediate reattack," said one. "Requesting clearance."

"Clear."

More of the remaining steel and concrete walls blew outward, shattering the trees surrounding the school. Only the front wall, facing the street, and random portions of remaining walls were still standing. Tatum saw more men stumble from the rubble, faces striped white and red with blood and dust. He was amazed anyone was still walking. Chain-link fence surrounded part of the school, and some of the dazed survivors jumped over it and ran into the neighborhood, darting between houses.

Tatum watched as the locals and Afghan soldiers chased them. Most ruthless among them were the Shia Hazaras. The Taliban had slaughtered them by the thousands when they seized the city in August 1998. Now it was the Hazaras' turn. They beat the fighters with shovels and rocks.

Back at the school, about thirty Taliban and Al Qaeda fighters had survived the strikes. They retreated into the rubble and hid in a corner of the building that hadn't collapsed. They would stay there for the next two days, exchanging gunfire with Atta's men, before finally surrendering—hungry, exhausted, somehow at the last choosing life instead of death.

Dean was pacing the safe house, waiting for Atta's return. The warlord was traveling somewhere in the city meeting with tribal elders and rekindling relationships that had been cut short three years earlier when the Taliban captured the town.

Dean felt terrible about Atta's men killed at the schoolhouse, but he also felt he had lost face with Atta. And if this was true, the mission could fail. Even though he hadn't bombed the schoolhouse, he was linked to it as an American soldier.

Back at the school, Atta's men were dragging the dead men out of the rubble and loading them into a truck. Soon the truck showed up at the safe house. Dean watched as some of the bodies were placed in simple wood caskets. Others were laid in the dirt under the trees.

One of Atta's men bent low over the bodies with pliers and heavy wire and bound the feet and hands together so the men would harden in rigor mortis positions that would fit in a coffin.

Dean's heart sank at the sight of the dead boys fifteen, sixteen years old. He had traveled with some of these fighters. He *knew* them. He resolved to make this right, somehow, with Atta.

The pensive, graying warlord appeared at the safe house later in the day, looking weary. Dean could tell he'd already heard the news.

Dean couldn't tell if he was upset. He approached Atta and apologized profusely for the deaths of his soldiers.

Atta stood listening and stroking his beard.

Dean walked over and shook hands with the commanders of the dead men and gave them a hug. He kept saying how personally sorry he was about the deaths, trying to communicate his contrition with every ounce of body language.

Atta asked to talk to Dean privately.

They stepped away from the other men. Atta looked him in the eye. He could see Dean was upset.

"These things happen in war," he said. "These nine people will not change the outcome of the war. It is very sad. But we understand."

Dean stared at him in disbelief, realizing what might have seemed coldheartedness on Atta's part was instead wisdom. After that, Dean felt a little better.

That night, some Mazar locals brought to Dean's door two Pakistani soldiers who'd been wounded in the aftermath of the school bombing.

Somehow, after running from the rubble through the city streets, they had escaped death at the hands of Mazar's angry citizens. Now they were Atta's prisoners. Dean looked at their hands, loosely bound with turbans. He wondered why anyone had even bothered to tie them up.

The prisoners glared at Dean and told him they had come to Afghanistan to wage jihad against Americans. Dean thought, *Roger that, and you'd kill me if you had the chance.* Medic Jerry Booker went to work cleaning the prisoners' wounds.

The first one was in his late teens, short and skinny. He had a pageboy haircut, thick and heavy. The kid reminded Booker of the movie actor Johnny Depp.

Booker asked him questions about the whereabouts of Al Qaeda operatives in the city. The kid said nothing. Like Dean, Booker saw hate in the kid's eyes.

He tapped his chest and heard a dull thud. He realized the boy's pleural cavity was filling with blood and air, and that this was putting pressure on the lung. He laid him down on an old wool blanket on the floor and picked up a pair of forceps from the duffel at his feet.

The kid grew wide-eyed, as if he thought Booker was going to hurt him.

One of Atta's junior commanders waved Booker away—*don't treat this piece of human crap*—but a senior commander told Booker to operate.

The bullet had entered the left shoulder and exited out the chest, also on the left side. The kid was panting, and he looked pale. Every breath magnified the pressure on his collapsing lung—a true sucking chest wound.

Booker had never treated one, but he knew what to do. In Special Forces medic training, he'd been told that in a pinch he could use an endotracheal tube, normally meant for insertion down the throat during surgery to allow a patient to breathe. But first he had to drain the chest cavity of some of the air and blood.

He numbed the left ribs with lidocaine, made an incision between them, inserted the forceps, and spread the bones. He then inserted a catheter into the chest. Booker could hear the trapped air rushing out of the tube with a small whistling noise, followed by blood flowing out onto the floor.

Atta's men were holding down the kid, who was screaming in Urdu.

Wasik, Dean's translator, told the Americans what he was saying: "They are trying to kill me!"

He yelled out that he wanted Allah to strike the Americans down. And even though the rib area was numbed, Booker knew the incision had to hurt like hell.

To fully drain the blood, he had to insert a larger tube in the kid's back. Again, he numbed the skin before shoving the forceps through the muscle. After the air and blood drained, he capped the hole in his chest with an Asherman Chest Seal, a dressing that works similarly to the seal on a bag of coffee. The seal lets out air, but it doesn't let any back in.

Little by little the kid's face relaxed as his breathing grew easier. He had finally realized the Americans weren't going to kill him.

Booker turned to the other prisoner—an older, bald man in a flowing blue smock and sandals. Blood was running down his face. His scalp was split, probably by the butt of an AK rifle. He was even more defiant than Johnny Depp had been, looking as if he would spit on Booker. The man's eyes darted about nervously as Booker studied his wound. Without provocation, the man erupted in a flurry of epithets. He shouted that he had also come to Afghanistan to kill Americans.

Booker, ignoring the outburst, debrided the scalp wound and carefully sewed it up. Afterward, he stepped back and looked the guy square in the eye. "By the way, dude," he said, "I'm an American."

When this was translated, the man studied Booker and grunted thanks. Then he was led away. Booker had no idea what would happen to him. Atta's men told him that the younger prisoner would be taken to a hospital.

When one of the senior Afghans in the room winked at Booker about this news, Booker thought, *Crap, why did we even fix these guys if they're going to take them out and slaughter them?*

A few days later, though, while touring a hospital, Booker spotted Johnny Depp, shaved and handcuffed to his bed. Booker was thankful he was healthy and alive.

The day after the bombing at the school, Dean and Nelson turned their attention to building peace in a shattered city. Still annoyed by Najeeb Quarishy's persistent questioning, Dean ordered the team to move out of the Quarishy household and into another safe house.

It was a propitious move. The new place was practically a palace by Mazar standards, with running hot water, working showers, and flush toilets. Dean set up guards on the roof and a secure perimeter. The team next started conducting what they called "bug hunts" in the city.

These involved driving through the city and visiting with locals, asking them if they were aware of enemy activity in their neighborhood. Each night when Dean climbed up to the roof of the safe house, he heard less and less gunfire. He wondered when they might be able to return to Fort Campbell. The question was increasingly on the minds of all the men.

As peace broke out, news reporters started rolling into the city. Alex Perry, who'd been hired by *Time* as a travel editor, had been trying to enter Afghanistan from Uzbekistan for nearly a month following the attacks in America. Perry had guessed early that the U.S. military would respond by invading Afghanistan, and he had waited out his chance to enter the country in a $25 a night guesthouse.

Perry had worked his satellite phone, calling Dostum, Atta, Mohaqeq, and their subcommanders, as they rode north on their horses with Nelson, Dean, and Mitchell. Perry filed his stories from an Internet café in Tashkent, increasingly eager to see the war firsthand.

"Are the Taliban Leaving Mazar-i-Sharif?" ran the headline of a dispatch he published on November 7, 2001 (two days after Essex, Milo, and Winehouse were overrun in their trench). No one in the press exactly knew the answer to the question. Reporters had been absent from the battlefield. The teams were operating in near total secrecy.

The following day, November 8, Perry filed another story that described his plight as covering "a war that nobody wants journalists to get to—not the Taliban, not the Americans, not the Uzbek border guards and not, one suspects, the Northern Alliance propaganda chiefs."

Hour after hour, Perry would reach wrong numbers and facile explanations while placing calls to contacts in Afghanistan, such as, "Sorry, Commander X is in a battle right now, can you call back later?"

The odd thing was that the person on the other end was sometimes telling the truth. Commander X really *was* in the field, fighting. Satellite phones and e-mail had multiplied and accelerated the lines of communications. As Nelson, Dean, and Mitchell used horses to ride up the Darya Balkh River Valley, theirs was the first war to be fought in an age of cheap, ubiquitous, and instant communication. This irony had not been lost on Nelson when he watched Dostum power his portable Thuraya satellite phone with a car battery he carried in a saddlebag.

Perry was frustrated by the ability to make easy contact while simultaneously being unable to make sense of the situation. Reporting at night from Tashkent was especially hard because his Afghan interviewees, often low-level commanders and Northern Alliance soldiers, were bombed out of their minds on hash and opium, the drug of choice among the Afghan lower class. The upper class drank vodka. Lots of it. His phone interviews often threatened to squib off into drunken blather and stoned silences.

He had spent hours trying to arrange safe passage into Mazar-i-Sharif when, finally, a Tashkent press chief invited several hundred journalists to the Uzbekistan border town of Termez, on the Amu Darya River.

Across the half-mile-wide current lay Heryaton, in Afghanistan. The press chief was desperately buttering up reporters to write about Uzbekistan's fabulous generosity in opening its border to deliver humanitarian aid to Afghanistan. He put a dozen reporters on the first barge of food aid with strict instructions to return with the ship.

When they arrived in Heryaton, Perry watched his fellow journalists dutifully filming the unloading of food from the barge.

Perry himself quietly slipped out of the port and caught a taxi that took him straight to General Atta's safe house in Mazar, thirty miles south. Perry, although a seasoned reporter, had never worked in a combat zone.

Yet he was surprised to find that he had become one of the first journalists to enter Mazar-i-Sharif after its fall. He felt he was on to something big.

As he sat at dinner with Atta—rice, raisins, onions, carrots, and boiled mutton—he met an enterprising and eager young man who spoke excellent English and harbored dreams of being a correspondent for CNN (or so he had told the last guests in his house, gruff American soldiers who moved out abruptly).

Perry hired the young man on the spot as his translator and "fixer," or guide. The young man introduced himself as Najeeb Quarishy.

The day after his arrival, Perry visited the Sultan Razia School. Five days after the bombing, Red Cross workers were still engaged in the grim chore of collecting the approximately four hundred bodies of Al Qaeda and Taliban soldiers.

Perry wrote: "The stench of death hung across the ruins. The team concentrated on intact bodies that could be lifted by the arms and legs. . . . Elsewhere, fire had reduced everything—furniture, clothing, people—to ash."

When Dean learned that reporters were in Mazar-i-Sharif, he took immediate, if simple, measures to avoid detection. Men and women toting laptops and cameras were pouring into the city, on the hunt for a story about the lightning-quick victory of a group of mysterious American soldiers on horseback, armed with lasers. Because the soldiers had left the States secretly, and because no reporters had been embedded with them (a concept that wouldn't come to the fore within Pentagon public relations for another two years), there had been a virtual news blackout. Nothing concrete was known about Nelson, Dean, and Mitchell—certainly neither their names nor their faces had been published in any kind of public media.

But the reporters were still largely stymied in getting a story. They spent their days hanging out in Mazar's only working hotel, talking mainly to each other about what might be happening in the street. It's difficult to imagine now, but in 2001, few reporters had any connection with locals on the ground—no ready-made network of contacts and translators, the same kind of supply lines the soldiers relied on to stay alive.

For Dean, whose modus operandi depended on his ability to blend in with local citizens and operate behind the scenes, the media was a new kind of digital-age meta-enemy. Within minutes, his face could be broadcast around the world, accessible to anyone with a television or an Internet connection. The thought horrified him. At the least, this disclosure could give the larger jihadi world the impression that the Americans had invaded and captured Afghanistan, creating ill will toward the entire postwar rebuilding of the country. At the worst, publishing his name, face, and location could get him killed. There was still a bounty on all of the soldiers: $100,000 for the body of each dead American soldier, to be paid by bin Laden's Al Qaeda network. Another scenario haunted the men. They feared some wannabe Al Qaeda nut job might show up outside their homes in the States . . . The possibility was too chilling to consider.

In short, having a reporter take your picture was like posing for your own Wanted poster. On bug hunts, the teams moved quickly and decisively through the city, keeping watch for anyone pointing a camera at them. If they were spotted, they left that area. They hid behind wraparound sunglasses, scarves, and Pakol hats, to avoid being recognizable in any pictures.

Several days after they'd moved into the new safe house, communications sergeant Brian Lyle and weapons sergeant Mark House were spotted in the bustling market area by a group of what looked like photographers.

Dean was at the safe house pulling security duty when his radio crackled to life. Lyle reported that the men were toting serious-looking cameras. Lyle and House had jumped in their truck and peeled away, but they couldn't shake their pursuers. Dean could hear the panic in their voices.

The photographers were swerving over the road, barely missing kids playing in the streets. Lyle told Dean they were afraid somebody was going to get run over.

"Drive by the safe house," said Dean. "Don't look at it—just drive and we'll see how many there are."

He saw their pickup race past, followed by a yellow sedan driven by three men pointing cameras out the windows.

Dean got back on the radio. "All right, lead them to an isolated area and box them in somehow. Stop the car, get out, and then bum-rush them.

"Tell them to stop photographing you, and that if they continue, tell them you're going to destroy their car.

"And, lastly, explain to them the whole Massoud incident," Dean added, recalling how the Northern Alliance leader Ahmed Shah Massoud had been assassinated two months earlier by Al Qaeda operatives posing as photographers.

Lyle and House turned down an alley, came to the end, wheeled around, and starting driving back out. They let the photographers' car pass and complete the same maneuver, and when it was behind them, Lyle wrenched the wheel and turned the truck across the road, blocking the path. He and House stepped out and ran at the photographers' car.

The men inside looked as if they might piss their pants.

"Don't take our picture," Lyle said, calmly. "It'll endanger us, and it endangers our families."

The photographers sputtered and cursed in French, "*Fuck you!* What about freedom of the press?"

But Lyle remained calm and repeated again what he had said. He saw that they were buying it. He knew that if he stayed calm and friendly, they would have little to react to.

The photographers talked over the request, and told Lyle they wouldn't take his picture—at least for now.

"Thank you," said Lyle.

He and House sped away, as anonymous as when they had left.

As Dean's and Nelson's teams ran bug hunts in the city, Mark Mitchell oversaw postwar operations from a second-story office in Qala-i-Janghi. The office was part of Dostum's spacious headquarters, which opened onto a long balcony overlooking the rose garden, placid groves of trees, and the cold, narrow stream flowing through the northern courtyard.

Across the courtyard, about 150 yards from the balcony, stood the 60-foot-high mud wall that separated the fort in two. In the southern courtyard, on the other side of the tall wall, several mysterious buildings sat among the stones and thorns, relics of the Soviets' occupation of the fort when they'd used it, like Dostum and the Taliban later, as a headquarters. One of the buildings was painted a startling pink, the flesh of the faintest rose, in stark contrast to the beige and gray of the fort's dry, mud walls. None of the Americans knew exactly what the pink building had been used for—Dean and Nelson believed it must have been a school because the floor measured about as much as a one-room schoolhouse in the States, about 75 feet on each side.

But it was the basement that gave one the willies. You entered by means of stairs standing about 50 feet off the side of the building itself. The doorway was made of stacked brick, with mud steps leading down 75 feet into darkness. The basement might have been a storeroom for ammunition as the walls and ceiling were several feet thick and reinforced with metal bars. Not one of the Americans, or even the Afghans, knew for sure. All stories seemed apocryphal, especially the one that the basement, which was dark and silent as a grave, had been a dungeon. The Americans had simply started calling the building the Pink House.

The rear wall of Mitchell's office in the northern end of the fort had been painted as a mural of an undersea fantasyland, with green fans of coral and exotically colored fish. Mitchell surmised that the Taliban hadn't destroyed the painting because it did not depict the faces of people or animals. Fish, he guessed, were A-OK with the Taliban.

Soon after moving into the fort, nearly everyone on the teams grew deathly ill. When fresh water had failed to arrive on a resupply drop, they'd taken their chances drinking from the local tap, and now everyone regretted the decision. Ben Milo was struck with a case of diarrhea that he felt would kill him (it would last nine days). He and the rest of his team were camped in three miserable rooms in the northern end of the fort, on the second floor. At one point, Milo grew so sick that he had to sleep in the bathroom, hurriedly scuttling between his sleeping bag and the dark hole in the floor a few feet away.

The very air in the fort seemed diseased. The men explored its maze of dank rooms and dark passageways, the halogen beams of their flashlights passing over rough mud walls that sprouted tufts of animal hair and straw—crude building materials—and revealed odd sights. Some rooms were filled floor to ceiling with small sticks; others were piled high with shoes; some seemed to have been the scene of horrific struggles, the floors tracked with heavy boots and dried pools of blood. After a half hour of wandering, the men would step out of a doorway and into another part of the fort without any idea of how they'd arrived there. The Magic Castle, some of them called it.

As part of postwar operations, Mitchell was busy preparing to dispose of the immense weapons and ammunition cache that the Taliban had left behind. Six Conex trailers hulked in the weeds in the southern compound. The rear doors of the trailers swung open with a screech. Inside lay hundreds of rifles, rockets, ammo, grenades, mortars, and BM-21 rockets, enough matériel to outfit an army. These were the Taliban's spoils of war after three years of occupying Mazar.

Mitchell marveled over the World War I rifles of French extraction, the Russian long guns, and the World War II–era machine guns. He picked up a British Enfield and admired the glint of its bayonet in the stormy half-light of the metal trailer.

A bayonet! It was stamped with a date: 1913.

On the day he was to cart the weapons away to a distant stretch of desert where they would be blown up, the ammunition cache started exploding on its own. The racket was incredible. At first, Mitchell wondered if they were under attack. He ran from his office to see what was the matter. (Mitchell would never discover the cause of this spontaneous explosion.)

It took about a minute to run from the north balcony to the southern courtyard. Passing though the tall gate in the middle dividing wall, he heard the small-arms ammunition—*crack crack crack*—and then the mortars—*shhhwwwooom*—followed by BM-21 rockets. The rockets banged around inside the metal trailers and sailed out the open doors, smashing into the interior walls of the fort, or sailing cleanly over and exploding outside in a surrounding field.

Mitchell retreated immediately. The fireworks lasted several hours. When he reconned the trailers, he saw that most of them had blown up and been incinerated. The insides were a cooked mess, consisting of twisted gun barrels and charred hunks of steel.

Several trailers were more or less intact. Mitchell wanted to take care of these, too, and he made a note in his green, hardback notebook to have this done as soon as possible.

If by some fluke the fort was attacked and fell into the hands of the Taliban, well, Mitchell didn't even want to imagine *that*.

Two days after they arrived in the city, the Americans received good news that the battle in other parts of the country was also going well. Northern Alliance soldiers had captured Herat, 60 miles west of Mazar, and Kabul, 150 miles to the south. Soon, other villages— Tashkurgan, Hairatan, Pul-e-Khumri, Taloqan, Bamiyan—fell, too. From Kabul, the Taliban had headed south to Kandahar, their spiritual base. In the north, they retreated to Konduz, 60 miles east of Mazar. The Taliban government was collapsing.

In Kabul, young men like Rocky Bahari, an amateur boxer and schoolteacher, were ecstatic to see the Taliban leave the city.

After the Taliban took charge of Kabul, at the height of the civil war, in 1997, there had been fewer robberies by highway bandits. Fewer explosions in the middle of the night. But Rocky had discovered that the security the Taliban had brought to the country came with a heavy price. One's very freedom.

Rocky did not understand why the Taliban made religion seem like a jail. They didn't let women go to school or work. If a woman's husband died, she had to beg on the street: "For the sake of God, give me money!" Often she and her children would starve.

In defiance to their rule, Rocky had refused to grow a beard. One day he was arrested on his way to the university where he was scheduled to take an exam. Rocky ended up spending a week in jail because he was clean-shaven. It had almost seemed humorous.

The Taliban executed women in the soccer stadium for sleeping with men who weren't their husbands. They cut off the hands of robbers. White-coated doctors would anesthetize them on the warm grass on the soccer pitch and do the operation in front of thousands of cheering people. This was disgusting. Beyond human comprehension. He believed that the future had to be brighter than the darkest past.

Rocky had lost his father, sister, and brother during the fighting with the Soviet troops. They had been riding in a taxi when it crashed with a Russian Army truck. Rocky had to go to work selling milk to support his family after that. With the Soviets gone and Afghans fighting each other, there had been little work in the city for any man, except to join the Taliban and fight the Northern Alliance.

Rocky sat in his small food shop all day, waiting for life to change.

One day, he was walking down a street in Kabul when he saw a jet from one of the warring faction's air force streak overhead. Rocky saw the jet drop a bomb on a nearby house, and he watched the house's roof fly off and go twirling across the street and crash.

He heard a rubbery sound—*plop*—and looked down at the ground.

At his feet was a woman's hand. It had fallen out of the sky.

It was lying with the fingers extended in the dirt, palm down, as if gripping the earth.

What Rocky noticed were the woman's fingernails. They were painted red with nail polish. She was wearing a ring. She was some-body's wife, he told himself. She had been somebody's mother.

Back at Qala-i-Janghi, Mark Mitchell heard stories about Taliban soldiers who'd "punished" children by cutting off their arms and legs to avenge their parents' attempted escape from the city of Taloqan, twenty-five miles from Konduz. The stories horrified him and made him think that any battle in those cities would be enor-mously challenging. In fact, as the Taliban's control of the rest of the country collapsed, it did appear that Konduz would be their last stronghold, which they would defend to their deaths.

On November 13, the British media reported that Osama bin Laden had a videotaped statement claiming responsibility for the attacks on American soil nine weeks earlier. "History should be a witness that we are terrorists," he said. "Yes, we kill [American] innocents."

Despite Pakistan's diplomatic reassurances that it was America's ally, Al Qaeda's ties to Afghanistan and Pakistan were increasingly clear. The CIA was picking up intelligence that approximately thirty-five Al Qaeda members, based in Pakistan, were planning to blow up the U.S. Consulate in Peshawar. Moreover, it appeared that

Al Qaeda planned to use Pakistan's nuclear capability for further attacks.

This news reminded Mitchell that the war was far from finished.

Four days after arriving in Mazar, Bowers convened a series of meetings at Qala-i-Janghi with Dostum, Atta, and the Hazara general, Mohammed Mohaqeq, hoping to lessen tension among the three warlords. He wryly referred to these as "the Paris peace talks," after the 1972 cease-fire discussions between North and South Vietnam.

Bowers realized that he needed to unite these men, bound together by the war, in common cause or risk a repeat of the tribal warfare that had erupted when the Soviets left. "This is your country, not ours," he began. "We will help you all we can. We have no desire to take anything, or to be responsible for your government." By the end of the meetings, the warlords had decided who would run the electric plant, the water system, and humanitarian efforts—the critical miscellany of daily life in a fifteenth-century town of shopkeepers, doctors, and teachers struggling in the twenty-first century.

Next, Bowers and the warlords turned their attention to Konduz, sixty miles away, in the country's eastern corner.

Konduz was a city under siege, a place of hellish spectacle. As intel reports trickled in, indications were that the city of 220,000 was boiling over with Taliban and Al Qaeda soldiers. The streets seethed with bandits, freelancers of no political stripe taking advantage of the chaotic situation. Soldiers broke into restaurants and stores looking for food. Families were killing their burros and eating them. Refugees were fleeing the city to camps outside of Mazar—or trying to. Men were shot in the back at checkpoints as they ran beyond the city limits.

Those who escaped spoke of atrocities committed by the Taliban, of citizens forced to run across minefields to the amusement of Taliban soldiers looking on. To prevent soldiers from abandoning their ranks, Taliban leaders ordered that the suspicious among them stuff their shoes with thorns.

Those citizens who managed to make it to refugee camps found conditions that were not much better.

Alex Perry, the *Time* reporter, discovered thousands of people living in stick tents draped with black garbage bag tarps on the outskirts of Mazar. Families of eight or nine were forced to share one blanket through the increasingly cold nights.

Perry saw that the refugees owned nothing except a few pots and pans. When the American soldiers pulled up in their pickups, men, women, and children swarmed the vehicles, saying, "Thank you, thank you," and making a hand-to-mouth gesture: *I'm hungry*.

Stu Mansfield, on Dean's team, gave one of his guards a pair of new boots to replace a pair of worn slippers. The guard thanked him profusely, and then gave the boots to a young boy who ran downtown, sold them, and returned with some money, which the man pocketed. Mansfield could only shrug, thinking that he had still done the right thing.

In an attempt to make the Taliban and Al Qaeda soldiers surrender in Konduz, B-52s trolled overhead, dropping bombs on fleeing trucks and tanks. At night, Spectre AC-130 gunships pounded the hillsides where the enemy forces were camped in trenches. In the dark, the specially outfitted planes flew at high altitudes and remained invisible to the enemy on the ground. The Spectre could shoot thousands of bullets per minute from Gatling guns and could fire artillery shells, barrels poking from the side of the plane. The enemy soldiers on the ground, believing they were concealed by darkness, often didn't know where the deadly fire was coming from.

On board each gunship, a weapons officer tracked enemy movements on a video screen. Human bodies resembled glowing grains of rice. The engines of trucks appeared as intense hot spots in the plane's heat-sensing optics. The weapons on board were aimed by means of a joystick.

The torrent of fire from the planes was relentless and accurate. Dostum had been delighted to discover that one of the Spectres' crew members was a female. He had been listening to his radio when he heard her voice amid the pilots' chatter. Dostum, who was standing alongside Nelson as the night attack unfolded, turned and asked, disbelieving, "Is that a woman?" Even though he professed an egalitarian streak, Dostum had met few women in positions of power in his lifetime.

"Hah!" he chortled into the radio, talking to the Taliban soldiers. "The Americans think so little of you that they have sent a woman to kill you!"

The Taliban shouted into their radios, cursing Dostum and the Americans.

"I will call her 'the Angel of Death,'" Dostum kidded the Taliban. The Taliban were apoplectic.

After several days of daily and nightly bombardment by B-52s and Spectre gunships, all roads in and out of Konduz were sealed. For the Taliban and Al Qaeda soldiers trapped there, they had one choice to make: surrender or die.

The trapped fighters fell into two camps: the "Afghan" Taliban who, if they did surrender, were allowed to join the Alliance or else return to their villages and pick up their lives. A man's newly sworn allegiance was his bond. These conscripts weren't even searched when they switched sides.

The other camp was comprised of "foreign" Taliban—soldiers from Chechnya, Pakistan, and China, and their hard-core brethren, Al Qaeda, with whom the Taliban had allied. Surrenders by these men were rare and often perfidious. An enemy soldier would walk up to a member of the Northern Alliance and blow himself up with a grenade hidden in his clothes.

"We are going to be martyrs," one of the hard-core fighters announced to the press. "We are not going from Konduz." The Northern Alliance soldiers decided they had no choice but to shoot "foreign" Taliban soldiers on sight.

There were approximately 3,000 of these committed soldiers in Konduz, and Bowers knew that his team's next move would be to rid this city of would-be martyrs.

Meanwhile, a faction of these otherwise suicidal fighters did entertain surrender—if they could keep their weapons and were offered safe passage from the city. When he learned of the proposal, Secretary of Defense Donald Rumsfeld in Washington, D.C., refused to accept it. He warned that if these particular prisoners were allowed to go free, they'd live to fight another day, and that this wasn't a palatable outcome.

"My hope," he told the press, "is that they will either be killed or taken prisoner."

In preparation for the battle in Konduz, Bowers and Mitchell's team moved their headquarters out of the squalid conditions of Qala-i-

Janghi to some place more hospitable. Over the previous ten days, Mitchell, like everyone else, had grown ill in the fort and he was glad to settle nearer downtown Mazar, into a modern five-story building known as the Turkish Schoolhouse.

The new headquarters offered a commanding view of the city. From the roof, Mitchell could see the Blue Mosque, located less than a mile to the west in the middle of town. Traffic streamed steadily past on the road fronting the school, piercing the air with the screech of brakes and the braying of car horns.

The Turkish government had built the boys' high school in the 1970s as a gift to the Afghan government. It had once been a gleaming monument to bureaucratic elegance with its rows of small, tinted windows and decorative, concrete facade. The Taliban had ripped the plumbing and electrical fixtures from the walls during their hasty exit twelve days earlier and now the building was a mess. But in comparison to Qala-i-Janghi, the place was clean and well lit.

At the same time, Nelson's team moved to a safe house owned by General Dostum located several miles from the fort. Spencer and his team hired local workers to paint the walls, fix the plumbing, and tile the floors. As if a spell had been broken, the health of all the men improved once they moved out.

Mike Spann and Dave Olson moved into the fifth floor of the Turkish Schoolhouse, along with several other paramilitary officers.

Mitchell, Bowers, and eight more Special Forces soldiers lived on the third floor, with the local interpreters on the fourth. The second floor became the operations and support center, occupying what had been, when the Taliban lived in the building, a mosque. Sensitive to the fact that he was using a religious space for other purposes, Mitchell was savvy enough to ask local Afghan clerics for permission to do so. Permission was granted and his Afghan colleagues in the building appreciated the gesture. *Brains before bullets*, thought Mitchell. *Outthink 'em so you don't have to outshoot 'em.*

The ground floor consisted of a tiled foyer, a large cafeteria, and a functioning kitchen that had running water and heat. Workers taped cardboard over the windows that had been smashed by the Taliban. Electricity was provided by gas-powered generators that broke down on a regular basis but were repaired by Afghan workmen using, it seemed to Mitchell, little more than a hammer and a piece of duct tape. He marveled at their good cheer in the war's aftermath.

Life in Mazar, it seemed, might begin to assume predictable rhythms. Master Sergeant Brad Highland of Dean's team sketched a diagram of a barbecue grill on a piece of scrap paper and Atta's men built one from spare metal they scrounged in the city. They butchered a cow and Highland next taught them how to grill hamburgers, which they'd never eaten.

Chief Warrant officer Cal Spencer, of Nelson's team, enjoyed watching the antics of the local children as they teased Garful, an aged Afghan who'd been assigned as one of the team's bodyguards.

Wherever Spencer went in the city, the children often trailed close behind. Spencer had to tell them that he was busy and that they should leave him alone. He was trying to be stern but this was hard. The kids wanted to shake his hand. Garful was less tactful. He would charge at the kids, waving a stick. "Leave the Americans alone! They have things to do!"

The kids would jump away and laugh, and Garful, looking over his shoulder, would wink at Spencer and smile. *Holy cow*, thought Spencer. *Garful's like us. He's making a joke.* Spencer and Garful started laughing.

But this familiarity aside, increasingly, out on the city streets, the teams found themselves thrust in the middle of family blood feuds that had been left to simmer during the Taliban occupation. And there were larger, more violent ethnic battles to contend with. Six of Atta's men were killed in a gunfight when they tried to prevent some Hazara soldiers from stealing a taxi driven by a Pashtun man. Even though the driver had done nothing to the Hazara soldiers, he happened to be a member of the same ethnic tribe as the Taliban, and this had awakened the Hazaras' bloodlust. Atta's men had prevented the robbery, but at great cost.

Such incidents were becoming more frequent. One faction of fighters kept asking Dean and Nelson to bomb individual dwellings, claiming that each was a Taliban safe house. Without exception, the men would discover that the house did not belong to a Taliban soldier, but to an old enemy of the group that had requested the air strike. Dean started calling this "attempted assassination by bomb." It was clear to both him and Nelson that the sooner they put the city back together, the sooner the Afghan soldiers might lay down their arms and reenter civilian life.

Amid the ongoing conflict and the wreckage of war, however, there were other signs of life, including desire itself. Wherever he went on the city streets, blond-haired and blue-eyed Staff Sergeant Brett Walden managed to draw a crowd among the women in their veils. They walked past him in their burkhas, upright and stiff as giant thumbs, faces hidden behind a fabric screen that fluttered in and out with their breath.

Walden was shocked when one woman beckoned him with a wave, *Come here*. He walked over and she lifted her veil, just a peek. Her beauty stunned Walden. He realized that he had stupidly assumed all the women were ugly just because he couldn't see their faces. This woman would have been stoned by the Taliban for this very behavior. Walden tried discouraging it. The team prided itself on being culturally sensitive and it was inappropriate for him to see any Afghan woman's face. But the problem only grew worse.

Some of the women started removing their veils completely and making eyes at Walden when they passed him on the street. Dean nicknamed him "Casanova." Walden blushed whenever he heard the word.

He also began to fear something bad would happen either to him or to the women he met in the street if he was caught looking at their uncovered faces. At the same time, the experience made him appreciate the plight of these women, many of whom had been beaten by their husbands simply for being female. He guessed this was their form of silent protest, a flirtation with a man from the outside world.

From his new office, Mitchell oversaw the rebuilding of the hospital and its water and electric plants, and supervised the disposal of unexploded bombs and land mines that still littered Mazar's airport and streets. Working alongside him was his friend Major Kurt Sonntag, who had helicoptered into Mazar from K2 after the Americans had ridden into the city. Major Steve Billings, who'd been running the teams' operations from the comfort of a desk back at K2, arrived, too, his worst nightmare realized: the war had gone on without him. With Billings was his buddy Master Sergeant Roger Palmer, who was equally glad to be entering the war zone, and an eager young captain, Paul Syverson, thirty, from Lake Zurich, Illinois.

None of these men had lately fired a weapon in a battle, if at all, and all of them were glad to have escaped the monotony of K2, which in the preceding weeks had grown by several thousand men. It was now a tented city populated by soldiers from all branches of the U.S. military, and they had been waiting for Mitchell, Dean, and Nelson to capture Mazar. Over the coming weeks, they would now enter the country in follow-on maneuvers. Just as they had planned, Mazar was a base from which the rest of Afghanistan could be controlled. The battle to win Mazar had proven decisive; but Mitchell, Dean, and the others all knew that the Taliban were still at large by the thousands, and that Mazar could be lost almost as quickly as it had been won.

With such uncertainty in mind, First Sergeant Betz, who had also flown in from K2 to staff the new headquarters, began fretting about the schoolhouse's security. The building was approachable from all sides. Betz worried about its close proximity to the busy street—an easy target for would-be car bombers.

Such an attack could deliver a devastating blow. Mitchell's feeling was: *Don't relax yet.*

With the temporary cessation of major military confrontations came the sudden yet murky game of strategic brinksmanship, with men, weapons, territory, freedom, and American dollars as the chips on the board. And soon Dostum made a move. Ever the wily diplomat, he lured a Taliban archrival, Mullah Faisal, deputy defense minister and commander of some 10,000 Taliban troops, into Mazar to negotiate the terms of his own surrender. If Dostum could achieve this surrender, he would decrease the enemy force he'd otherwise face in Konduz, and add more soldiers to his own ranks. He had set the date for his movement to Konduz as November 25, so time was of the essence.

On November 21, eleven days after he'd driven the Taliban army from Mazar, a gloating Dostum welcomed the dour, defeated Faisal back to Qala-i-Janghi. Faisal roared through the fortress gate in a convoy of about forty vehicles and five hundred men, all of them heavily armed. They pulled into a parking lot inside the fortress walls and Dostum's men instantly surrounded them.

As each side stood with its guns trained on the other, the portly

Faisal, his head wrapped in a black turban, and dressed in a brown sport coat, marched up to Dostum's chambers to discuss the terms of surrender. He was trailed by several dozen of his own gunmen, who took up positions around the room. The room was tense.

As a show of Dostum's strength, helicopters from the U.S. 160th SOAR, the same unit that had delivered Nelson and Dean into the country, patrolled overhead as the two warlords talked over tea and biscuits, with Bowers listening quietly in the corner. Dostum leaned on Faisal to order the Al Qaeda soldiers in his midst to surrender. Many of the local Taliban were already eager to give up.

At all costs, Dostum wanted to avoid urban combat in Konduz. Like the Al Qaeda fighters who'd been trapped at Sultan Razia, the enemy soldiers inside Konduz were fanatical and opposed to surrender; subduing them would mean another bloody slaughter.

For his part, Faisal wanted passage from Konduz to Herat, about sixty miles west of Mazar, where pockets of sympathetic Taliban still had not surrendered. The usually defiant mullah had a proposition for Dostum: he would pay the Uzbeki warlord $500,000 if he and his fellow Taliban were allowed to travel safely to Herat.

After some discussion about the overall plan, the enemies looked each other in the eye and shook hands. They were agreed.

The two men emerged from Dostum's quarters and announced to the Taliban and Afghan soldiers gathered below that nearly 13,000 Taliban would surrender in Konduz.

"The fighting of the Taliban is finished in Afghanistan," Dostum happily proclaimed. "They are preparing to surrender to us." Ever the diplomat, he carefully described the surrender not as a moment of defeat for the Taliban but as a chance for all warring Afghans to be reunited.

Dostum realized he was about to vastly enlarge his army, as long as he could convince the enemy fighters to join his side. The Taliban in Konduz owned valuable tanks, weaponry, and a fleet of beat-up Toyota trucks. That he would be $500,000 richer for the deal had to seem a bonus.

The Afghan Taliban would be allowed to join the Northern Alliance or return home, to their farms, businesses, and families, as long as they didn't have an affiliation with Al Qaeda. The fate of the approximately 3,000 Arabs, Pakistanis, Chechen, and

Chinese—the "foreign" Taliban—would be decided later. Dostum would eventually sort them out, he announced, deciding who was an Al Qaeda terrorist and who was not. But for now, they would be safe.

Faisal and his men loaded up in their Toyotas and delivery trucks and shot off from the fort in a cloud of dust. Bowers watched them go. He didn't like the feeling he had about the meeting. Dostum, it seemed, was suffering an attack of hubris. (Bowers would later be surprised to learn that Dostum also believed he might convince Al Qaeda and foreign Taliban fighters to surrender to him.)

During the meeting, Bowers had found Faisal to be vague about his plans to carry out the surrender in Konduz, and he distrusted the warlord (Bowers was also unaware of Faisal's cash payment). But there was nothing he felt he could put a finger on. And besides, there was the pressing matter of Konduz to turn to.

After his meeting with Dostum, Faisal informed his soldiers that any Taliban who put a picture of Massoud or Dostum on the windshield of their vehicle would be allowed safe passage into Mazar and on to Herat. Faisal didn't mention that at first they would be held as prisoners, and that the local Taliban would be freed, while the fate of the foreign ones was as yet undecided. The foreigners were also led to believe that they likely would be allowed to continue to Herat. Faisal misled them in order to get them to comply.

His double cross went further, in the later estimation of Max Bowers: with Dostum pacified by the surrender, Faisal planned an attack on Mazar itself. His payment of money to Dostum had bought him the time and space to accomplish this.

Taliban fighters located in Balkh, to the northeast of Mazar, were awaiting word from Taliban command when to attack. Further attacks would come from Taliban soldiers in Konduz, and from those still hiding in Mazar.*

If they could retake Mazar, then the Americans' battle for the north would be lost. And with it, the battle for Kabul, and thereby the rest of the country. The Americans would soon be mired in

---

*While Bowers believes that Faisal was an architect of the Taliban's perfidious surrender, he also suggests, as do other U.S. soldiers, that other "rogue actors" in the Taliban organization had a hand in the double cross.

conflict through a long, bitter winter. Without Mazar, they wouldn't have access to the Mazar airfield and nearby Freedom Bridge to Uzbekistan, with which they planned to bring humanitarian and military supplies into the country.

Without Mazar, they would lose everything.

While the surrender negotiations were under way, Diller had worried that he'd be left out of the fight in Konduz. Bowers, again, seemed to have a sore spot for Diller's team. Sam Diller wondered what kind of action he himself would see if he had to stay behind in Mazar. Not much, he guessed. He couldn't imagine sitting out a fight.

To remedy the situation, he knocked on the door of the ranking officer on the ground, Admiral Bert Calland, who had arrived in mid-November as a high-ranking liaison with Alliance forces. The ultimate decision about who would be going to Konduz would be Calland's. Diller knew he was going over Bowers's head. The move could end his career.

Calland was looking through some papers when Diller entered. He looked up. "Yes?"

Diller removed his boonie cap and said, "Sir, there's a fight coming up in Konduz, and we want to be there. I've heard that we might not get to go.

"Nobody put me up to this," he went on. "Captain Nelson didn't ask me to talk to you. So if there's gonna be any heat, I'll take it."

Calland, obviously impressed with Diller's initiative, smiled and thought a moment. "You won't be taking any heat, Sergeant. I'll see what I can do."

After that, Diller and the team were placed on the list to go to Konduz.

That night, First Sergeant Dave Betz drove to a helicopter landing zone north of the city to pick up some soldiers newly arrived from K2. These men would help form the support staff at the schoolhouse. Betz was relieved to be receiving more men to supplement the skeleton crew. He was looking forward to seeing his buddy Captain Kevin Leahy. Leahy had been riding out the war back at K2. Now he walked up and slapped Betz heartily on the shoulder. "Ol' Sarge, how the hell are you?"

The two men had worked in adjoining offices back at Fort Campbell, and Leahy remembered the day the news of the attacks in New York City had come over the radio, when he had worried about his brother who worked for a brokerage firm on Wall Street. Betz had been a steady presence as Leahy waited for the news that his brother hadn't been in the World Trade Center when it collapsed.

Not everyone could fit in the van already stuffed with gear and weapons, so Leahy volunteered to stay at the landing zone with eight other soldiers while Betz drove eight British SBS (Special Boat Service) soldiers back to the city. The Brits had arrived to help Mitchell secure the Turkish Schoolhouse in the aftermath of the war. Betz was glad to see them, too.

About an hour after Betz left, Leahy watched red tracers streak across the horizon, maybe a quarter mile away. That was odd. Leahy had been told when boarding the helicopter that he shouldn't expect to meet any resistance when he landed. He ordered the other men to don their night vision goggles and find cover among the ruined walls of some nearby mud buildings. U.S. Army Staff Sergeant Jason Kubanek hadn't been part of an active combat team, like Dean Nosorog's or Mitch Nelson's, but he was hungry for a fight. Maybe now, he thought, he was going to get it.

Leahy heard the angry roar of a vehicle approaching, tires skidding to a stop at the edge of the landing zone. Three men in desert camo—Leahy could see they were Special Forces soldiers—jumped out. Captains Paul Syverson, Andrew Johnson, and Gus Forrest had been ordered by Betz to come back and pick up Leahy and the remaining men.

"They started firing at us at a checkpoint!" yelled one of the soldiers. "That was an ambush. They're coming after us right now!"

That explained the tracer fire, thought Leahy.

"Let's move the vehicle to the buildings," said Leahy.

"If we do that," said one of the soldiers, "somebody could see we're here because of the taillights."

"Okay," said Leahy, "then break the taillights!"

He waited and nobody moved. Then he realized that some of the guys were hesitant to take a swipe at the truck. *Like it was war, but you couldn't break anything,* Leahy thought.

"Like this," he said. And he raised his rifle butt and smashed one of the lights. Somebody quickly smashed the other.

These guys were staff people, not warriors who had recently seen battle, Leahy reminded himself. In fact, he counted himself among them. But he could feel his own fear as he reminded these guys of the seriousness of the situation.

"Can somebody get comms with the schoolhouse?" he asked.

"Sir," said blue-eyed, blond-haired Ernest Bates, a thirty-five-year-old sergeant first class from the Midwest who was fiddling with a radio he'd drawn from his rucksack. "Sir, I can't raise the schoolhouse."

Leahy knew he had to get a message to Betz explaining why they hadn't returned already.

Bates pulled another transmitter from his bag. "Sir, using this, we *can* send a distress signal to K2," he suggested.

Leahy considered it. "Nah, 'cause we're not in distress, yet. Listen, everybody has got to relax."

Back at the Turkish Schoolhouse, Betz was, in fact, wondering why Leahy hadn't yet returned. Impatient, he grabbed another sergeant, Bob Roberts, and they jumped into a vehicle, Betz driving like hell for the HLZ.

Shortly, Leahy looked up from behind his pile of rubble and watched as a van ripped over the desert road, heading toward him.

"Is it ours?" he shouted out.

"We think it's Betz," one of the men answered.

"Keep an eye on it," said Leahy. He told everybody to be ready to hammer it with automatic fire.

The door flew open and a stout man silhouetted by headlights bounded up, sputtering. *Betz.*

He walked up to Leahy. "Hey, sir! *What the hell!* What are you guys doing?"

Leahy snapped. "*What the hell?* I'm out here and we've got guys getting *shot* at!"

And then they heard something falling toward them from overhead. Around them the ground thudded. *An air drop,* thought Leahy. *We're standing in a supply drop zone!*

They scattered to the edges of the gravel field and listened as the night erupted with shouts. There were people out there, scurrying around in the dark, tearing at the bundles of food and blankets.

"Does anybody know who those people are?" asked Leahy.

"Locals," said Betz. "They've been raiding our supply drops."

"We're going to recon the crowd and see," Leahy decided.

He turned to Sergeant First Class Roberts. "Get in and drive. And put the headlights on."

Leahy, Betz, and Roberts drove toward the crowd.

Leahy watched as the looters stood up, arms laden with bandages, blankets, and boxes of MREs. He stepped out into the headlights and held his hand up in peace. *Salaam alaikum*, he said. Peace be with you, brother.

Leahy could see the looters were nervous, fingering beat-up AKs.

Betz couldn't figure it out. These were local citizens, with some Northern Alliance soldiers in the mix. What was making them so nervous?

Leahy decided that he and Betz and several others would make a run for the Turkish Schoolhouse. They would have to leave the rest of the group behind until daylight. Leahy wanted to get word to Mitchell that they had arrived and were all right. He wanted to know: *What in hell is going on?*

PART FIVE

AMBUSH

Mazar-i-Sharif, Afghanistan

November 24, 2001

Probing close to the city limits, but unseen, unheard, and completely unexpected, a force of six hundred armed Taliban warriors, tired, filthy, and wavering on the verge of total physical collapse, had halted in the cold desert sand, about two miles from Leahy's position at the ruins.

It was these men who had made the local Afghans so nervous.

They had walked and driven several days from Konduz. These were some of the foreign soldiers who had been tricked by Mullah Faisal into believing that General Dostum would afford them safe passage to the nearby ancient city of Herat, west of Mazar, one of the Taliban's remaining strongholds.

Six large trucks sat idling on the highway. Among the Taliban fighters was a young American so tired he was unable to speak. He squatted in the dust and made a fireless camp, waiting for dawn.

Through the night, the Americans in the city could hear gunfire erupting, growing in intensity, as if there were a battle brewing that no one had told them about. At about 3 a.m., Betz and Leahy back at the schoolhouse were still trying to sort out why the locals at the landing zone had been so jumpy. At the same time, Dean was fitfully dozing in his safe house when a team member burst in and shook him awake.

The team member had gotten up to go to the bathroom and

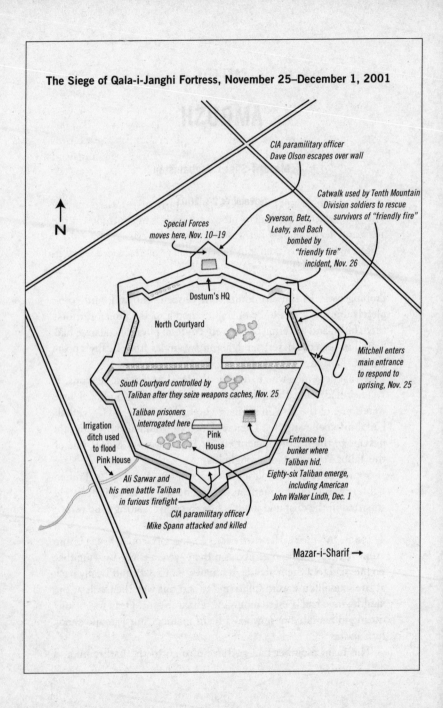

The Siege of Qala-i-Janghi Fortress, November 25–December 1, 2001

N

CIA paramilitary officer Dave Olson escapes over wall

Catwalk used by Tenth Mountain Division soldiers to rescue survivors of "friendly fire"

Special Forces moves here, Nov. 10–19

Syverson, Betz, Leahy, and Bach bombed by "friendly fire" incident, Nov. 26

Dostum's HQ

North Courtyard

Mitchell enters main entrance to respond to uprising, Nov. 25

South Courtyard controlled by Taliban after they seize weapons caches, Nov. 25

Taliban prisoners interrogated here

Pink House

Irrigation ditch used to flood Pink House

Entrance to bunker where Taliban hid. Eighty-six Taliban emerge, including American John Walker Lindh, Dec. 1

Ali Sarwar and his men battle Taliban in furious firefight

CIA paramilitary officer Mike Spann attacked and killed

Mazar-i-Sharif →

had seen a light on in an office often used by Atta in an adjoining room.

The man had thought it was unusual that Atta should be there so late. Dean jumped up from his sleeping bag, which was laid out on a large pillow on the floor; he dressed and pulled on a brown Army T-shirt and his hiking boots, and rushed across the courtyard to the general's makeshift office.

Atta looked up in surprise as Dean entered.

The usually dapper warlord had also apparently dressed quickly and hurried to his office. His brown smock was misbuttoned at the neck and his trademark hat was askew over one eye.

Dean saw the stricken look on the man's thin face and asked in pantomime, *What's wrong?*

Atta had unrolled a map of northern Afghanistan on a table and he bent over it, pencil in hand, and traced a line along the road from Konduz to Mazar.

"Taliban and Al Qaeda!" Atta said.

He ran his finger over the road leading from Konduz. He then tapped the map with the pencil around the city of Mazar and said again, "Taliban and Al Qaeda!"

"Here?" asked Dean, pointing at Mazar.

Atta shook his head. "Yes."

Dean stood up. "You're shitting me, sir," he said in English, forgetting for a moment that the warlord couldn't understand him.

"You mean the Taliban are here? In the city?" said Dean, pointing at the map.

"In Mazar, yes," said Atta.

Dean's worst nightmare—the nightmare of all the teams—had come true.

After arriving back at the schoolhouse, Leahy bounded up the stairs to the operations center, where he met Major Kurt Sonntag.

He explained the ambush and nervousness of the looters and that his remaining men were back at the HLZ, waiting for pickup. He asked Sonntag why the friendly Afghans should be so nervous.

Sonntag said he didn't know.

No one at the schoolhouse knew what was unfolding outside the city's limits.

Leahy saw CIA paramilitary officer Garth Rogers walk down the cement stairs from his office on the fifth floor. He did not look happy. "We got some Taliban at the wire."

Leahy and Sonntag shook their heads, not understanding.

"About six hundred of them," the grizzled officer said. "They want to surrender. They're out by the airfield."

Leahy and Sonntag also immediately wondered how they would maintain control of 600 Taliban when most of the American combat fighters and Northern Alliance soldiers were in Konduz.

Dean's and Nelson's teams, accompanied by Atta and Dostum, were scheduled to leave for Konduz within several hours, while Leahy had been assigned to stay at the Turkish Schoolhouse. Major Sonntag would oversee the schoolhouse staff, while Major Mitchell would coordinate the movement of the soldiers patrolling the city. Leahy set aside his worries for a moment, and asked if there was any danger in driving back to the landing zone. No, the CIA officer told him. He should be safe.

The Taliban had confined themselves to the area near the airfield. In fact, they were asking to be taken prisoner there.

Sonntag, Leahy, and Betz headed out the door.

Leahy jumped up in the back of the pickup and Betz gunned it. The news of the Taliban had unnerved Betz. Suddenly the war seemed incredibly close. Leahy had to grab hold of the .50-caliber machine gun, mounted on a tripod on the truck's bed, to avoid being whipped around as Betz tore out of the drive.

As they approached the landing zone, Betz saw a roadblock up ahead, a handful of men with weapons standing in his headlights. He slowed the pickup, tires crunching on gravel. Leahy stood in the back, manning the machine gun as the truck glided ahead.

The events of the last six hours were untangling themselves for Leahy. He realized that the Northern Alliance had known of the Taliban's arrival earlier in the night and that this news had spooked them. They had started seeing demons in every shadow. That's why they had fired on the second truck that had come to pick him up.

At the same time, the Americans hadn't realized they had just run a checkpoint during a curfew imposed to keep the Taliban off the streets. The fact that news of the Taliban's arrival hadn't been passed on any earlier to anyone at the Turkish Schoolhouse, including the CIA officers on the fifth floor, reminded Leahy of how volatile his new world was. The intel had become trapped in some eddy in the rapport between Dostum and Atta—it was impossible to tell how.

Betz rolled down his window so the kid at the checkpoint with the AK-47 could see his face and that he was an American.

"How ya doing?" said Betz.

The kid was fingering the rifle trigger, looking uneasy.

Betz still hadn't stopped the van. They were already rolling past. Betz was ready to stomp the gas. His eyes locked with the kid's. The young soldier finally waved them through.

It was at that moment that Betz knew that something was going wrong with the war as they had been fighting it. The kid seemed to know something he didn't know, and couldn't know, and wouldn't know, until it was too late.

What he'd seen in the kid's eyes was *fear*.

When Betz arrived moments later at the HLZ, the men still hiding behind the mud walls were drained from the stress of standing the whole night with guns popping off around them. Staff Sergeant Jason Kubanek realized that the rock his foot was resting on was actually an unexploded artillery shell. He looked around and saw ordnance every five feet—they'd been standing all night in a huge field filled with explosives. It was a miracle no one had been blown up. Now it was safe to get in their vehicles. Kubanek was relieved when they pulled up to the schoolhouse in Mazar. He could now rest easy.

At Atta's safe house, Dean marched next door to Mark House's room. "Hey, man, get your ass out. We got Taliban in town!"

House rolled over. "What?"

"Wake up," said Dean. "We got *bad guys* in the city."

House dressed in seconds. The day before, he had played a game of *Buzkashi* at the Mazar field with some local riders. The game involved two teams on horseback attempting to drag a headless goat into the other's goal. House couldn't believe that just hours before they had been playing a game to celebrate their victory, and now the Taliban seemed to have returned.

He grabbed his rifle and loaded into a truck outside the safe house door.

House had pointed to the stock of his weapon with his daughter's name—*Courtney*—written with white paint. He patted the gun and said a prayer as they sped to the surrender site.

John Walker Lindh had left Konduz so quickly, clinging to the splintered side of a truck, that he'd taken no food, water, or adequate clothing. The desert nights had been cold and black, the days dry and blistering.

He was afraid he was dying. As the hours ticked by, the memories must've pulsed in his head. Of reading the Koran in a madrassah near a shopping mall in California. Of his mother and father and their divorce. The divorce had changed everything—the family had dissolved.

His feelings about his father were complicated. There were men in this camp who would call his parents infidels. His father had announced about a year earlier that he was gay. Abdul had left the permissive environment of Marin County, California, and traveled to one of the strictest madrassahs in Pakistan, where the thousands of sentences of the Koran had been like iron in his mouth, sure and unbending. He had memorized half of them.

Sitting in an Internet café in Peshawar, Pakistan, he had written: "I am Suleyman Lindh, eater of much wheat crop, drinker of much buffalo tea." And then he had packed his sleeping bag and left for Afghanistan.

That was five months ago, another life. Now Abdul Hamid had come to surrender in Mazar-i-Sharif. He wanted to go home to California. He did not want to die.

When he learned that airplanes had been crashed into buildings in the United States of America, he was dismayed. Why had innocent people died?

Abdul feared men like Dostum as much as he did the hard-core Al Qaeda fighters surrounding him. After his surrender, he hoped that Dostum wouldn't line them all up and shoot them.

Following behind General Atta in his convoy of six trucks headed to the surrender site, medic James Gold pulled his vehicle to the side of the highway, parking beside a primitive wooden arch spanning the bombed-out blacktop.

Next to the arch stood a rough-hewn guard shack. Dean, sitting beside Gold, studied the shape-shifting mass of Taliban through his binoculars; they were about a mile away, across the red tabletop of sand. Some of the men's turbans flashed in the sun, shiny and black.

That's a lot of bad guys, thought Dean.

Dean wondered if there were Al Qaeda fighters mixed in the bunch. He studied the men and saw various ethnicities—Pakistani, Chechen, Arab—and the grimace of what he imagined were Afghan farmers, shopkeepers, and doctors conscripted to fight the jihad.

Dean knew that these last soldiers hated the "foreigners" and given a chance, they'd probably gun them down. Dean got out of the truck and started following Atta up a trail cut into a dune, about 100 feet high, overlooking the plain and the prisoners in the distance.

"Walk where I walk," said Atta, turning around. "This place is mined."

At the top, Dean could see that the desert stretching before them was littered with the rusted hulks of Russian tanks and jeeps left over from the Soviet war with Afghanistan.

He asked Atta what this section of Mazar's outskirts was called. Atta said it had a colorful name: "the grave of snakes."

The hilltop was dug with fighting positions—these, too, had been left behind by the Russians—from which Dean saw you could easily defend the guard shack.

Next to the shack was a bent piece of metal that could be lowered across the road as a gate. Dean figured the road around the gate was mined as well, and that whoever controlled the hilltop controlled traffic on the road.

Atta called one of his men over. He touched the courier by the shoulder and gave him instructions.

The man took off, trotting down the dune to the pickup. He fired it up and drove toward the Taliban.

Dean watched him drive for maybe thirty seconds, through his binos.

The man carried no white flag, only a beat-up AK-47 with some cartridges slung across his chest in a battered leather belt. Dean watched as several Taliban fighters moved forward to meet him.

The men talked for several minutes. And then Atta's courier started driving back.

He trudged back up the dune, out of breath.

"They want to keep their guns," he announced.

Atta shook his head. "No. They may not."

The man headed back down the dune, to the waiting truck.

Atta turned to Dean. "This may take a while," he said.

Dean was standing on the hilltop overlooking the highway when he heard Lieutenant Colonel Bowers at the Turkish Schoolhouse on the radio talking to Dostum, as the warlord drove to the surrender site. Dean believed that someone needed to figure out what to do with these prisoners in Mazar, and then get to Konduz as soon as possible. The problem was, it seemed to Dean, that no one was in charge.

Dean believed Atta now had control over the surrender, as he and his men had arrived first at the scene. Dean also expected Dostum to demand an equal share in the spoils of the negotiations, principally any of the local Taliban soldiers willing to give up and switch their allegiance.

Dean picked up the radio and called Stu Mansfield back at the safe house, saying, "We've got to get the guys together out here." Dean told Mansfield they would need more guns on these prisoners once they turned themselves in.

He then tried raising Bowers on the radio to report the situation as he saw it, with the hope that he would relay this news to Dostum and tell the warlord to back off.

But as he was talking, the radio went dead. Dean rapped it with his hand. Nothing. It had suddenly malfunctioned. Now he was cut off from both his command at the Turkish Schoolhouse, and his team at the safe house.

At about this time, Bowers piled in his vehicle and headed for the surrender area.

Everybody was about to converge.

Chief Warrant Officer Stu Mansfield had been at the safe house cooking a turkey, of all things, when he'd gotten the call to meet the anxious captain. He'd reluctantly left the bird in its roaster pan and got into a truck and started driving. Mansfield, who back in Tennessee ran a real estate business out of his home and lived a quiet life fishing and golfing when he wasn't soldiering, wasn't easily rattled. He was a bit rattled now. The idea that all of these Taliban had showed up in Mazar was bad news.

Riding with Mansfield were Sergeants Walden and Lyle. Lyle had an extra radio. Dean used it to call Mitchell back at the schoolhouse and told them what Atta had explained to them, that he had these Taliban who wanted to surrender, and they were located on the highway about twelve miles east of the city.

From the hilltop, Dean watched as Dostum and his entourage of vehicles appeared on the horizon and roared to a stop at the bottom of the hill.

Atta explained to Dean how this Afghan surrender would work. "These men are not going to surrender right away. You cannot surrender"—Atta snapped his fingers—"like that. If you do, you will lose face."

Dean watched as Dostum and Lieutenant Colonel Bowers got out of their vehicles. Behind them were more people . . . with cameras and notebooks . . . *Oh my God,* Dean realized. It was the press.

Dean saw about a hundred reporters in all. They were following Dostum! Dean and his team were trapped on top of the hill. Dean was convinced his face was about to end up on the front page of a dozen newspapers.

Seeing this, Atta announced that he wanted to meet with Dostum and tell his rival that he had the situation under control.

Dean decided he had better follow the warlord, making sure any friction between Atta and Dostum didn't develop into a fight. With trepidation, they descended along the narrow road down the dune, careful to avoid the land mines that Atta warned were planted alongside.

When they reached the bottom, Dean pulled his scarf up around his nose and stood apart from the horde, hoping that no one was taking his picture.

He watched Atta and Dostum conferring, each of them gesturing to the other, *"I'm in charge."*

Dean realized that there was no way to ensure that there weren't any Taliban fighters in the crowd—men who had stayed behind in Mazar in hopes of slipping back into civilian life unnoticed. Now would be a perfect time for someone to whack two warlords at once. Dean understood the surrender as a complete security nightmare.

He could hear the swarm of cameras going off: *Click-clack, click-clack.*

Locusts, thought Dean. Little metallic mandibles just chewing at his anonymity, his safety. His mission.

Dean was relieved when the meeting ended without a dustup. But he was sorely disappointed when he saw that Dostum was going to come up the hill after all.

One of the reporters shoved a camera inches from Dean's face and he batted it away. He heard a cacophony of voices, French, German, Spanish—everybody in the world was here to report this surrender. Dean and Gold hurried back up the hill.

Dean watched as Dostum followed and occasionally stopped to point in the direction of the surrender, which nobody could actually see with any clarity with the naked eye, as the cameras clicked and whirred.

Dean thought, *He thinks he's General George S. Patton.* Part of him marveled at the general's savvy.

Dean looked up at the B-52 circling lazily overhead, sketching a hairline contrail. The bomber had been called on station by Air Force combat controller Malcolm Victors, who had set up his radio gear on the hilltop.

The Taliban sitting on the highway could look up and see the jet and know that they were being watched by some very serious firepower. Victors had plotted the Taliban's position and was ready to call in a strike at a moment's notice.

Now, thought Dean, the only thing remaining was to get the Taliban to actually hand over their rifles and surrender. Dean looked at his watch. They had planned to leave for Konduz hours ago. They were running seriously behind.

Two hours later, the surrender was over. Dostum had agreed to the Taliban's demand that they be held at the Mazar airport. While there wasn't a fence on the airstrip to contain the prisoners, they would have to run a half mile in any direction before reaching safety, by which time they'd be gunned down by guards.

From his hilltop lookout, Dean turned to Mark House, his weapons sergeant, standing beside him.

"You know what? They're not really searching these guys."

When the Taliban told Dostum they were surrendering, this declaration was accepted as it always was on the battlefield: at face value, as inviolable.

"I don't like the looks of it," said Dean.

"What can we do?"

"It's their surrender." Dean was right. The entire success of the campaign rested on the idea—and the reality—that this was the Afghans' war. To change gears now, in the middle of this confusing series of events, would be easier said than done.

It appeared to Dean that about every fifth Taliban fighter was getting a cursory pat-down from a Northern Alliance soldier.

Dean watched as the Taliban men's rifles were confiscated and piled on the bed of a large delivery truck. Soon there were hundreds of guns lying in a pile.

Back at the Turkish Schoolhouse, Mitchell had been listening to the negotiations on the radio. He hadn't thought using the airfield was a good idea, but, on the other hand, there wasn't a jail big enough to hold all of these prisoners. And then his radio popped to life.

More news was coming in. Dostum had suddenly changed his mind.

The prisoners were now being taken to Qala-i-Janghi, where Dostum had decided they would be held securely within the walls of the fortress.

No, thought Mitchell.

He recalled the weapons, rockets, RPGs, and rifles, all that ammo . . . still stored at the fortress. He and his men had never finished blowing them up.

After the Taliban surrender at the airfield appeared to be complete, Dean and his team came down from the hill and got in their trucks, waiting to leave for Konduz.

It was now late, about three o'clock in the afternoon. They would now leave directly from the surrender site.

Sitting in his idling pickup, Dean gazed over at the newly surrendered prisoners sitting only about fifty feet away in their own vehicles.

Dean was shocked to see the look in their eyes. They did not look defeated.

Two of the Taliban trucks pulled alongside Brian Lyle and Mark House, who also thought the detained soldiers looked like men prepared for battle.

Atta ordered his men to load up.

The plan was to get into positions around Konduz, a half-moon perimeter around the eastern rim of the city, about twelve miles from the center. From there, they would receive prisoners and call in air strikes to further convince the Taliban to surrender. A good plan, perhaps. But if they should be needed back in Mazar, where the six hundred prisoners were to be housed in the fort, it would be a long trip back.

At the Turkish Schoolhouse, several hours after Dean and Nelson had left for Konduz, Major Kurt Sonntag was at his desk when he looked up and saw the newly surrendered prisoners drive by.

Well, I'll be damned, he thought. *That's a first.*

The prisoners were still driving their own vehicles, headed to the fort. Sitting beside each of them, though, was a Northern Alliance soldier with a gun pointed at the prisoner's head. It looked almost comical.

Except why had they stopped outside the headquarters?

Sonntag could see the prisoners staring at the schoolhouse as if they were casing it, as if they were trying to memorize a way inside.

Upon hearing that the prisoners were heading to the fortress, Sonntag had not had a hard opinion one way or another, except that Major Mitchell had registered his displeasure about the plan. But both men knew that Dostum would be in charge of these enemy fighters, as they would be garrisoned in his quarters. Mitchell and Sonntag would stand by on an as-needed basis.

But Sonntag hadn't counted on them stopping outside his building. He realized that he didn't know if the men with their AKs pointed at the drivers were really Northern Alliance soldiers. The whole thing could be a setup for an ambush.

He called for Sergeant First Class Betz, who flew into action.

"Anybody who looks American, stay indoors," he ordered. A handful of soldiers took positions around the cafeteria, their M-4 rifles pointed at the street.

Mitchell came down from the second-floor office, worried. *Why have these guys surrendered? You just don't turn yourself in. . . .*

There were prisoners hanging on the sides of some trucks; the truckbeds were piled high with men. Mitchell couldn't believe these men were being brought straight through town. He realized they'd have to drive past the Blue Mosque and the market square, where they could bolt from the trucks.

The convoy was led by two Northern Alliance vehicles, and two or three more were bringing up the rear, but to Mitchell's way of thinking, this certainly wasn't enough to contain an outbreak, should it happen.

As the trucks sat idling, Mitchell watched as one of the prisoners was pulled from a truck by local men and disappeared behind it. There, on the ground, he was stomped to death by local citizens and Northern Alliance soldiers who'd come rushing to the scene.

The trucks moved on. Mitchell and his men inside the Turkish Schoolhouse breathed their relief.

Later that night, a civilian truck roared through a security road-block on the street outside the schoolhouse, and the nervous Afghan soldiers guarding it opened fire. The truck sped away, unhit, as more gunfire broke out around the school. Mitchell's men ran for cover. Some of the windows were shattered as machine-gun fire raked the building. No one could tell who the shooters were. The fire seemed to be coming from indeterminate directions, as if the attackers had been waiting to pounce. And then the shooting stopped.

Mitchell learned that some of the fighting might have been between Atta's and Dostum's soldiers. He didn't know why they were fighting each other. The situation worsened even more when Mitchell learned that one of Dostum's lieutenants had been blown up in a suicide attack at the fortress. As the prisoners were unloaded from their trucks in the southern courtyard, a video-camera man, working under Dostum's direction, was filming the event for posterity. The man was nervous to be surrounded by so many enemy fighters. These scowling, filthy men did not look like they wanted to surrender.

"This is a very bad situation," the man kept saying as he video-taped the scene. "A very bad situation."

One of the Taliban soldiers beckoned Dostum's lieutenant over, and when he approached, the man pulled a grenade from his clothing and pulled the pin.

The cameraman watched through his video camera as a small "pop" appeared in the background of his frame, followed by a puff of smoke. At least two men lay dead on the dirt.

In retaliation, the guards quickly herded the prisoners into the basement of the Pink House. For good measure, they threw a grenade of their own down one of the square-shaped air vents in the brick foundation. The grenade came tumbling inside from overhead and exploded.

When he heard this news, Mitchell feared that the city was spinning out of control.

At the schoolhouse, Betz had already enacted the highest security alert, ThreatCon Delta, with soldiers posted on the roof and in the windows, guns poised.

At one point, someone had pulled up to the schoolhouse in a van and then the driver had leaped out and run away, not to be found.

Betz and Mitchell believed the vehicle was rigged with explosives, and so they called in a team to dispose of the bomb. But after cautious study, the team discovered that the van hadn't been rigged with anything.

It felt like a warning: *Next time.*

When Major Leahy crawled into bed that night, he thought of the last phone call he'd had with his wife back in the States, several days earlier.

"Please be careful," she had said.

"I'm the support company commander," Leahy had said. "What could possibly go wrong?"

The following morning, the prisoners were led by twos up from the basement of the Pink House and set in the dirt courtyard on the west side of the building. Their hands were tied loosely behind their backs with their scarves.

The Arabs came first, followed by the Pakistanis, and then the Uzbeks. Spann and Olson hoped to methodically debrief each of these men in hopes of mining a precious nugget about Osama bin Laden's whereabouts, and Al Qaeda's plan for future attacks, anywhere in the world.

Before them were several hundred angry, filthy faces. As the CIA paramilitary officers marched up and down the ranks of seated prisoners, some of the men cursed the Americans: "We have come here to kill you."

Spann bent down and looked at one young man, whom he suspected could speak English. He was wearing a British-style commando sweater—there was something about the kid that piqued his curiosity and, finally, his ire.

The kid wouldn't speak.

What Spann didn't know was that the young man was afraid that if he did talk, this would mean he could be singled out as a defiant foreign terrorist.

In the basement the night before, many of the prisoners had at first worried they would be killed in the morning by Dostum's men. Even though one of his guards had thrown the grenade down into the basement, wounding several men, the prisoners had later received news that they would be able to continue their voyage to Herat, as had been agreed in the negotiations between General Dostum and Mullah Faisal.

Still, the Uzbeks, in particular, were concerned. Because they shared Dostum's ethnicity, they believed he would be especially harsh with them. He would torture them, and send them back to Uzbekistan, where the anti-Islamic government would execute them. As the Uzbeks saw it, they had nothing to lose. All of the prisoners also had shared the story of what they'd heard had happened to some Taliban prisoners when they found themselves trapped in a school in Mazar after Dostum captured the city. The place had been called Sultan Razia. The Taliban inside had been bombed and killed by U.S. warplanes.

Many in the young man's group believed that the basement was as far as they would travel in this life. They fingered the grenades and pistols they had secreted in their clothing. It might be possible to attack their guards in the morning. They would probably be killed, but they would also be martyrs.

Sitting on a carpet in the dirt, the young man facing Mike Spann wanted no part of this plan. He wondered who the fair-haired, blue-eyed man was. The officer yelled at him to hold up his head so he could take his picture. He might be a Russian. He appeared to be working for General Dostum.

Neither the blond-haired man nor his friend—a tall, broad-shouldered man in a dark *shalwar kameez*—had protested when Dostum's guards kicked and punched at the prisoners. Any man who talked without permission had been beaten.

No, he would keep his mouth shut, his head down, and pray that he would live.

Spann bent close to the prisoner. He wondered if he was British because of his European-looking commando sweater.

"What's your name," Spann snapped. "Who brought you here to Afghanistan?"

The young man stared at the ground.

Spann gazed at him, looking for an angle, some way to get the kid to talk.

"Put your head up. Don't make me have to get them to hold your head up. Push your hair back so I can see your face."

Spann raised a camera and took a picture.

"You got to talk to me," he said. "I know you speak English."

Nothing.

Finally, Spann stood up. He would stop the questioning, for now.

His partner, Dave Olson, approached with a handful of passports he'd confiscated from the prisoners. Olson wanted to know what Spann had learned, if anything.

"I was explaining to the guy we just want to talk to him, find out what his story is," said Spann.

"Well, he's a Muslim, you know," said Olson. "The problem is, he's got to decide if wants to live or die, and die here. . . . We're just going to leave him and he's going to fucking sit in prison the rest of his fucking short life. . . . We can only help the guys who want to talk to us."

Spann addressed the young man, "Do you know the people here you're working with are terrorists, and killed other Muslims? There were several hundred Muslims killed in the bombing in New York City.

"Is that what the Koran teaches?" Spann went on. "Are you going to talk to us?"

"That's all right, man," said Olson. "Gotta give him a chance. He got his chance."

Spann decided that he was finished with the prisoner.

A guard led the tall, frail-looking man back to the main group of prisoners, who were seated in orderly rows on the west side of the Pink House.

The prisoner sat. He had made it. He hadn't talked. And they hadn't killed him, yet.

Awhile later, an explosion suddenly rang out on the east side of the Pink House, filling the tall, bricked entrance that led into the basement with dust and smoke.

An Uzbeki prisoner started up the stairs when he'd suddenly produced a grenade from his clothing and tried tossing it up the stairs at a guard.

He missed and the live grenade rolled back toward him and blew up. He now lay in a heap, dead on the stairs.

Out in the courtyard, hell had broken loose.

Screaming *Allah Akbar*—God is great—some of the prisoners jumped up and swarmed over the guards, killing them instantly with grenades or weapons they pulled from their clothes.

Mike Spann started running. He headed across the courtyard, east, as if trying to get inside the Pink House itself, where he might take cover and fire at the prisoners with his AK-47. He had one magazine of ammo and two pistols—one of the handguns secreted in his jeans waistband.

The firefight probably wouldn't last long, but he could take a lot of the prisoners with him. And maybe he could hold out long enough for fire support to arrive by air, or by ground, from the Special Forces soldiers back at the Turkish Schoolhouse. It had all gone so wrong so quickly.

Three days before, he had called his wife in the States and told her he had new prisoners to process. He told her that he loved her. He missed their two daughters and their young son. He said he couldn't wait to hold the little boy. Shannon rarely expected to hear from Mike, if at all, and she was delighted to hear his voice. But this phone conversation had somehow, almost mysteriously, it seemed, left her in tears. She had hung up feeling that something terrible was going to happen to Mike.

Before falling asleep, she had prayed, "God protect Mike from having to see too much craziness today."

Now, as Mike ran, he felt arms tackle him from behind, and he fell in a storm of flesh. Fists and feet rained on him. He reached to one of his pistols and managed to shoot several of his attackers. But still they punched and kicked.

One of the prisoners walked up and fired twice, point-blank, at Mike Spann.

Mike Spann was dead.

At the sound of the grenade exploding in the stairway, Dave Olson had turned to see what was the matter.

After conferring with Mike about the prisoners, he had decided to return to Dostum's headquarters in the north half of the courtyard, and he'd reached the middle wall when he saw Mike fall under a crush of bodies, all punching at him.

He turned around. He was shocked to see that all of his guards were dead.

Olson fired his pistol as some of the prisoners ran at him. Others were running across the courtyard and picking up weapons that had been dropped by the dead guards. They started firing at Olson.

One man ran straight at him until Olson dropped him with his pistol. He tried firing again but he was out of bullets.

So he ran.

He sprinted across the courtyard, scanning left and right, seeing up on the fort walls that some of Dostum's and Atta's soldiers had scaled the heights at the first sound of gunfire. Now they were shooting down at the rioting prisoners. The volume of gunfire was deafening.

Olson was joined by one of Dostum's soldiers running alongside him, both of them pumping their knees high and straining for speed as bullets whizzed past their heads.

Olson was worried they'd be shot by the men up on the walls if they mistook them for Taliban fighters. He relayed the fear to his new companion, saying they needed to get out of this courtyard, quickly.

They headed for a wall, hoping it might screen them from the gunfire.

"Allow me," said the man, jumping over the barrier first and making sure that no enemy fighters were hiding behind it.

"All clear!" he yelled back.

Olson couldn't believe the man's daring. He shuffled over the chest-high wall and continued running.

Reaching Dostum's headquarters building, he bolted straight inside and up the stairs, where he ran smack into a short, wiry, blond-haired man who was standing on the stair landing.

Arnim Stauth, a German television reporter, looked up in surprise. He had taken cover on the landing when he heard the gunfire erupt several hundred yards away, in the southern courtyard. He had no idea what was happening in the fort. He had driven to Qala-i-Janghi early in the morning from Mazar to film some International Red Cross workers as they monitored the well-being of the newly arrived prisoners.

He looked at the tall, dark-bearded man now standing before him. He appeared out of breath, confused, dazed even. The man had been holding his pistol by the barrel, instead of the grip, as if he'd forgotten how to carry it, and he kept trying to put it back in its holster and missing.

Stauth asked him what had happened.

Olson answered in a dry, shaky voice. It was probably now that it occurred to him that his covert status as a CIA paramilitary officer had been suddenly, and permanently, blown.

He explained to Stauth that the prisoners had rioted in the southern courtyard, and he thought one American was dead. He wasn't sure. They moved to another room in the headquarters, deeper inside, where Olson surprised Stauth by asking if he could borrow a phone.

Stauth guessed, because the man was wearing a nonstandard "uniform"—a long *shalwar kameez*—that he was a CIA officer, and he found it hard to believe that he wasn't carrying some means of communication. He offered Olson his satellite handset.

Olson now had another problem. He didn't have a number to call at the Turkish Schoolhouse.

So he dialed the U.S. Embassy in Tashkent, Uzbekistan, which patched him through to U.S. Army Central Command at MacDill Air Force Base in Tampa, Florida, which then routed Olson to the Turkish

Schoolhouse. The urgent call had traveled some 20,000 miles before it arrived on the fifth floor of the CIA's makeshift headquarters in Mazar-i-Sharif.

There, a gruff, bushy-bearded fellow officer named Garth Rogers received the news. The seemingly bulletproof shield that the CIA had been traveling under during the entire mission had been pierced. It was almost inconceivable that this could be true.

Rogers hung up, shocked.

Back at the schoolhouse, Mark Mitchell had been receiving reports throughout the late morning about steady but unexplained gunfire occurring somewhere in the city. Did he know anything about this? No, Mitchell said. All was shipshape, so far as he knew.

Earlier that morning, Admiral Bert Calland, along with U.S. Army medical surgeon Craig ("Doc") McFarland, had been touring area hospitals and surveying their suitability. McFarland had been appalled by the conditions. The outdated X-ray machines were barely powerful enough for the simplest tests, and the anesthesia equipment looked to be 1950s vintage. Almost none of the officers had antibiotics or bandages. McFarland saw he would have his work cut out for him as the chief medical officer in Mazar.

As they returned to the Turkish Schoolhouse, McFarland thought he heard heavy gunfire coming from the vicinity of Qala-i-Janghi. He found this strange. He knew Mike Spann and Dave Olson had gone over there to question prisoners.

As soon as he got to Mitchell's office, he asked about the gunfire. He told Mitchell that it had sounded like a real battle was going on at the fort.

Kurt Sonntag was getting the same kind of queries by radio from Special Forces soldiers traveling in the city. "Hey, there's gunfire out to the west. What's going on?"

"Does it sound like anything?"

"No, just a lot of intermittent fire."

Finally, Sonntag got a call from an officer at K2, in Uzbekistan. "We've got a report that there's, you know, two Americans dead in Mazar."

Sonntag said he hadn't heard anything about this. He thought it must be some mistake.

Just then, Mitchell got a message that he had an urgent visitor wanting to see him. Into the office burst one of Dostum's friends, a civilian named Alam Razam, dressed in an ill-fitting suit coat and tie. Razam was in a panic.

"There's been a terrible incident," he said. "You must come. I think there are dead Americans. The prisoners have seized control of the fortress. You've got to come right now!"

Kurt Sonntag had entered the office with Razam, and Sonntag and Mitchell looked at each other, disbelieving.

Mitchell and Sonntag struggled to maintain their composure. Mitchell had lived in the fort longer than any of the new arrivals to Mazar, and knew most of the maze of rooms and passageways.

"Look," he said, "I'll volunteer to take some guys over there." He wondered if maybe just a few unhappy prisoners were causing a big, loud disturbance.

Several minutes later, more bad news came, though this message was grimmer. Garth Rogers, the CIA officer, came down from the fifth floor and announced that Mike Spann was probably missing or dead.

Mitchell saw Rogers had taken the news hard. Mitchell respected him. Rogers wore a Harley-Davidson ball cap and wraparound sunglasses, and resembled an outlaw biker more than a veteran CIA officer. Mitchell wanted him in the search and rescue team. Rogers readily agreed, suggesting they should leave immediately.

Sonntag interrupted him. "Let's make a plan," he said. "We have to be careful here." He saw Rogers was agitated and itching for a fight.

Sonntag dragged a chalkboard into the operations center, and he and Mitchell sketched out a floor plan of the fort. They marked Spann's last known position around the west side of the Pink House, set in the southern courtyard.

They agreed their mission was to find Dave Olson, who was reportedly still alive, and to search for Spann and retrieve his body if, in fact, he'd been killed.

After they were done, Mitchell squinted out the window. They had about three hours of daylight left, at best. They had to hurry. Once darkness fell, it would be impossible to tell friend from foe in the fort. And the Taliban would likely take the opportunity to escape under the cloak of night.

Sonntag estimated about 50 Northern Alliance fighters were positioned at the fort, with maybe another 100 scattered in the city who would be available for a fight. Among their own forces, Mitchell had 8 men he could count on as trigger pullers. A thin string of men to confront a fortress filled with 600 enemy fighters.

Mitchell rustled up the 8 British Special Boat Soldiers, who had arrived in Mazar the night before at the harried landing zone. He didn't even know these new soldiers' names, except for one, Stephen Bass, a U.S. Navy SEAL attached to the SBS in an intergovernment exchange program. Complicating matters, the SBS would not be allowed to fire their weapons at the enemy unless they were attacked first. This was because they had not received their orders for rules of engagement, which instructed them when and where they could fire at enemy fighters. Mitchell reasoned that self-defense would be a reasonable justification to shoot back, and he was pretty sure they were about to drive into a serious firefight.

In addition to Garth Rogers, Doc McFarland, and the eight SBS personnel, the fifteen-man rescue force included two Air Force lieutenant colonels, attached to the Defense Intelligence Agency, who had happened to be visiting Mazar. Mitchell considered these men to be "military tourists"—a name given to officers who like to fly into war zones and sightsee, even though they are not officially part of the action. Now these Air Force officers were going to get a close-up view.

Still, Mitchell was glad to have them along. One of the men spoke Russian, which could be helpful. All of the interpreters working at headquarters had left with the rest of the fighting force for Konduz, and Mitchell realized he didn't have any other way to communicate with the locals. He also brought along two trusted sergeants: Pete Bach and Ted Barrow. He would need them both. Barrow was an excellent weapons specialist, and the steadfast, good-natured Bach would run communications.

When he was in the regular Army fifteen years earlier, Bach had been stationed in Alaska, and he'd trained to get in shape for the Special Forces qualification course by asking his wife to drive him far into the wilderness, and then running home for twenty miles through hip-deep snow.

None of the men had worked with each other as a combat team, but they did share common training. Mitchell hoped this expertise would be enough of an edge to help them survive in the fighting.

Having solved the problem of fielding a team, he then had to figure out how they would travel to the fort. In addition to interpreters, Nelson and Dean had also taken almost all of the headquarters vehicles to Konduz.

Mitchell had to dig through a pile of car keys, and by trial and error, find a spare vehicle that would actually start.

Finally, thirty frantic minutes after they had received the call that Mike Spann was missing, the team was heading to the fort.

Mitchell drove at breakneck speed.

The Land Rover's suspension was mushy, and the heavy vehicle drifted over the road. Mitchell had trouble shifting with his left hand and steering with his right in the European-style interior, and he felt that any minute he might careen out of control.

The midday traffic in Mazar was heavy, and he was forced to weave in and out of lines of cars, honking the horn constantly, and barely slowing down at intersections.

Rogers rode in the passenger seat, calling out approaching obstacles.

"Goat cart!"

Mitchell: "I see it."

"Donkey cart!"

"Okay."

"Man with a load of bricks!"

"Got it."

As Mitchell drove, he worried that they'd be ambushed by prisoners who'd escaped from the fort and who were lying in wait, parked in vehicles of their own at some intersection. As they approached the bazaar in downtown Mazar, a tractor-trailer pulled out in front of Mitchell and he had to slam on the brakes. Traffic backed up behind him and he nervously checked the rearview mirror. Rogers scanned the area in front of them.

This is it, thought Mitchell. *Now we're going to get whacked.*

After several tense minutes, the long truck finally pulled ahead and Mitchell shot off again, tires squealing.

• • •

As the urgent call went out about the fighting in the fort, Dean in Konduz got word that Mike Spann was missing and that Dave Olson was trapped somewhere inside the fortress walls. Dean's satellite phone rang: it was someone back at the Turkish Schoolhouse in Mazar asking him if he had all of his men with him.

"Just checking," said the caller. He said that the details were sketchy, but definitely something bad was going down in the city. The call sounded like it was coming from the fort.

Dean thanked the caller and next got Brad Highland on the radio back in Mazar. He, along with teammates Martin Graves and Evan Colt, had stayed behind to run logistics from the safe house.

"Listen, man," said Dean. "Get yourself ready to move to the Turkish school. Don't mess around. The city is heating up."

Highland began packing up sensitive items—maps, radios, reports.

Dean next got Mark Mitchell on the radio. Like Dean, Mitchell at that point was just grasping the enormity of the fight raging at the fort.

"I've got guys inside Mazar," said Dean. "You need to know they're there before you get crazy." Dean meant that he didn't want his men getting fragged in a bomb strike aimed at enemy fighters.

Then he asked Lieutenant Colonel Bowers if they would be redeployed to Mazar to take part in the fighting. Nelson, who was standing nearby, wanted to know, too.

The question was a reasonable one, given how closely the men had worked with one another. But Bowers, worried by the fact that the situation in Mazar had suddenly spiraled out of control, now feared that the same could happen to them at Konduz. He reminded his men that they were about to take charge of thousands of angry, armed Taliban and Al Qaeda prisoners. No, they would not be leaving immediately for Qala-i-Janghi.

Nelson wanted to get in the fight at Mazar. Mike Spann had been his friend. Now he wanted revenge.

Approaching the fort, Mitchell could hear the *rat-tat* of gunfire and the sonorous hiss of RPGs launching inside the tall mud walls.

Smoke trails whizzed upward from the interior and corkscrewed overhead in the blue sky before smashing into the farm field around Mitchell's Land Cruiser.

Several hundred prisoners had climbed out onto the walls and were firing down into the entrance drive. Rounds were hitting around the vehicle as Mitchell ran the gauntlet and came to a rocking halt inside the fort walls, next to a guard shack.

Cowering near the shack were forty Alliance soldiers, all of them terrified by the gunfire. Mitchell ignored them. They'd have to come to their senses and reorganize as a fighting force on their own.

Time reporter Alex Perry had been driving near the Qala-i-Janghi Fortress when he saw something out the taxi window that made him want to stop. He tapped the driver on the shoulder and said, "Let's drive over there."

He pointed to black plumes of smoke drifting above some distant, towering walls.

The fighting had been raging for forty-five minutes when Perry arrived, pulling right up to the front gate, oblivious to the danger inside. He knew something bad was happening, something big . . .

He was standing beside the taxi listening to the gunfire when a white Land Cruiser came up the gate, running through a hail of fire pouring down from a high wall above. A large group of armed men piled out and one of them hurriedly walked up and asked Perry what he knew about the fighting inside. *These guys don't know what's going on*, thought Perry. He realized he was the first person to have arrived from outside the fort after the fighting had started.

Perry told them that the greatest volume of fire seemed to be coming from the southern courtyard, which was to their left, on the other side of the ten-foot-thick mud walls.

The soldier thanked him and trotted off. Perry was left wondering what he should do next. This was his first assignment as a war correspondent and he understood he had hit the jackpot. He didn't know if this made him lucky or doomed.

Inside the guard shack by the front gate, a set of cement stairs led up to the fortress wall. Mitchell ordered Stephen Bass and the rest of the SBS personnel to the top. He wanted them to spot where the Taliban fire was coming from within the huge southern courtyard.

They charged up and set machine guns on tripods behind the long, four-foot-high mud wall. Swinging the gun barrels, they started raining fire into the southern courtyard. Mitchell saw they were killing a lot of the prisoners.

However, the Northern Alliance soldiers who'd arrived at the fort and now lined the walls were another matter. They were firing into the courtyard by holding their battered AKs over their heads and doing something the Americans called "spraying and praying." This method of fire was wildly inaccurate. It was difficult to tell if any of the rounds were hitting anyone. Oddly enough, that was the point of this method of fire. According to their beliefs, which Mitchell had studied, if one of the Afghans killed a man on purpose—that is, if he deliberately aimed and shot someone—then the dead man's soul would become the killer's responsibility. If, on the other hand, a man was hit by accident—by an errant bullet that had been fired wildly—then it was God's will that the man should have died. Mitchell hoped to hell God was looking down and urging the Alliance fighters on.

Mortars kept popping up from some horse stables built against the wall below him, and Mitchell couldn't pinpoint the exact position of the launches. He considered jumping onto the stable roofs and creeping along them until he found the mortar position. He could chop a hole and drop a grenade inside. But he then realized that the jump itself would probably send him straight through the thick mud roof and land him in the stables, a clear suicide mission. Reluctantly, Mitchell called off the plan.

Sonntag at the Turkish Schoolhouse had arranged for bomber aircraft to be over the fort by 4 p.m. Mitchell looked at his watch. He had only one hour to rescue Dave and look for Mike before finding cover in preparation for the bomb strike. Mitchell radioed to Sonntag that he was moving along the north wall of the fort and heading to Dostum's headquarters, Dave Olson's last known position.

He next told the communications sergeant to get comms up and track the status of the incoming air support. They were going to need those bombs. It was already clear to Mitchell that there was no way he and his fourteen men could match the volume of gunfire pouring out of the fort. They were going to have to bomb the Taliban into submission.

As they crept along the base of the wall, heading to Dostum's headquarters, Mitchell ran into one of the warlord's subcommanders, a frazzled-looking man covered in dust who had clearly borne the brunt of the battle's beginning.

Mitchell didn't have a radio that operated in Dave's frequency. The man offered his walkie-talkie. "Here," he said, "I have been talking with Baba Dave." Mitchell called Olson and soon he heard the CIA officer's voice.

"They've exploded some kind of bomb," Dave said. "They killed the guards, and they control half of the fort."

Despite the bad news, Mitchell was immensely relieved to hear Olson's voice. He was alive.

"What about Mike?" Mitchell asked.

"The last I saw him," said Dave, "he was fighting hand-to-hand." He paused. "I don't think he made it."

"We're coming for you," said Mitchell.

He and the team pressed ahead, wary of being shot at by Taliban who might be positioned above them on the wall. They hadn't gone far when the position they had just left exploded in a shower of fire and dust—a mortar had fallen directly on the spot. Mitchell didn't believe this was targeted fire, but that wouldn't matter much if you got hit.

Sooner or later, however, the Taliban would likely get lucky. Mitchell remembered the piles of weapons in the open Conex trailers. Now he heard reports from the Brits at their machine-gun position along the wall about 100 yards behind Mitchell that the Taliban had found the makeshift armory as well as a garage-size mud building located near the middle wall, and could be seen lugging all kinds of weapons—rifles, ammo, mortars, even BM-21 rockets—across the southern courtyard to intensify their attack.

As fast as they could load the tubes, the prisoners were launching mortars, trying to bracket the long, broad facade of Dostum's headquarters. And they *had* started to hit the building. Chunks of plaster and concrete were careening off the wall and spinning into the northern courtyard.

After about twenty minutes of scurrying along the base of the fort, Mitchell and Rogers now stood looking up at a sixty-foot-high parapet at the northwest corner. (The remaining team members had dispersed along the wall, with Sergeant First Class Bach keeping in radio contact as Mitchell and Rogers moved ahead.)

At the base, Mitchell was joined by one of Dostum's trusted men, Commander Fakir, who had made his way to the fort shortly after the fighting had started. Fakir had ridden against the Taliban in horse charges during their battle en route to the city, and he faced this next challenge of breaching the fort wall with the same fearlessness.

Digging the toe of his scuffed shoe into the mud wall, Fakir began scampering up, securing each toehold with a thrust of his leg. He nimbly made it up to and over the parapet that ran along the top of the walls of the entire fort, and stood.

He then unwrapped his long green turban and threw one end down to Mitchell, who was standing about thirty feet below him.

Fakir gave the cloth a shake: *Hurry, grab hold.*

Mitchell reached up and clenched the fabric, knotting it around his hands, and began climbing as Fakir hauled on the other end, bracing himself against the parapet.

Mitchell would later recall that he felt like he'd entered a violent telling of the fairy tale "Rapunzel." He quickly reached the top. Fakir repeated the process with Rogers, and the three men climbed the rest of the way, another thirty feet above, scrabbling on all fours up the crumbling wall.

Getting down on his belly and studying the scene, Mitchell could see bullets peppering the front of Dostum's headquarters, which only several days earlier had been his home.

He picked up the radio and called Dave Olson. "Where are you?"

Olson and Mitchell figured out that he was somewhere in the offices directly below Mitchell, about thirty feet down, through mud and timbers.

"Are you hurt?" asked Mitchell.

Olson said he was fine. But he said he was concerned about his security. RPGs were smashing into the balcony, rocking the place. The Taliban had figured out how to fire the long, heavy BM-21 rockets without the aid of a standard launchpad. They were aiming these like giant bottle rockets at the headquarters building. The explosions shook the fort's walls.

Because of the volume of gunfire, Mitchell decided that he couldn't move any farther without risking getting shot. He would have had to dash about 150 feet across the hard-packed roof of Dostum's headquarters, run down a long ramp leading to the floor of the north courtyard, and then rush up a set of stairs to get inside the building itself. He thought the run would be too risky.

About three hundred yards away, in the south courtyard, he could see Taliban prisoners darting out from behind some blown-up vehicles to fire their weapons at the building. The prisoners hadn't spotted him, he hoped. But he was about to let them know that America's bombs had arrived.

He scanned the fortress floor for any sign of Spann's body, a glimpse of denim jeans or a black sweater, a shock of blond hair among the dozens of dead bodies littering the southern courtyard. Mitchell didn't want to drop any bombs if Mike was still alive in the courtyard. But he couldn't see anything that resembled his friend. He decided to call in the air strike. With any luck, this would quiet the enemy gunfire and allow them to reach Olson, still hiding inside Dostum's headquarters. As a large, well-armed group, they then could try to make their way safely out of the fort.

Bach called Mitchell on the radio and informed him that he had an inbound F-18 jet, ready to drop.

Mitchell called Olson and told him. "When the bomb hits, get over the north wall. Use the explosion for cover."

Bach then came back and informed Mitchell that the pilot was insisting that he drop at an altitude of 20,000 feet.

Mitchell was incredulous. At that height, the pilot might hit anything.

"We're Danger Close," he told Bach, meaning that he, Rogers, and Fakir were located within 300 yards of the target, in peril of being injured or killed by the explosion. He needed the pilot to fly as low as possible to minimize this risk.

Bach came back on, "He'll go to eighteen thousand feet." Bach wryly explained that the pilot feared getting shot at by ground fire.

Shot at? thought Mitchell. *I can tell him about that.*

News of the ongoing fight at Qala-i-Janghi continued to reach Nelson's team at Konduz. Nelson was still itching to get back to Mazar. Diller was of a different mind. There was nothing they could do for Spann—they couldn't help him now. But what boggled Diller was hearing that Dean's team members hadn't been asked to go into the fort and smoke these dudes out.

Diller believed a trained-up team like Dean's would have been a useful asset. In addition, another team, run by a Fifth Special Forces Group soldier named Don Winslow, had recently entered the city, ostensibly to link up with the Hazara warlord General Mohaqeq, a plan that hadn't yet taken effect.

Winslow was hot for the fight. He had started carrying a camping hatchet on his pants belt as a sign of strength, explaining to Mohaqeq's violent, aggressive warriors that he used it in battle to hack men to death. The Hazaras thought the American carried some serious mojo. Yet Winslow wasn't going into the fort, either. He was stationed at the Turkish Schoolhouse as part of the defensive force protecting the headquarters.

In fact, members of the other teams hadn't been tasked with fighting in the fort because the fight had erupted quickly, and as events kept rolling over Mitchell and Sonntag, they believed that they were winning. Further, the connecton between Mitchell's command and control unit and Nelson's and Dean's team, had been ad hoc at best and nonexistent at worst, as evidenced by the accidental bombing of Dean's men by Nelson's team at the Sultan Razia school earlier. In short, Mitchell correctly believed that he had under his command, in the faces of his own men, the firepower and experience to put down the insurrection. If he were to be forced to storm the fort, then he could call the reserve team members. But for now, he would try to kill the men inside with bombs—something far less risky than entering the enormous fortress and clearing it room-by-room, hand-to-hand, guns blazing.

Looking at the Pink House, Mitchell figured that the building held the largest concentration of prisoners. Using his range finder, which shot a laser at an object and then displayed its distance in a viewfinder, he calculated the building's GPS coordinates.

He relayed these to Bach, who then radioed them to the F-18 pilot.

"Do you have a visual?" Bach asked.

"Roger, I see the building," said the pilot.

Mitchell cleared him to unleash the bomb.

The explosion rocked the ground near Mitchell. When the smoke and dust cleared, he glassed the area with his binoculars and saw that the bomb had fallen about 100 yards away from the Pink House.

Mitchell asked the pilot to correct the coordinates, and the second bomb hit the Pink House squarely, collapsing a portion of the roof and sending chunks of mud and concrete tumbling down the interior stairwell, suddenly illuminating a portion of the main chamber. Inside, the several hundred prisoners who had retreated to the building cowered in its corners. The room filled with smoke and dust.

Even more prisoners were packed into the narrow hallway and on the stairs leading up to the outside. Terrified by the explosions, they braced for another attack. Some of them had begun to think of surrendering, walking out of the basement and turning themselves over to Dostum's men. They might be executed, but surely their chances were slimmer if they remained in the basement.

The foreign Taliban in the group shouted at the huddled men that no one would surrender. They would fight in the fort to their deaths.

Mitchell ordered three more bombs dropped on the building, followed by two others that exploded along the southeast wall, where some of the prisoners had sought refuge in the horse stable.

Mitchell was dismayed by the effect of the strikes. The gunfire rising from the Taliban positions scattered around the courtyard only seemed to increase after each explosion. He realized these guys were not going to give up easily.

In the intensity of the fight, he hadn't noticed that darkness was descending on the fort. The vastly outnumbered American team would have to leave soon or risk being ambushed by prisoners as they made their way back to the gate and their vehicles.

Mitchell tried raising Olson on the radio but this time he came up cold. Either he'd been wounded or killed by gunfire, or by the Danger Close explosions of the air strikes. Or maybe he was simply out of contact for less serious reasons. Either way, Mitchell now couldn't be sure of Dave's location. He realized he had to go find him.

"We've got to get out of here," cautioned Rogers.

Mitchell nodded. He called Bach.

"Drop one more," he said. He ordered this one on the Pink House. They would make a dash over the wall under cover of the air strike.

As soon as the explosion went up, and along with it a scorching fireball, they slipped down the back wall to the desert floor. Once there, Mitchell and Rogers stripped off their heavy load-bearing vests. Keeping their pistols and M-4 rifles and a radio, they were able to scale back up the wall to a window they'd discovered leading into Dostum's headquarters.

Mitchell slid inside and yelled for Dave, running along the hallway, swinging doors open, as the walls started shaking under the impact of more RPGs.

Mitchell shook his head, disbelieving. Even after all the firepower he had dropped on them, the prisoners were still able to pound at the building.

In one hallway, he stumbled upon a group of frightened Afghan men who'd been trapped inside when the shooting started. Speaking in Dari, and pointing frantically down the hall, they pantomimed that Olson had managed to get out down a back wall when the bombing began.

Mitchell was relieved, but then decided that he needed to check for himself. He couldn't trust Dave's safety to secondhand information like this. He and Rogers searched more rooms, but still they came up empty-handed. At the main gate, they rejoined the rest of the team, where Mitchell asked if anyone had seen Olson.

No one had.

Mitchell still wasn't sure that his friend had made it out of the fortress alive, but at the same time, he could hear more gunfire growing in intensity inside the fort. The Taliban were standing just on the other side of the thick mud walls, about 150 yards away.

Even the nearby village of Deh Dedi had fallen into chaos. Some of the homes abutted the fort's southeast quadrant, and the settlement had become a free-fire zone, with Northern Alliance soldiers and escaped prisoners roaming from house to house, shooting at each other. Most of the village's inhabitants had fled to Mazar.

Mitchell realized it was time to leave the area. And quickly.

On their way back to the Turkish Schoolhouse, Mitchell learned by radio that a quick reaction force (QRF) of twenty-four Tenth Mountain soldiers was landing at a dirt airfield about a mile away, nearby Deh Dedi. Mitchell thought the QRF's arrival was an excellent idea. These soldiers could provide extra security at the school-house, and they had trained to act as a search and rescue team.

Sonntag had requested their deployment when, at one point during the course of air strikes, he'd been unable to reach Mitchell by radio. Sonntag worried that maybe he'd been captured, or killed. (In fact, Mitchell had only switched radio frequencies from Sonn-tag's to a "strike" frequency needed to talk with pilots.) Believing that a majority of his security force at the schoolhouse might have been eliminated, Sonntag had called K2 and asked for the QRF.

Sitting on his cot in his cramped tent at K2, Tenth Mountain Division soldier Eric Andreason, nineteen, from Florida, had jumped up excitedly at first word that he and his buddies were going into battle. They'd hurriedly packed and soon were standing on the tarmac, as Nightstalker mission commander John Garfield, who, a little more than a month earlier, had helped insert Nelson's and Dean's teams into the country, addressed the group of twenty-four young men.

Garfield was concerned by the worried looks on their faces. These men were some of America's most eager, highly trained infantry soldiers, but they seemed unsure of being inserted in the middle of a raging firefight. Some of them, Garfield saw, were paralyzed with fear.

Garfield, perpetually witty and affable, didn't help matters when he addressed the young crowd just before takeoff. He was keyed up and not thinking (he would later realize) when he said, "Many of you won't be coming back from this mission. We're going in hot. And there will be shooting."

For Garfield, this was a way of mitigating the terror of approach-ing combat. He remembered what it felt like to be shot at the first time, when he was a young, regular U.S. Army soldier, and then later as a covert Delta Force operator.

The first time, all he could remember was being angry. And then he managed to focus on simply shooting his weapon. But for the Tenth Mountain soldiers, Garfield's speech felt like a pronounce-ment of doom.

Some of them already in the helicopter started throwing up. *What have I done?* Garfield wondered. He watched as one soldier broke away from the helicopter. Garfield could tell he wasn't running *to* anything—he only wanted to get the hell away from the bird. Garfield, who was maybe fifteen years the kid's senior and out of shape, started jogging after him.

He caught up with the tall, lean soldier, and still running alongside him, talked calmly. "Hey, where are you going?"

The soldier looked around, as if finally realizing what he'd done. "Nowhere," he said.

"Then do you mind if we walk?" Garfield laughed.

"No, that's okay."

The kid stopped running. Garfield was glad. He was out of breath. He put his arm around him. He knew the young man was not a coward; he was just scared, and he'd given in to that emotion.

"Your men need you," said Garfield. "They need you to be with them."

He pointed at the three large Chinook helicopters sitting on the tarmac, their long rotors spinning.

"How about we walk back and take charge of this situation?"

The soldier nodded.

Once at the helicopter, he strode up the ramp and spoke confidently to his fellow men about the need for all of them to psych themselves up for the battle. He took his seat along the wall. Garfield was proud, and relieved.

This was going to be a hairy landing in an active combat zone. He needed these guys to be on their game. He still believed that many of them wouldn't make it out alive. Including himself. Garfield decided he would go down shooting.

Now, at the Deh Dedi airstrip, Mitchell helped load the Tenth Mountain soldiers into vehicles and the convoy picked its way along the dark streets to the Turkish Schoolhouse.

News of the fighting at the fort had driven many of the city's inhabitants indoors, fearing that the Taliban would soon be returning.

At about 6:30 p.m., Mitchell stepped down from the Land Cruiser before it had even stopped outside the schoolhouse. He still didn't know if Dave Olson had made it out of the fort alive. He feared the worst. He ran inside and asked the first soldier he met about the CIA officer.

Just then, he looked up. Olson was coming down the stairs. Mitchell relaxed. Olson, it turned out, had arrived at the schoolhouse about a half hour earlier. Mitchell asked again about Mike Spann.

Olson's face grew dark. "I don't think he's alive," he said.

Mitchell felt a wave of exhaustion sweep over him. On the second floor in the operations center, he and Sonntag convened an emergency meeting to review the day and plan their next attack.

They knew that about 50 of Dostum's men had been killed so far, and another 250 had been wounded. An estimated 300 Taliban, from the initial group of 600 prisoners, lay dead inside the fort. As Sonntag pieced the situation together, he believed that the prisoners had rioted at Qala-i-Janghi in order to force the Americans to rush to the fort's aid, leaving the city unprotected and vulnerable to attack from the Taliban soldiers now stationed to the north. The American forces would be caught from behind and decimated. The Taliban forces would then join up, re-arm themselves from the armory kept in the Conex trailers, and sweep east to Konduz, trapping Bowers, Nelson, and Dean between the large Taliban and the Al Qaeda armies dug in at Konduz.

After the meeting, another frightening bit of intelligence came into the schoolhouse, this time from one of General Mohaqeq's Hazara soldiers. The news was that a gathering force of Taliban soldiers was preparing to attack the city.

This information seemed to corroborate some intelligence Mitchell had received earlier in the day, that at least several hundred more Taliban soldiers had escaped from Konduz and parked themselves northeast of Mazar, about fifteen miles from the schoolhouse.

When he'd heard this, Mitchell immediately radioed Sonntag. Sonntag was already on the case. He'd ordered an AC-130 Spectre gunship to fly above the enemy troops and lock them in place at their desert location. The plane's crew had orders to shoot if the enemy force moved on Mazar.

Mitchell and Sonntag could still hear gunfire and explosions coming from the western side of the city. Mazar-i-Sharif was no longer safe. All of them resolved that they would not retreat.

The following morning, November 26, Mitchell was lying on his belly on top of the middle wall, on the fort's eastern edge. The wall, dividing the fortress into its northern and southern courtyards, gave him an excellent vantage point. Looking down and to his right, some 200 feet away, he could see three Taliban fighters, in ragged gray and green blouses, their black trousers flapping as they ran, firing their AKs and spraying the northern walls. One of them dropped a mortar down a tube set up on a tripod.

Mitchell heard the *thwoop* sound and watched it launch.

It landed wide, but pretty soon more mortars were dropping around him and he feared the guy was going to get him bracketed and smoke him.

Looking to his left, across the northern courtyard, Mitchell saw Captains Paul Syverson and Kevin Leahy, along with Sergeant First Class Pete Bach, Air Force combat controller Malcolm Victors, and Sergeant First Class Dave Betz 300 yards away, also taking mortar fire. This was coming from another position that Mitchell couldn't locate.

Bach was on the radio talking to the F-18 pilot overhead, trying to get him to drop his bomb. Betz, Leahy, and Syverson were crouched down and firing their M-4s into the southern courtyard.

Mitchell lowered his binoculars. *These guys just won't die.* He had no idea how any man could have withstood yesterday's bombardment. As he looked down into the fortress, at the blood soaking the dirt floor, he saw hundreds of suicidal Taliban and Al Qaeda fighters firing rifles, rockets, throwing grenades.

If these fighters escaped the fortress, if they broke through these walls and escaped into the scrum of Mazar-i-Sharif's streets, it might take the entire winter to recapture the lost ground. If they escaped, Mitchell reasoned, it might be impossible to hold off the larger enemy force.

He thought of his daughters, his wife, whom he loved to tell, "I may not be pretty, Maggie, but I'm dependable." He wanted to get home.

Up on the northeast parapet of the fortress, Paul Syverson heard the rounds hit the mud wall behind him, generating little puffs of dust, as if the wall were being stung by hundreds of invisible bees. The pilot overhead was demanding that Syverson relay his position coordinates, which neither Bach nor he wanted to do. Bach keyed the mike so the pilot could hear the gunfire.

"Hey, would you hurry this up!" he yelled into the radio.

A mortar round landed not fifty feet away. The explosion was loud and Bach yelled, "Goddamn it!" He keyed the mike again and said, "You hear that?"

Both he and Syverson were worried that if they gave their position to the pilot, he'd mistake it for the target's.

"Repeat," said Bach. "You will not drop without coordinates!"

"Roger," said the pilot. "I need your position before I drop."

It had already occurred to Syverson that they could be overrun.

Down below in the southern courtyard, scurrying in the grove of poplar trees, were several hundred Taliban bent on killing him. None of the enemy wanted to be taken alive. They had all decided to go down shooting. Syverson had heard the stories about what the Taliban did to its prisoners. Being strung up by your guts from the barrel of a tank was not preferable to emptying your 9mm at some hardened Pakistani before he killed you.

Syverson finally nodded at Bach and said, "All right, give 'em the coordinates."

Bach relayed their position. Ten thousand feet up, the pilot typed into the keypad the target designation for the JDAM.

The long, heavy bomb awoke, its electronic brain—a GPS— whirring with this newly inserted information. It dropped from the jet and began flying to its target.

Mitchell heard the countdown over the radio. He closed his eyes and opened his mouth to relieve the overpressure that he knew the bomb would create on detonation.

The mud floor shook beneath him. He felt the air being sucked out of the fortress. The noise was so intense that a thick silence draped over him, and in this quiet he opened his eyes and was staring directly at the southern courtyard, where the mortar position had been.

It was still there. The courtyard was untouched. Mitchell turned to see the mushroom cloud blossoming over Syverson's position. He later learned the reason for the errant hit: one of the pilots overhead had transposed two digits of the intended target's coordinates.

At the last second, Syverson had seen the bomb coming in, accompanied by a sound like the crack of dry kindling. The air flashed. The bomb flew straight into the mud parapet that he and the four other soldiers were sitting on, penetrated, and kept going.

Buried some twenty feet in the dry, caked mud wall, it finally exploded, throwing a nearby Northern Alliance tank high in the air, flipping it. The tank landed on three Afghan soldiers and they lay dead beneath its green iron turret, their thin brown legs sticking out from underneath.

Syverson himself had been thrown airborne, punching and kicking within a gray cloud of dust. He didn't feel himself land, but he heard it—a *whumpf* that set every bone in his body ringing with a hollow, angry tune.

Sergeant First Class Dave Betz shook his head after he landed, trying to clear his thinking. He'd been catapulted some sixty feet, and at the end of his arc he felt he was falling down a cool, dark cave. He smelled the dead animal stench of burnt hair and realized it was his own. The blast had scoured his face. Miraculously, only his eyelashes and mustache were gone. His nose hurt where his Oakley sunglasses had melted; they, too, had disappeared. He looked at Bach and laughed—and even that hurt. Bach's hair was sticking straight up. His face was powdered completely white with dust; only the whites of his eyes were visible, wet, with blood running from the corners in thin lines. Bach's ears were bleeding as well. Betz reached up and touched his own and found blood there, too. Betz was amazed at the impact he had endured. He turned and looked back up at the parapet, where he'd been standing when the bomb landed.

Then Betz looked down at his waist. The lower half of his body had disappeared. He panicked and started digging at the pebbles and sand. His legs emerged, then his knees. The blast had buried him up to his belt in rubble. He realized that even the shirt on his back had been blown off.

Betz looked around and couldn't find Syverson. He heard Leahy moaning. He turned and saw Kevin in a heap at the bottom of the fort wall, which had collapsed inward in a pile of pebbles and sand. Kevin was twisted like a doll. He looked as if two fingers had reached down from the sky and shaken him until he broke. *Oh, man, not Kevin.*

Kevin Leahy was his buddy. Betz looked after him like a mother after a wayward son. His father-in-law was the commanding general of all U.S. Special Operations Forces operating in the world. Who was going to make that phone call?

"Uh, General Brown? Your son-in-law was blown up by one of our own bombs."*

Betz wiggled out of the loose grip of sand and chunks and stumbled over to Kevin. Then he started screaming for help. Everything was moving in slow motion. He couldn't tell if hours had passed since the explosion, or minutes. He stood there screaming for someone to help his friend.

It was nearly noon. Betz worried they were about to be overrun by Taliban soldiers who could use the chaos of the moment to easily charge them. In a matter of minutes, the military campaign that was America's answer to the attacks that had terrorized New York and Washington, D.C., had gone terribly wrong.

Mitchell looked across the fort and tried to see through the towering cloud of dust just what had happened. Shrapnel and debris whizzed past his head as he stared through his binoculars.

"Okay," he wondered, "where did this thing hit?"

The entire northeast end of the fort was obscured by the dust cloud.

He grabbed the radio and called, "Syverson, this is Mitchell. Over." Silence.

"Syverson, this is Mitchell. Over."

Nothing.

Mitchell began to think that maybe Syverson had been so close to the explosion that his eardrums had burst, and he'd been deafened.

He called again. "Syverson, this is Mitchell. Over."

And then, more urgently, "Any station, do you copy? This is Mitchell. Over."

One of the Tenth Mountain soldiers, located at the wall along the highway, replied, "Go ahead, Mitchell."

Relieved, Mitchell asked, "What can you see? Do you have contact with anybody?"

---

*In the errant bombing, four British SBS soldiers, with whom U.S. Navy Seal Stephen Bass was deployed, were also wounded. At least eighty Afghan soldiers suffered serious injury.

The soldier then delivered the devastating news: the bomb had hit Syverson's position.

About this time, the dust was clearing, and with his binos Mitchell could see a couple of guys rising up out of the rubble. They stood and started wandering around, dazed. He swallowed hard at the sight.

Then he picked up the radio and ordered everybody to rally at the main gate. He grabbed his gear and ran.

Betz looked up through blurred eyes. Coming for him and the other wounded men were nine soldiers from the Tenth Mountain Division. They had been crouched at their position behind a wall along a nearby road when the bomb hit. As the gray cloud boiled upward, they'd been shocked to hear the cries, "Oh my God, we may have killed the wrong people!" rise up among some other Special Forces soldiers standing there.

"That was wrong, that was very, very wrong," yelled some of the Afghans.

Wounded men—Afghan and some British SBS soldiers—were already stumbling from the fort's entrance, walking dazed like zombies, hair sticking up, faces white.

Sergeant Jerry Higley, twenty-six, looked up and saw the sky darken as the debris cloud drifted the several hundred yards toward their group. A loud, thwacking rain began to fall as shrapnel and bits of the exploded building wheeled through the dusty air and hit around them.

"It looks like we're in for a bit of a scuffle," joked Higley's friend, Specialist Thomas Beers, twenty-two, from Pennsylvania.

They quickly piled into a minivan and sped toward the fort, jumping out at the front gate and scaling a set of stairs leading up to the wall's pathway, where, after a breathless run through Taliban gunfire, they found the bleeding and broken Dave Betz.

When they arrived, Betz didn't know where he was, or even *who* he was. The soldiers, having swallowed their earlier fears, scooped up stocky, muscular Betz and the others and walked with them arm-in-arm, dodging the furious gunfire, along a narrow catwalk on the top of the wall. Their movement was slow going. The Taliban prisoners were taking the opportunity to pound the blown-up position. Machine-gun fire zigzagged across the crumbling dry walls.

Private First Class Eric Andreason, who the day before had been leisurely reading a novel in his tent back at K2, thought for sure they'd be shot. Dazed and in shock, the wounded men sometimes tried running away, heading back down the path to the bomb site. Andreason and his buddies chased after them, ducking as bullets smacked the wall at their back. Andreason found the roar of the shooting relentless. Yet he was suddenly bathed in a soothing calm. Finally, he and the others struggled down a steep set of stairs near the main gate, where Doc McFarland was waiting.

Kevin Leahy was near death and had to be resuscitated several times. All of the men had broken bones and concussions, but miraculously, they were alive. Syverson had suffered, among other wounds, a broken hip and back. Combat controller Malcolm Victors was badly burned on his face, his beard and hair singed. He felt like he really had died and miraculously come back to life. Sergeant First Class Pete Bach, blood still running from his ears and from the corner of his eyes, wandered away from the aid station and found a truck with the keys in the ignition. After fumbling with the door for several minutes, too dazed to figure out how to open it, he managed to get inside and drive himself back to the Turkish Schoolhouse, where he walked into the lobby, shocking everyone with his macabre, cartoonlike appearance. Everyone at the schoolhouse, hearing of the errant bomb strike over the radio, had thought that all of the guys had been killed. (After fully recuperating from their wounds, all five men would receive the Purple Heart and Bronze Star and return to active duty.)

As he headed toward the wounded men, Mitchell struggled under the sixty pounds of gear he'd stuffed in his load-bearing vest. The endless rope climbs, bench presses, and gut-wrenching twenty-mile runs back at Fort Campbell were paying off. Mitchell felt like collapsing on the ground, but pushed himself ahead one stride at a time.

He tried raising the bombers on the radio. He had to stop them from dropping any more bombs. "Shasta one-one," he said, using the plane's call sign. No reply. And then in the melee's confusion, he couldn't remember the call signs of any other planes, so he blurted out: "Any U.S. aircraft, any U.S. aircraft, cease fire, cease fire, we have friendly casualties on the ground."

Finally he heard a voice on the radio.

"Roger, this is Shasta one-one, I understand that you have U.S. casualties on the ground. Ceasing fire."

And then the pilot asked Mitchell if he wanted him to stay in a holding pattern overhead, ready for another air strike.

Mitchell considered this, then said sadly, "Negative. We are done for the morning."

He had to gather everyone including Betz, Leahy, Syverson, and Bach, and retreat to the schoolhouse. There, they would make a new plan to destroy the fort once and for all.

Mike.

He thought about Spann, lying somewhere in the courtyard.

We have to find Mike.

Finding Spann was now his mission.

As the battle raged, Najeeb Quarishy caught a taxi in Mazar and rode out to Qala-i-Janghi to watch the mayhem. He stood outside the fortress on the main highway along with several hundred other men, many of them armed. Every few minutes, another taxi would arrive from town, and another soldier would get out, leaning in to retrieve his long RPG tube from the backseat, pay the driver, and then trot off across the farm field leading to the towering fortress.

Najeeb could feel the explosions of the bombs and mortars in his chest. The air crackled overhead with the passing of stray rounds, like a thousand snapping whips. He wanted to go inside and see for himself what was happening, but knew it was too dangerous. He was glad to see the Taliban receiving so much punishment from the Americans' bombs.

Inside the city, Nadir Shihab, who had been beaten severely by the Taliban and whose father's house had been dug up by vengeful Taliban soldiers, had climbed to the roof of a friend's house and watched the spectacle in the distance, the sky flashing like lightning as the daylight faded. The sound of the bombs falling rattled the windows in his friend's house. The sky was filled with black smoke.

He saw a big plane, a B-52, fly overhead, very high, and then it dropped its guided bombs and the fortress exploded. After a while, he went home. Nadir was sitting in his own house when suddenly the walls exploded around him.

A rocket had flown from the fortress, one of the hundreds that were exploding on their own in the fires burning in the courtyard, and crashed through the roof of his house, several miles from the fort. Nadir was the only one home, and when he regained consciousness in the rubble, he knew he was lucky to be alive. He was bleeding. He felt around at his back and found it wet. He tasted blood in his mouth. His legs were cut with shrapnel. He couldn't walk.

He lay in the dust and fallen chunks of mud. Then his father rushed into the house and carried him outside to a car, and drove off to the hospital in Mazar. The Afghan doctors bandaged him and sent him home later that night. Nadir smiled at the damage he knew the bombs must be inflicting upon the men who had tormented him during their occupation of his city.

News of the errant bomb strike soon flashed across the airwaves in the United States. Karla Milo, Ben Milo's wife, heard it while she was up early readying their three kids for school. She was standing at the ironing board in the living room when footage of the explosion flickered on the television. Almost simultaneously, the phone rang in the kitchen.

It was a family friend of theirs, the wife of one of Ben's buddies from his days in the regular Army, wanting to know if he'd been hurt in the bombing.

Karla froze, standing there in her bathrobe, the iron poised in midair over the board.

"I don't have a clue," she told the woman, and hung up.

Then the phone rang again and didn't stop as the rest of the team wives called each other, asking, "What have you heard? What do you know?"

They didn't know anything. The wait over the next several days was excruciating. Usually, when something bad happened, when someone was killed or wounded in a training exercise or a training deployment overseas, it took at least a few days before anyone got an official account of what had happened. This was because the command wanted to positively identify who had been hurt, or who was now dead, before notifying the family. Not that the wives weren't able to figure out who was dead or hurt before this official confirmation. Karla just wished that Ben would call. The phone looked enormous as it hung in silence on the kitchen wall.

After the errant bombing of Syverson's position, Sonntag and Mitchell decided to level the south courtyard with cannon fire from an AC-130 Spectre gunship. The two officers felt they were running out of time. The Taliban had to be smashed *now*. The enemy soldiers remained too sheltered in the various rooms, stables, and storehouses that dotted the courtyard. And they were able to re-arm themselves too easily with the weapons and ammo lying around in the various nearby caches.

The U.S. attack was planned for that night.

Meanwhile, the Northern Alliance launched a daylight assault and began battling their way into the south courtyard, yard by yard. Often, the fighting was hand to hand. Ali Sarwar and several other men had started from the secure north end and crept along the narrow path lining the fortress wall, engaging in vicious firefights along the way. At one point, one of Sarwar's men stooped to reload in front of a small six-inch opening in a mud wall and was shot in the stomach by a Taliban soldier. The man looked up in surprise and went tumbling down the wall into the fort. He lay at the bottom, moaning.

Ali had to leave him behind as they pressed ahead.

He and his men spent the night hiding in a house in the nearby village of Deh Dedi while the Spectre gunship pounded the fort. The Taliban's main store of weapons and ammunition was hit by one Spectre cannon shell, and it sent a fireball shooting several hundred yards in the air. The explosions could be seen by some of Sonntag's men standing on the roof back at the Turkish Schoolhouse, eleven miles away.

The following morning, Ali and his men arrived at the southern end, behind the main parapet, the Taliban's remaining stronghold. They hacked their way through a mud wall into a small room—a makeshift guard shack that overlooked the wide parapet. They hacked quickly because inside the room were Taliban soldiers who were firing across the fort into the north courtyard, and they wanted to take them by surprise.

The Taliban soldiers turned to see Ali and his men standing

before them in a hole that had suddenly opened in the wall. Ali's men gunned the Taliban down and occupied the position, eventually moving out to the lip of the parapet, where they had a clear view of the remaining Taliban forces below in the south courtyard.

The fighting grew so intense that Ali and other Alliance soldiers who joined them dragged the dead bodies of fallen soldiers in a pile and used them as barricades from which they could safely fight.

After several hours, near dark, Ali and the rest of the Alliance force, now numbering several hundred men, were able to move down the sloping ramp into the south courtyard itself.

From one of the horse stables, fifty yards to the east, across grassy, open ground, a Taliban started charging at the line of Alliance soldiers, screaming in rage, a war cry that was interrupted by a fusillade of gunfire. Still, the man continued running, a grenade clenched in one hand held high over his head. Ali could see the man jerk as the bullets hit him. But he would not drop.

He managed to reach one of Sarwar's men, stopping to heave his grenade. The heavy, metal object sailed through the air and struck the soldier on the head, blowing up and killing him instantly. The Taliban soldier was finally cut down by a last, long volley of fire.

Ali's men walked another 100 feet through waist-high grass, when suddenly some fifty Taliban soldiers jumped up from hiding on the ground and opened fire, their AKs held at their hips, spraying bullets.

When the shooting stopped, some 150 Taliban and Alliance soldiers lay dead. Bitter smoke and the cries of bleeding men drifted over the suddenly quiet courtyard. Somehow, Ali was unhurt. He cautiously thought, as he had back in the canyon when confronting the Taliban there, and surviving, that his day to die still had not come. He continued moving through the fort, fighting and winning.

Shannon Spann was visiting her parents' house in California when there was a knock at the front door. Standing outside was a "grief team" of five CIA employees. One of them was the paramilitary officer whom Mike had asked to be the person to deliver any bad news to Shannon.

Shannon listened as the man spoke, but she could barely believe he was telling the truth. Mike *couldn't* be dead.

The officers explained what had happened at Qala-i-Janghi, telling her that her husband had been shot. Shannon immediately thought of Mike's two young daughters and their own six-month-old son. The girls' mother, Spann's first wife, was dying of cancer and wasn't expected to live much longer. Shannon resolved to remain steadfast, even when one of the girls later started crying and asked, "Who's going to teach Jake the daddy stuff?"

By November 28, Ali Sarwar and the rest of the Alliance soldiers had captured the southern courtyard. The area was shaded by pine trees whose shaggy trunks had been shattered and pockmarked by bullets. Severed tree limbs lay on the ground. Nearby stood a tiny mosque, its smooth white walls defiled by hundreds of bullet holes.

The Pink House, too, was pocked by gunfire, and thousands of tiny, pink chunks of concrete were scattered on the hard ground.

That morning, General Dostum and his troops returned to reclaim the fort as their own. Dostum had driven back from Konduz the night before with now-prisoner Mullah Faisal in tow, along with Dean, Nelson, and Bowers.

Freelance video cameraman Dodge Billingsley, who had been inside the fort filming much of the battle, had been asleep in his hotel in downtown Mazar when he heard Dostum's armored column roll past along the main street, shaking all the windows in the building. Billingsley got out of bed and went to the balcony. He counted about six troop carriers, several large cargo trucks, and an assortment of Toyota Land Cruisers. He guessed that the warlord would not be happy when he finally reviewed the damage the Taliban had inflicted upon his headquarters.

At the fort, Dostum grabbed Mullah Faisal, the Taliban general with whom he had negotiated the ultimately perfidious surrender, and shoved him along on a tour of the destruction.

Also in tow were hundreds of news reporters and television cameras, capturing Dostum's furious glare as he pointed out the physical damage, as well as the piles of dead bodies strewn everywhere. The stench of death was overpowering.

As the soldiers and reporters walked around the southern courtyard, a number of Taliban prisoners were still trapped in the basement of the Pink House. None of the Alliance or American soldiers knew exactly how many were there. Estimates ranged from a handful to several hundred. No matter the number, it was clear to everyone they were refusing to surrender.

In the basement, the prisoners sitting near the air vents and holes that had been punched in the foundation could hear the anxious, upbeat voices of their guards, who believed they had killed most everyone below. In fact, the prisoners were still armed with rifles, mortars, and grenades and they were considering rushing up the stairs and attacking.

Up top, Dostum angrily asked Faisal, "Did you know anything about this?"

The Taliban leader shook his head, no.

Dostum, surveying the human carnage, told one reporter that he was "sick of death." Soon, the Red Cross would begin hauling away seemingly innumerable bodies that would be buried in a mass grave outside the city.

Back in his headquarters office, Dostum held court, looking beyond the messiness of the battle to his political future. He explained that he wanted to build a new country. Women should be educated, he said, and he insisted that they would have representation in the new government. He further explained that he wanted to spend more time with his family.

Back in the basement of the Pink House, the last Taliban fighters waited and considered making one final attack.

In an effort to get the remaining prisoners to surrender, the Alliance soldiers began to throw rockets and grenades into the basement through jagged holes they'd hacked in the brick foundation. Periodically, men also walked up and jammed their gun barrels into the holes and emptied their clips.

Still, there was no sign of surrender.

At one point, several foreign aid workers had started walking down the long stairs to the basement and were shot by machine-gun fire. After they were bandaged up, all concerned decided that no more such trips would be attempted. Dostum's men next poured fuel oil through the holes and then set fire to it.

Down below, men were burned alive. Others pressed against the wall, away from the licking flames.

Still, no surrender.

Finally, Ali Sarwar found a shovel and ax and started digging a trench all the way across the southern courtyard.

A small stream entered the fort at the southwest corner; in prosperous, pre-Taliban times, the water had been used to irrigate fertile plots of corn and cucumber that dotted the fort. Sarwar was now diverting the stream so that it flowed to the Pink House.

Someone in his group hacked another hole in the brick foundation, which opened at ceiling level in the basement below. Just then, a prisoner pointed a gun barrel out of the new hole and fired, hitting some of Sarwar's men in the ankles. They crumpled and had to be quickly dragged away to an aid station.

Soon, however, the current of muddy, brown water was lapping against the building and then slurping through the hole in the foundation. Freezing water started filling the basement.

After a day, just thirteen Taliban soldiers had emerged from the basement. The water had risen inside to nearly seven feet, causing the men inside to dog paddle as their heads brushed the ceiling. A majority of them who were wounded or sick drowned. The room became a slurry of human body parts, excrement, even raw horsemeat, which some of the men had been eating, having hacked pieces off some of the dead animals lying in the courtyard.

Seeing that the basement was now flooded to a depth of about eight feet, leaving two feet of air space, Ali Sarwar and his men shut off the water supply. As the water drained through the porous soil floor below them, the surviving men, freezing and delirious, took a vote as to whether they would surrender.

They could either die in this stinking hole or they could die in the sunlight above, on firm ground. At least, by surrendering, they would have a chance of living. To remain in the room was to surely tempt being flooded again.

So it was that four days after Mitchell had tried leveling the Pink House with fusillades of bombs, eighty-six wet and miserable Taliban prisoners walked up from the basement, blinking in daylight, many of them, for the first time in seven days. John Walker Lindh was among them. Shortly after the fighting had begun, he'd been shot in the leg and lay in the courtyard as the battle raged around him. Finally, he was dragged by others into the basement. There, he had feared for his life, both at the hands of the soldiers firing from above, and from his fellow prisoners hunkered in the dark room. For the first several days of the battle, any talk of surrender was silenced by Lindh's hard-line comrades in the group. Lindh feared getting shot in the back if he made a break upstairs, to what he hoped would be freedom.

He felt he had no choice but to remain in the basement and hope for the best.

As the fort was secured, Ben Milo finally called his wife, Karla, back in Tennessee. He had been shaken by the errant bombing at the fort as well as the savagery of the battle within its walls. Ben knew Karla would be worrying about him, given all the extremely bad news.

She knew it was Ben calling because the words "Ft. Campbell" lit up on her phone's caller ID at home. To make the call, Ben had phoned the Fort Campbell switchboard, which then directed the call to their home in the Hammond Heights neighborhood on the Army post.

She didn't even say hello. "Tell me you're okay."

"I'm fine."

"Is the team okay?"

"Yes." Karla could tell he was being tight-lipped. She wanted him to say more.

"Was it the battalion?" she asked. Meaning: Were you guys hurt in the bombing at Qala-i-Janghi? Had *he* been hurt?

"I can't say."

Because of security measures, he wasn't able to tell her that he hadn't even been in the vicinity of the bombing. They talked for maybe five more minutes and then Ben said he had to hang up, that he had to go to work. The lag in their conversation caused by the satellite link had been horrifying. Ben's voice had sounded like he was talking to her from underwater. She had ended up interrupting half of his sentences. But still: *he was alive.*

She was relieved, but she had no one outside the immediate group of Special Forces wives to share the news with. Ben's being in Afghanistan was officially still secret (even though news reports had noted their presence in the country). In fact, several days earlier, President Bush had come to Fort Campbell to eat Thanksgiving dinner with the regular Army troops, but as far as Karla knew, none of the Special Forces wives was invited.

Hey, what about us? My husband's over in Afghanistan fighting, too! Karla thought.

Then she went back to ironing the clothes and getting the kids ready for school.

Spann's body was retrieved and delivered to the Turkish Schoolhouse, where it was stored, sealed in a blue vinyl bag, in one of the kitchen's overlarge chest freezers adjoining the cafeteria.

The body bag was wrapped in an American flag that had been presented by Mitchell's fellow teammate Martin Homer. Homer had carried the flag into battle as a Special Forces soldier in the first Gulf War and it had accompanied him on this last campaign. He had been honored to give it to Mike. The body was driven from the schoolhouse to the Mazar airport, where two Chinook helicopters were waiting to escort it to K2. At the airport, the flag was neatly folded and presented to one of Mike's CIA colleagues, who sat in a seat on the helo and held it tightly. Helicopter pilot Greg Gibson saw that the man was nearly in tears. Spann's body was loaded onto the lead bird.

Sitting behind it, in his own aircraft, Nightstalker pilot Jerry Edwards, who in the last month had learned during a phone call with his wife that he was going to be a father again, watched the honor guard solemnly unfold on a video screen in his aircraft. He had turned on the helo's FLIR radar, which picked up a grainy image of the soldiers as they walked up the lead bird's ramp with Spann's remains strapped to a body board.

Edwards and his crew had been shaken by Spann's death, even though not many of them knew him well, if at all. But his death had diminished their own sense of invulnerability. They hadn't wanted to miss the ceremony, which they learned would take place in the dark (the only time when the Nightstalkers liked to fly missions), and Edwards felt lucky to have figured out how to use the FLIR to witness the moment. He sat strapped in his seat reflecting on his own mortality and the impending birth of a new baby.

At almost the same time, CIA director George Tenet publicly announced Mike Spann's death.

"Mike was in the fortress of Mazar-i-Sharif, where the Taliban prisoners were being held and questioned. Although these captives had given themselves up, their pledge of surrender—like so many other pledges from the vicious group they represent—proved worthless.

"[Mike's] was a career of promise in a life of energy and achievement. A precious life given in a noble cause." Spann's death soon dominated the news. President Bush issued a statement of condolence.

In his hometown of Winfield, Alabama, neighbors plied his parents' house with food and sympathy. Many of the townspeople were shocked to learn that Spann had worked for the CIA. A next-door neighbor would remark that Spann never seemed to be home much. On Main Street, a black bow hung on the front door of the real estate office of Johnny Spann, Mike's father. Flags flew at half mast at the post office.

Mr. Spann held a press conference at a neighbor's home: "He was a cherished son, he was an amazing brother, a devoted father, and a loving husband. Our family wants the world to know that we are very proud of our son, Mike, and we consider him a hero." Mr. Spann was near tears as he spoke.

News of a CIA employee's death is usually kept a secret, with a gold star being placed afterward on the memorial wall at Langley's CIA headquarters to honor the deceased. In the case of Mike Spann, he bore the grim distinction of being the first American to be killed in the first war of the twenty-first century, and George Tenet as director of the CIA took the unprecedented decision to make Spann's death a matter of public record. Complicating any idea that his death might have been kept secret were the reporters who had been at the fort chronicling the uprising. Tenet had no other choice than to disclose Spann's death in the melee.

The effect was that Mike Spann became an overnight hero, an accolade he would have shunned while alive. He became a figure of everything that a nation felt was honorable about itself, namely, selfless sacrifice while at war with a new and mystifying threat.

The further effect was to make Shannon Spann the face of a nation's grief as it met this new threat. The publicity over Mike Spann's death tore away the veil of secrecy that had covered her own employment as a CIA officer, and she found herself thrust before the television cameras, and reporters camped outside her home in Manassas Park, Virginia.

• • •

News of the Battle of Qala-i-Janghi started to make headlines around the world.

Perhaps most shocking for Americans was the discovery of John Walker Lindh hunkered in the basement of the Pink House, as well as the death of a CIA paramilitary officer. Who even knew America had such men fighting in Afghanistan?

Back in Marin County, California, Lindh's mother, Marilyn, logged on to the Internet when she heard news that an American had been discovered in Afghanistan. She had a sinking feeling it might be John, and this feeling was confirmed when she happened upon a photo of him at the fortress.

Yet Marilyn was shocked. She had never expected John to get into any kind of trouble.

"He knows we wouldn't have approved," his father, Frank Lindh, told a reporter, one of many who would soon be descending upon the Lindh household.

Marilyn felt that John must have been brainwashed to have been tangled up with the Taliban. "He would freeze" when confronted with danger, she said. "He's totally not streetwise."

For his part, Lindh was spending his days under close guard in a second-story room in the Turkish Schoolhouse. But he had not been identified immediately. In fact, after emerging from the Pink House basement, wounded and dirty, he had been loaded onto a truck headed for Sheberghan, seventy-five miles west. At Sheberghan, he and the other prisoners would be housed in a large prison there.

As the truck was about to leave, a reporter for *Newsweek* named Colin Soloway had happened to pass by and peer inside, curious about a filthy, Western-looking figure he saw hunched there in the gloom. Soloway had come to Qala-i-Janghi, like hundreds of other reporters, to report on the battle and Spann's death. But now his interpreter had approached him and said that an American was on board one of the trucks. Soloway found the bed of the vehicle filled with moaning, bleeding men. He was amazed that any of them had survived the bombs that had been aimed at the fortress. The young man in question was dressed in a black tunic and blue sweater. He was covered with dirt. His face looked as if it had been burned.

One of the truck's guards tapped Soloway and motioned again at the prisoner. He insisted that he was an American.

"Are you American?" Soloway asked the young man.

"Yeah," he replied.

This struck Soloway as unusual—indeed, shocking. He asked Lindh his feelings about the attacks in America, which had taken place some eleven weeks earlier. Did he support this action?

Groggy, speaking slowly, Lindh replied, "That requires a pretty long and complicated explanation. I haven't eaten for two or three days, and my mind is not really in shape to give you a coherent answer."

Soloway asked again for his opinion.

"Yes, I supported it," Lindh said.

Accompanying Soloway was translator Najeeb Quarishy, who earlier in the month had hosted Dean at his home, and a French photographer named Damien Degueldre, who recounted the surreal encounter this way:

> Earlier in the day, one of the survivors of the fighting in the fort had come up from the basement [of the Pink House], asking if their lives would be spared if they surrendered. The local Afghan commander agreed and they started coming out, eighty-six in all.
>
> I was at the Red Cross office when they received a radio call saying survivors had been discovered. I tried catching Alex Perry [of *Time*], but couldn't find him. Then I caught a taxi and drove straight to Qala-i-Janghi. That's where I met Najeeb, who said there was an American within the prisoners.
>
> Lindh was sitting in the front of a truck with some of the others. Being an American, Soloway engaged in conversation with Lindh while I stood filming nearby. Lindh wasn't too keen to be filmed. I didn't really want to interview him, as I felt these guys were coming back from hell. They looked completely in shock. I felt the visual of this young American with his long hair and beard, and his face covered in dirt, was really amazing. The image was talking for him.
>
> I still believe he was in shock when he answered Soloway's questions. It is true that he answered "yes" when he was asked if he was approving of the September 11 attacks, but he didn't focus on this matter or extend the discussion. He was embarrassed.
>
> I also perfectly remember his answer when he was asked why he

came to Afghanistan. He said it was the most ideal Islamic republic in the world, the *purest*.

That same night, Colin Soloway's interview was out on *Newsweek*'s website. When we finally made it to Shebergan prison [the next day], it was too late. The U.S. Special Forces had taken him.

In fact, after Lindh had arrived at Sheberghan, one of Dostum's aides had talked with the young man and discovered that he could speak English. He immediately informed Dostum that there was a captured American in his midst.

As medic Bill Bennett was treating the dehydrated, starving Lindh, the moment was recorded by video camera (a CNN freelancer, Robert Pelton, conducted an interview during the medical exam). And once the video appeared on television, John Walker Lindh's odyssey from California to the Qala-i-Janghi Fortress became worldwide news.

Among those watching the broadcast of the medical examination were Frank and Marilyn Lindh. They hadn't seen or heard any sign of their son for seven months. They had believed he was in Pakistan, where he'd told them he had been studying Arabic at an austere, demanding madrassah. The sight of him on television, shirtless and emaciated-looking, crushed them.

Soon, the crank phone calls and death threats started to come in: "Great parenting job."

"You should be shot with the same gun used to shoot your son."

Frank Lindh, a lawyer for Pacific Gas & Electric in San Francisco, had been proud of his son's disciplined study of Arabic and Islam. But he had never imagined John was in Afghanistan, fighting with the Taliban. When he saw his son on television, he broke down crying.

Not long after Mike Spann's death, Shannon received a package from one of Mike's former colleagues. It was a series of tattered, charred pages from his journal.

The morning he was killed, Mike had parked their truck in the fortress and walked with Dave Olson to the southern courtyard, to interrogate the prisoners. Inside the truck was a journal he'd been keeping. In the air strikes that followed, as Major Mitchell fought to quell the uprising, the truck was blown up.

Shannon now held what was left of the charred journal.

"One thing has troubled me," Mike had written, as he rode north to Mazar. "I'm not afraid of dying, but I have a terrible fear of not being with you and our son . . . I think about holding you and touching you. I also think about holding that round boy of ours. . . ."

Five days after leaving Afghanistan, Spann's body arrived at Andrews Air Force Base by military transport. Shannon, along with Mike's family, went on board to view the coffin in private. The casket was then carried to a waiting hearse, while Shannon and the family watched tearfully. Johnny Spann, Mike's father, was consumed by grief.

Mike, trying to prepare his father for the dangerous, clandestine nature of his work, had always warned: "Daddy, don't ever believe I'm dead until you see my body."

Shortly before Mike's burial in Arlington National Cemetery, Mr. Spann visited his son one last time. He stood at the casket and gazed down. He touched Mike's forehead. He then lifted his boy's head, bent down, and looked.

He saw two holes in Mike's skull. He lowered the head back down on the satin pillow and looked at his son's temples. There was a hole on each side. He figured the bullets had entered there, and exited at the two wounds he saw at the back of the head. He touched his son's arms and legs and studied them. They were bruised. He wondered if Mike had been tortured. The wounds told him nothing.

Back in Mazar, Mark Mitchell could not help but remember the very moment that the fighting in Qala-i-Janghi had stopped. He had surveyed the smoking rubble, a dusty M-4 in his hand, barely able to stand after fighting for his life for seven straight days.

Dead and dying men and wounded horses had littered the courtyard, a twitching choir that brayed and moaned in the rough, knee-high grass. The horses killed first had swollen to twice their size. The fort's mud floor was spongy with blood.

Mitchell's buddies were equally wrung out. With every nerve jangling, Major Kurt Sonntag stumbled upstairs in the schoolhouse and told Doc McFarland, "Gimme something." Whatever it was, Sonntag was soon out cold and slept for a day. The schoolhouse was filled with men who'd been pushed beyond their limits.

They had come within minutes of losing the entire war in Afghanistan. That was the scary thing, thought Mitchell. That was the really scary thing. And yet they had prevailed. Mitchell had been part of "the first U.S. cavalry attack of the twenty-first century," a feat trumpeted in news photos showing bearded Americans riding across a sunlit Afghanistan plain. "It was as if warriors from the future had been transported to an earlier century," General Tommy Franks later remarked.

On November 14, 2003, Mitchell found himself standing uncomfortably at a podium at Special Operations Command back in Florida. That day, USA Today had carried the headline "Soldier to Receive Honor for Valor in Afghanistan."

"Maj. Mark Mitchell doesn't like to talk about what happened in Afghanistan in late November 2001," ran the story. "But the Army thinks it's a pretty big deal."

Mitchell stood at attention as General Bryan D. Brown, commander of Special Operations Command, pinned his uniform with the Distinguished Service Cross for "extraordinary heroism." The audience was a mix of family and colleagues from the covert world of special operations, America's brain trust in the war on terror. Mitchell's parents had arrived from Milwaukee, Wisconsin; Mike Spann's wife, Shannon, sat near the front, fighting tears. Mitchell's wife tended two fidgeting daughters who knew their father better as someone who read them *Goodnight Moon* and then was gone for months out of the year.

Emceeing the ceremony was a former Delta Force commander, Brigadier General Gary Harrell, who had survived the harrowing 1993 fight in Mogadishu, Somalia. Harrell described the ninety-six hours of fighting that Mitchell and the men of Fifth Special Forces Group had endured as "the most intense urban combat conditions to date in Operation Enduring Freedom." General Brown described Mitchell's task as "mission impossible."

"The fortress [was] nearly impregnable," recounted Brown. "Seventeen warriors attacked on what—on a good day—would have [involved] thousands of soldiers."

Balding, smiling, looking more like a high school biology teacher than a stone-cold killer, Mitchell was suddenly the most famous soldier in America.

Pretty good, he thought, *for a kid from Milwaukee who had nearly quit Special Forces a year earlier.*

A handful of men had gotten on wild horses and defeated the Taliban.

Mitchell had to laugh. He remembered the day when he couldn't even ride a horse.

EPILOGUE

After the victory in Mazar and the Battle of Qala-i-Janghi, combat operations in northern Afghanistan began to wind down. The Taliban were beaten, and the Northern Alliance was in control. Conventional U.S. Army and Marine forces soon arrived in large numbers and began the long-term effort to secure the country and hunt down Osama bin Laden and Al Qaeda.

The epic success of the Horse Soldiers, as they were dubbed, was stunning, by both historical and contemporary standards. The campaign is, in fact, a template for the way the present war—and future ones—should be fought. Instead of large-scale occupations, we should rely on small units of Special Forces who have proved it's infinitely more effective to work with a country's soldiers and citizens at eye level. The SF soldiers believe that we must now resolve the root problems plaguing Afghanistan; they are uniquely trained and equipped to do just that.

At the time of the capture of Mazar-i-Sharif, there were fewer than fifty U.S. military personnel like Nelson and Dean on the ground. They accomplished in two months what Pentagon planners had said would take two years. In all, about 350 Special Forces soldiers, 100 CIA officers, and 15,000 Afghan troops succeeded where the British in the nineteenth century, and the Soviets in the 1980s, had failed. Between October 19, 2001, when Captain Mitch Nelson's team landed in Afghanistan, and the early months of 2002, when Nelson and other Special Forces teams left the country for good, the United States would eventually spend a mere $70 million to defeat an army of 50,000 to 60,000 Taliban fighters. The story of the Horse Soldiers was not unlike a Western, featuring high-tech lasers rather than six-shooters, fought on horseback.

"It's as if the Jetsons had met the Flintstones," Special Forces soldier Ben Milo said to me.

The political victory proved just as overwhelming as the military one. "To this day Al Qaeda still considers the campaign their largest, most destructive defeat," Dean explained. American military planners and Afghan leaders like General Dostum and General Atta had taken great pains to make sure that the war remained the Afghans' war, and not one of American occupation. Local Afghans hailed Nelson and Dean as liberators. In 2001, as hostilities began, the Taliban had little base of support among local people. And when their leadership began to collapse, they had nowhere to retreat. They couldn't blend into the countryside and fight an insurgency from there.

By entering Afghanistan with a small force, and by aligning themselves with groups that once had been battling each other and pointing them in one direction at the Taliban, U.S. forces found robust support among Afghans. They proved the usefulness of understanding and heeding, the "wants and needs" of an enemy, and the local population that may support it. Awareness is the soldier's number-one tool in his kit, beside his M-4 rifle. To win wars against enemies like the Taliban, which are often stateless in their affiliation, you adapt.

You eat what they eat, sleep where they sleep, and think like they think. The information and insight gained from this was the essence of the Special Forces soldiers' training and experience.

As the Chinese general Sun Tzu wrote in the sixth century B.C., "To win 100 victories in 100 battles is not the acme of skill. To subdue the enemy without fighting is the acme of skill." Sergeant First Class Sam Diller, part of Nelson's team, carried this aphorism in a notebook, and he read it often while atop his mountain lookout.

The success of the mission was "about as perfect an execution of guerrilla force as could be studied," reflected the commander of U.S. Special Forces at Fort Bragg, Major General Geoffrey Lambert. Unfortunately, said Lambert, "It may never be repeated."

His words would prove prescient.

On December 10, 2001, Nelson and his men had to suddenly leave their safe house in Mazar—and Afghanistan. They would not return as soldiers riding at Dostum's side. For all intents, their time in Afghanistan was done. The order was as shocking as it was abrupt. Leave? Without saying goodbye to anyone? To Dostum? It was unthinkable. Dostum, when he heard the news, was hurt and angry. But Nelson and his team had no choice. They'd been summoned to fly to K2 and meet with Secretary of Defense Donald Rumsfeld, who, riding a tide of popularity in the United States, wanted first-hand to hear their tales of war.

They had just a few hours to pack. They needed a hot shower, clean clothes, and haircuts, but there wasn't time. That was okay, the order said. *Come as you are.* The SecDef wanted to meet the men who captured Afghanistan.

Ben Milo had picked up a bayonet at Qala-i-Janghi as a souvenir. He polished off the dirt and decided he would give it to Rumsfeld.

Cal Spencer wanted anybody but the easily excitable Milo to present the gift to the SecDef. Milo was the soldier who, after he had almost been overrun in a firefight with the Taliban, jumped up in his foxhole and gave the enemy fighters the finger.

"Just make sure you watch what you say to him," said Spencer. "Can you tell a story without swearing?"

"I could try," said Milo.

"Well, just try," said Spencer.

They landed at K2 and set foot back on the ground that they had tried so hard to leave six weeks earlier, worried that they'd never see combat. In their absence, a veritable tent city had sprung up. Thousands of soldiers and aircraft and vehicles buzzed along paved and graveled streets with names like Broadway, Main, and Lexington, all marked with road signs. The men in camp who recognized them watched warily as they passed, as if Milo and the rest had stepped from the pages of a fantastic spectacle.

They entered a large tent near Colonel Mulholland's headquarters. Donald Rumsfeld was sitting inside at a table, waiting.

Milo gave him the bayonet, shook hands, and found himself tongue-tied. This was the secretary of defense of the United States of America. Ben awkwardly stepped back into the line of men. Rumsfeld broke up the formality of the moment.

"A lot of people say you're heroes," he said.

The guys shook their heads.

"You were doing your job," said Rumsfeld.

They agreed that they were.

Rumsfeld explained that many people in the United States were surprised that such a small number of men were able to accomplish so much, so quickly.

The men said it had been the Afghans' victory, and that they had simply helped. And they meant this. They weren't being coy.

The meeting ended as quickly as it had begun, after about half an hour.

On December 20, 2001, Nelson and five members of his team briefly returned from K2 to Mazar for a difficult, tearful farewell ceremony with Dostum.

The general, along with many of his commanders and troops, had wanted the men to stay as they tried rebuilding the country.

By the middle of January 2002, practically all of the Horse Soldiers were headed home. Upon their return, reporters clamored for the story of these men on horseback, galloping under fire through an Afghan sunset; but after a brief period of notoriety on television and in newspapers, they fell silent, true to their motto as the Quiet Professionals. They didn't give interviews. They didn't write books.

Among the many things the team didn't discuss was John Walker Lindh, though they had been shocked by the discovery of the young man in the fort during the battle.

When Sam Diller had heard that an American had been found among the prisoners, he was incredulous.

"An American *what?*"

"American Taliban."

Diller told his fellow teammates to stay away from the man. "Do you all want to go to court?"

Diller understood full well the implications of discovering an American citizen in the middle of a bloody, defiant crowd of Taliban and Al Qaeda prisoners. Diller figured that Lindh would be heading to court, and so would any U.S. soldier who had contact with him.

Nelson told Lindh in his makeshift jail cell on the second floor of the Turkish Schoolhouse, "You're safe here. But if you jump out of this window, you're probably going to break your leg." The message was: *stay put*.

"What's going to happen to me?" Lindh kept asking.

"Man, I don't know. That ain't for me to decide."

On December 7, 2001, a week after walking out of the Qala-i-Janghi basement, Lindh was transferred to Camp Rhino, a U.S. Marine base in Kandahar. He was blindfolded, stripped naked, and strapped on a cot that was placed inside a metal trailer. At night, the desert temperatures plummeted. Loud music was played outside the dark, nearly airless container. Periodically, someone would bang on the metal walls and yell insults.

Lindh had announced that he wanted to speak with a lawyer. His FBI interrogators told him that he was certainly entitled to a lawyer, but, "as you can see, there are none here in Afghanistan." Frank Lindh had in fact hired a San Francisco attorney, James Brosnahan, to represent his son. Brosnahan contacted the FBI to say that the family had retained him, but this information wasn't communicated to John in Afghanistan.

On January 15, 2002, Attorney General John Ashcroft, in a nationally televised press conference, announced that John Walker Lindh was being charged with aiding and abetting the Taliban in Afghanistan. The charge carried the possibility of life in prison.

A week later, Lindh returned to the States, setting foot on American soil for the first time in seven months, this time in handcuffs.

On February 3, he entered the federal courthouse in Alexandria, Virginia, where he pleaded not guilty to the ten-count indictment against him.

Afterward, Shannon Spann told CNN correspondent Deborah Feyerick that she would prefer the death penalty for Lindh, if he were to be convicted.

She had sat in the courtroom thinking about her husband. "Mike's life is all about taking responsibility. I wanted to come today to see if John Walker Lindh will take responsibility for what he has done," she told Feyerick.

When the pleading was over, Spann's parents and Shannon left the courtroom and walked to the elevator.

Frank Lindh was also leaving the courtroom at the same time. Suddenly, Lindh spotted the Spanns and walked quickly toward them.

Standing before Johnny Spann, Frank Lindh held out his hand in greeting.

Spann stood with his hands in his pockets, unmoved.

"I'm sorry about your son," said Frank Lindh finally. "My son had nothing to do with it. I'm sure you understand."

Johnny Spann turned and walked away. He later remarked of the meeting with Lindh, "I should have taken out some of my revenge on him."

Johnny Spann blamed Lindh for Mike's death. Of course, Lindh had not actually been the person who fired the shots that killed Mike Spann. But he had aided the enemy, and for the Spann family he was the face of that enemy, perhaps easy to hate but not to understand.

On July 15, 2002, shortly before his trial was to begin, Lindh's lawyers and prosecutors agreed to a plea arrangement. Lindh would plead guilty to "contributing services to the Taliban and carrying explosives in commission of a felony." Government prosecutors, on the other hand, would not have to drag into court disclosures about classified operations and CIA officers in Afghanistan. As far as the government was concerned, this was a good deal.

For Lindh's lawyers, it was the best outcome they felt their client might get, given prevailing attitudes about Lindh. He had become known as a traitor to his country—an "American Taliban." The deal also meant that the charges against Lindh that he had conspired to kill Spann, and provided material support to the Taliban, were dropped. Lindh would not be held accountable for the death of Mike Spann.

Johnny Spann was in Winfield, Alabama, at the drive-through lane of a fast-food restaurant, when his cell phone rang with the news of Lindh's plea. He had to pull over and collect himself. He was upset that Lindh's role in his son's death would never be investigated.

He didn't understand why Lindh hadn't asked Mike for help. He felt that Mike would have done anything to help Lindh escape from the prison. More than ever, he held Lindh responsible for his son's death. Spann also felt that justice for his son's death had been sacrificed in order to prevent U.S. soldiers and CIA officers from testifying in court. Lindh's plea bargain meant that the trial was over.

He later told a television reporter, "We never thought he would get less than a life sentence."

On October 4, 2004, Lindh was sentenced to twenty years in prison. After the sentencing, Frank Lindh told reporters: "John has no bitterness. He has never expressed the slightest bitterness about any of the treatment that he suffered. . . . Never once did John ever say anything against the United States. Never once, not one word.

"John loves America and we love America."

Today, Lindh is serving his sentence in a federal prison in Indiana. He will be released in 2022, when he is forty-one years old.

Since 2004, his father has petitioned the U.S. government to commute his son's sentence. His requests thus far have been denied.

As for Shannon Spann, on the cold, gray morning of December 10, 2001, two weeks after he'd been killed at Qala-i-Janghi, she buried Mike at Arlington National Cemetery, in an area where Ira Hayes, one of the Iwo Jima flag raisers, had also been laid to rest. Mike would have been happy with the serendipitous choice of this resting place.

When he lived nearby in Manassas Park, he had enjoyed walking through the cemetery and studying the headstones. Even as a boy during family trips to Washington, D.C., he liked to walk among the headstones.

This bored a sister, who told him, "Mike, let's go! They're all the same!"

"No, Tanya, they're not," he had said. "There are stories behind them."

At a memorial service in Alabama several days earlier, one of Mike's daughters had written him a letter that read: "Dear Daddy, I miss you dearly. Thank you, Daddy, for making the world a better place." The letter was placed in his casket. The church was packed with five hundred of Mike's neighbors, his Sunday school teacher, former football teammates, a former Marine buddy. American flags flew at half mast around town, and Christmas lights had been hung up to read: GOD BLESS AMERICA.

Now, at Arlington, the crowd of two hundred people, including George Tenet, listened as Shannon addressed the group with a eulogy about Mike. At one point, she spoke to Mike directly: "Darling, if you were here today, I would tell you that I love you with every part of who I am, and I would thank you for giving me the greatest honor of my whole life, and that was to be called your wife."

She would later write of the emptiness that Mike's death had left in her life. "There are times when I just lie on the floor and say, 'God, why did this have to be part of your plan? I miss my husband so much.'" After the service, she approached the casket, kissed her hand, and then placed her palm on top.

"Farewell, my love," she said, barely above a whisper.

Captain Mitch Nelson came home to a hero's welcome in his home state of Kansas, at the state legislature in Topeka.

Standing next to his wife and their new baby daughter, who had been born while Nelson was riding out of the Darya Suf Valley during the first horse charges, Nelson listened as a resolution was read acknowledging that he had been "instrumental in the liberation of over fifty towns and cities; for the destruction of hundreds of Taliban vehicles, bunkers, and heavy equipment; and the surrender, capture, or destruction of thousands of the Taliban and al-Qaeda."

"It was an extremely challenging situation," said Nelson, typically understated. "You never quite knew how things were going to turn out.

"I'm not a hero," he added. "The men of my detachment are the heroes." He explained that General Dostum had called him "my brother." Nelson had been honored by this statement.

Dostum, too, had complimented Nelson on his skill as a warrior. "I asked for a few Americans," said Dostum. "They brought with them the courage of a whole army."*

Cal Spencer's wife, Marcha, had decorated their team room with red, white, and blue balloons and streamers. Propped up on a desk was a large, white dry-erase board on which one of the wives had written WELCOME HOME!

---

*Chief Petty Officer Stephen Bass also received recognition; he was awarded the Navy Cross Award for "extraordinary heroism" during his mission to locate Mike Spann and Dave Olson.

Spencer and some of his teammates had walked in, dead tired from the three-day flight, but smiling. Spencer appreciated Marcha's effort, but all he really wanted was to go home and have a beer.

He dropped his bag on the dingy linoleum floor and took the scene in. The ceiling tiles were still sagging, and they were still water-stained from the leaky roof overhead. It was good to see things hadn't changed. The fluorescent-lit hallways still had the same damp, chalky smell.

In the absence of a bona fide public homecoming celebration, like the kind the guys in the regular Army got, the wives had wanted to make this moment special. They had set out cookies and SunnyD, a nonalcoholic fruit drink (officially, beer wasn't allowed in the team rooms).

"Honey," said Spencer, "let's go home."

On the way back, he and Marcha held hands. She was driving. It had been so long since he'd driven on a civilian street, in a country not at war, that he felt it was better Marcha take the wheel. He didn't trust himself. Any loud noise, such as the sudden slamming of a door, made him jump.

Once they got home, one of the first things they did was to have a huge fight, which seemed to clear the air. After that, they got along just fine.

Ben Milo came home in the middle of the night on January 15, 2002.

It had been a long wait for all of the wives at Fort Campbell. First, they would be told by the command that the guys would be coming home on a particular day, then that day's return was inexplicably canceled. Many of the wives had spent the night waiting in the team room, drinking coffee and eating the cake they'd set out for the men.

Karla finally gave up trying to anticipate Milo's arrival. But she still went through the normal rituals that all of the wives performed before a homecoming. She made lists of tasks. "Monday, I'm going clothes shopping. Tuesday, I'm going to Sam's Club to get food. Wednesday, clean the house. Thursday, I'm going to mow the grass . . ." Karla ended the week by giving the three kids baths, combing their hair, cutting their nails. Everything had to look perfect: *We were okay in your absence. It is okay that you were gone. We love you.* After several weeks of false alarms, she was sleeping when the phone rang and jarred her awake at five in the morning.

"I'm home."

It was Ben. She breathed a sigh. It was all she could manage.

"Could you come pick me up in the parking lot?" he asked.

She threw on some clothes and ran into the kids' room. "Dad's home and I'm going to go get him!"

She asked their eldest son to watch the smaller ones and drove like mad to the church parking lot a mile away. She expected to find him surrounded by a crowd of people and that they would run toward each other and hug, like in the movies.

Karla pulled into the lot and Milo was standing under a street-light with his duffel bag at his feet. He was alone. He looked tired. He looked up when she pulled in. He gave her a smile as she got out, and they hugged and he kissed her, and then he said quietly, "Let's go home."

They got back in the car and drove in silence. There was so much to say and so few ways to say these things. How do you catch up on four months of anything, let alone four months of being at war? *It will take time*, Karla thought. She looked over at him and she could see that he had aged.

She could tell Ben had learned something, but hadn't liked the experience of learning it.

The couple had just gotten into bed when the kids came running in and started jumping up and down. "Dad's home! He's home!"

The two youngest children didn't know who Milo was. They had been just one and two years old when he left. Milo bent down and picked them up and gave them a kiss. He knew he'd have to reintroduce himself. And that, too, would take time.

Over the next few weeks, Karla wondered when she and Ben would start fighting over the little things needed to run a household. The other wives had told stories about their husbands who had come home from deployments and insisted on taking charge of the household, as if their wives had been waiting to be told what to do.

With Ben, Karla felt none of that belligerence. He pretty much split the parenting fifty-fifty. He asked Karla for advice when it came to handling a discipline problem with one of the kids. He wanted to know how she'd handled things in the past. He seemed to think first and then open his mouth.

At the end of January, they celebrated their fifteenth wedding anniversary. They went to visit Milo's parents in Chicago, and his mom and dad watched the kids and they took a hotel in the city and went out on the town. She would catch Ben thinking and wonder what it was that made his face so troubled. Over time, he told her: *the killing*.

Had it been right? Karla guessed he had shot his weapon before, but that this was the first time he had killed someone.

Karla didn't feel they were a strictly religious family, but she saw her husband struggle with the fact that he had killed someone. He talked about being in the trench with Essex and Winehouse, and being overrun, and shooting at people.

He talked with his sister a lot about these feelings. Karla figured it was easier this way for him to unburden himself. He told his sister that he had killed people and he didn't know how many. He wanted to know if God would forgive him. He knew that he had belonged in Afghanistan, and that the Taliban soldiers would have killed him if he hadn't shot them first, but still the killing troubled him. He figured it was something that he would carry with him for the rest of his life.

As for Major Mark Mitchell, his wife threw a Christmas party when he returned home to Tennessee. Also invited were the guys who'd been blown up at the fort, many of whom Mitchell hadn't seen since last glimpsing them through his binoculars moments after they'd come tumbling back to earth, bleeding and dusty.

Mitchell, like all of them, felt lucky to be alive. He reflected on how, exactly, he had survived. His Special Forces training had taught him to ignore pain and mental exhaustion. But it had taught him something more important, and complex: *to think first and shoot last*. He was able to set aside his ideas about how he thought Afghans, Pakistanis, and Saudis might act in a given situation, and he listened instead to what they were saying or watched what they were doing in response to a question or turn of events. He had been trained to see the world through other people's eyes.

He knew that if he asked an Afghan man to do something for him, and if the man answered, *"Inshallah"*—God willing—what the man really meant was, I don't want to do this for you. This was an important insight to have if what you had just asked was whether or not the man could provide soldiers for a next day's battle.

Such an insight required Mitchell to ask new questions of himself, such as, "What can I do for him, so that he will do something for me?" His weapon would stay in its holster; coercion rarely worked in attempting to gain the cooperation of a population already terrified at the hands of an oppressor like the Taliban.

Shortly after the battle, Mitchell called his wife in Tennessee. "Did you see what happened?" he asked. "I was the ground commander during that fighting." Maggie could tell he was overwhelmed. He kept trying to explain to her the enormity of the battle. He sounded frustrated and tired. Months later, even as he prepared to receive his Distinguished Service Cross at the ceremony at Central Command, he still hadn't grasped the battle's impact. One of his uncles, a World War II veteran, explained to Mitchell that the ceremony would bring back memories of the fighting, and that not all of them would be welcome. And he was right.

What had happened during those days suddenly seemed as vivid and real as ever, and Mitchell started reliving it. He smelled the dead bodies again, heard the roar of the bombs. It was not pleasant and he tried to put a lid on it. As he stood on the stage, he looked up, sought out Maggie in the audience, and smiled. She beamed. He felt he'd come home, finally.

Soon, however, he had to leave again.

In February 2003, along with most of the soldiers of Fifth Special Forces Group, a number of whom had fought in Afghanistan, Mitchell deployed to Iraq. Some of the men in Special Forces were not fans of this war from its beginnings but, as true and loyal soldiers, they fought there.

And they died.

Bill Bennett, the jovial, good-natured medic who had ridden with Sam Diller and yodeled country and western songs from the saddle, was shot and killed in a firefight in Ramadi on September 12, 2003.

Ten months later, Captain Paul Syverson, who had been blown up in the errant bombing in Qala-i-Janghi, was hit by an enemy mortar that came flying into his base camp in Balad. Syverson had been walking along and talking to a friend as they made their way to the camp's mess hall.

And Brett Walden, whom the Afghan women had found hand-some and who had been embarrassed by their flirtations, was killed in Rubiah on August 5, 2008, while carrying out a convoy mission.

Before deploying to Iraq, Master Sergeant Kevin Morehead, a medic who had been attached to John Bolduc's team (Bolduc had been the Fifth Group soldier tasked to K2 to survey its building), paused to reflect on the injuries two close friends had survived in Afghanistan.

"I would never in my wildest dreams have said, 'Mike [a team-mate], you're gonna get your right arm blown off. And Corey, you're gonna get shot in the gut.'"

At that time, I was visiting with Morehead in his team room at Fort Campbell. Strewn around us were radios, weapons, and Iraqi tank manuals, which men like Morehead were studying before they headed into Baghdad.

"To me," said Morehead, "that's the amazing part of the story. You're sitting here talking to me, and maybe you'll look back and say, 'I knew that guy once—that's the guy that got killed on infil, goin' into some other place.'"

Morehead was killed in the same firefight alongside Bill Bennett.

When I had asked Morehead how this premonition that he might be killed affected his daily life, he said: "I'll tell you honestly, it doesn't, because I believe in God, and I believe in America."

As many of the same men who had ridden horses against the Taliban in Afghanistan were deployed to Iraq, the worst fears of the Afghans who had fought with them against the Taliban were confirmed.

Nadir Shihab, whose house had been blown up by the Taliban during the fighting at Qala-i-Janghi, told me, "We were very happy, sir, after the Americans came to Afghanistan. *Very happy*.

"The Taliban was a bad regime. And now we want to have secu-rity. We want to rebuild our country once again, sir." He wondered if the Americans would stay and help with this enormous task.

Commander Ahmed Lal, the subcommander who'd fought alongside Nelson, later echoed his same concern to a U.S. government official. "We are men," Lal had said. "If we give our hand to someone, we will be with him until the last drops of our blood."

Lal, who had sacrificed much while fighting the Taliban—he had not seen his family, who were living in exile in Iran, in over five years—felt a bond with his American counterparts, one that he did not want to see broken.

In May 2003, when Ambassador Paul Bremer, the director of reconstruction and humanitarian assistance in Iraq, "fired" the Iraq National Army by disbanding it, I received a phone call from a soldier who'd fought in Afghanistan alongside General Dostum.

"We just lost Iraq," said the officer, referring to Bremer's decision.

This was well before pundits or reporters had begun making similar assessments about America's fight in that country.

I asked what he meant.

"Ambassador Bremer has sent 500,000 young men home with their weapons, after we've bombed their country. They're angry. In the end, they won't be on our side."

The Special Forces officer was correct. Instead of assimilating, and working with, this former enemy army, the Americans had driven it underground, where it mutated into a potent insurgency.

That this officer had anticipated this outcome was not surprising. During the course of my research, it became clear that solutions to problems such as these likely would be found in the ethos of the Special Forces community. Just as a painter would not study light without studying shadow, or as a composer would not consider one tempo without its counterpoint, we would do worse than to study war-fighting as practiced by these men in order to study and create peace. Wars, as the earlier military thinker Carl von Clausewitz pointed out, are not fought to kill people; they are fought to effect political change. They are violent, expensive, and represent one of the universe's great rifts in the social contract. To study peace, then, is, de facto, to study war. Any political or social movement, of any stripe, that does not grasp the degree to which these opposites are actually twins is fruitless.

In this way, the story recounted here is also a flag raised against the brute visage of fundamentalism, in all its forms, here and abroad. The book is, I hope, an account of religious and cultural hubris and misanthropy. What struck me during my research was learning the degree to which violence had often been a third or fourth choice in resolving conflict. Indeed, some men in this book never fired their weapon, even when doing so would have put an "end" to a problem. Instead, the crisis of a particular moment was fixed by crouching in the dirt with a stick, opposite the "opponent," and scratching out a solution. This method, though time-intensive, can be far more effective and lasting than kicking down doors, guns blazing, a more usual (albeit often incorrect) perception of a modern soldier's modus operandi.

In short, the story recounted here seems to exist in another time, before news cycles filled with stories about abuses and sometimes confused military thinking. Because the Horse Soldiers were so underequipped when they deployed, and because, in fact, the United States was not prepared at all to fight a war in Afghanistan, these warriors landed in Afghanistan and comported themselves with the nuanced awareness of anthropologists, diplomats, and social workers. They had realized that their deployment was historic. Indeed, Special Forces, and America, had never fought a war in just this way. These soldiers were loath to offend any customs of local people or to appear as hegemonic imperialists. This demeanor is drilled into them during their grueling training. As Major (formerly Captain) Dean Nosorog recently told me, it's precisely this ethos that accounted for their success.

In Iraq, America would be perceived by the local population to be the invader, the heavy-handed imperialist. In Afghanistan, the Taliban, especially the "foreign" Taliban soldiers, bore this unlucky distinction. But as more men and money and lives were lost in Iraq, the Taliban regrouped in Afghanistan, feeding on the growing discontent of villagers who did not see the promise of a post-U.S. victory bringing a new, prosperous future.

Millions of dollars of aid poured into the country, yet the funds remained bottled up in Kabul, a city now teeming with nongovernment workers and diplomats afraid of Taliban attack in the countryside. This is unfortunate, because the rural and remote regions are often where the goods and services are needed most. Car bombings, kidnappings, ambushes, all tools from the war in Iraq, are now endemic in Afghanistan.

In light of these developments, U.S. officials have gone so far as to announce a new willingness to deal with moderate elements of the Taliban organization, an enlightened move that men like Dean would approve of. "We weakened their leadership," he told me, "by asking them to defect and join us. They started fighting each other. Their organization collapsed from the inside as well."

In the parlance of guerrilla war, we will have to "get down in the weeds" and also work a diplomatic magic from there. The reality of working with and uniting disparate factions can be fraught with traps, such as when, after the Battle of Qala-i-Janghi, Dostum's men were accused of sequestering hundreds of Taliban soldiers in truck containers and suffocating them, a timeworn method of murder also practiced by the Taliban.

At present, in early 2009, the Taliban once again control large portions of Afghanistan, and to subdue them, the U.S. government has promised to commit greater forces in the entire country. At the same time, Pakistan is becoming less politically stable, a development that affects both Afghanistan and America's ability to defeat the Taliban. The clock is ticking.

Afghanistan president Hamid Karzai, the saying goes, is really just the mayor of Kabul.

On February 11, 2009, however, even that seemed in doubt. Suicide bombers, reportedly supported by Pakistani fundamentalists, shocked Kabul's four million residents by attacking the Ministry of Education, Directorate of Prisons, and the Ministry of Justice, killing, according to the *New York Times*, "at least 20 people and wounding 57." These attacks took place just several hundred yards from Karzai's presidential palace.

People would later ask Master Sergeant Pat Essex, "Did we do the right thing by fighting in Afghanistan? Do you think we made a difference?"

Essex felt he could reply, "You won't be able to say today or tomorrow if it was the right thing. You're gonna have to go back to Afghanistan in ten or fifteen years from now and say, 'Was this right?'"

He believed that it was.

In Memory of U.S. Army Fifth Special Forces Group
Soldiers Who Died in Operation Enduring Freedom and Iraqi
Freedom

Dustin Adkins, Specialist
William Bennett, Sergeant First Class
Jason Brown, Staff Sergeant
Nathan Chapman, Sergeant First Class
Jefferson Davis, Master Sergeant
Gary Harper Jr., Staff Sergeant
Aaron Holleyman, Staff Sergeant
Matthew Kimmell, Staff Sergeant
Paul Mardis Jr., Staff Sergeant
Ryan Maseth, Staff Sergeant
Kevin Morehead, Master Sergeant
Daniel Petithory, Sergeant First Class
Brian Prosser, Staff Sergeant
Michael Stack, Sergeant Major
Paul Syverson III, Major
Ayman Taha, Staff Sergeant
Michael Tarlavsky, Captain
Benjamin Tiffner, Captain
Brett Walden, Sergeant First Class
Justin Whiting, Staff Sergeant
Daniel Winegeart, Specialist

ACKNOWLEDGMENTS
AND SOURCES

I would like to thank the following people for their assistance and support as I traveled in the United States and Afghanistan conducting interviews, gathering research, and touring key sites. In all, I conducted approximately one hundred interviews with pilots (helicopter and fixed wing), soldiers, civilians, and family members, and consulted material such as news and magazine articles, books, scholarly papers, soldiers' journals, monographs, "after-action reports," hundreds of photographs (taken by soldiers in battle), and various detailed documents describing the movements of the Special Forces soldiers in Afghanistan, as well as interviews with their Afghan colleagues.

Much of my research is drawn from my own interviews with the principal people involved, as well as the observations I made while visiting key locales in Afghanistan. Because of this, I have been able to look at the campaign of the Horse Soldiers from many important perspectives—the ground, the air, and the living rooms of the families left behind as their fathers and husbands deployed to Afghanistan. In short, I was able to speak with someone (and in most cases, multiple individuals) involved in nearly every element of the campaign.

I would also like to express my thanks to the soldiers and members of the defense community who met with me, but whose names they've asked me not to disclose. (Some appear below using their pseudonyms.) All military ranks listed here, unless otherwise noted, are contemporaneous with the time frame of this story.

First, I want to especially thank Major General Geoffrey Lambert (now retired) for opening doors for me within the entire Special Forces community, both in the United States and Afghanistan. His astute insight into the battle, as well as the mind of the Special Soldier, was invaluable. Likewise, I want to thank then–Fifth Group Commander Colonel John Mulholland (now general) for his hospitality whenever I visited Fort Campbell, and acknowledge as well the help of Sergeant Danny Leonard and ad hoc press officer Major James Whatley. And in Afghanistan, Colonel Jeffrey Waddell provided much appreciated support as I moved around the country.

Brigadier General David Burford, Colonel Charles King, and press officer Major Rob Gowan were invaluable whenever I visited Fort Bragg. At MacDill Air Force Base, Deputy Public Affairs Officer Ken McGraw, of U.S. Special Operations Command, helped as I arranged interviews. The following individuals also extended a welcome in the course of my travels: Colonel John Knie, Colonel Warner "Rocky" Farr, Command Chief Warrant Officer Lawrence Plesser, Major Gary Kolb, Major William Owen, Major Christopher Fox, Lieutenant Colonel Kent Crossley, Carol Darby, Colonel Robert L. Caslen, Tommy Bolton, Colonel Manuel Diemer, Major Christopher Miller, Colonel John Fenzel, Barbara Hall, Marie Hatch, Jim Ivie, Gabe Johnson, General Mike Jones, Major Rich Patterson, Major Scott Stearns, and Kevin Walston.

I also want to acknowledge the hard work and scholarship of Dr. Charles H. Briscoe, U.S. Army Special Operations Command historian, at Fort Bragg, who provided counsel and support, and pointed me in the right direction to invaluable information concerning Afghan soldiers and American personnel on the ground. I also appreciate the support I received from archivist Cyn Harden, and U.S. Army Special Warfare Center and School historian Dr. Kenn Finlayson. Dr. Briscoe, along with a team of writers—Richard L. Kiper, James A. Schroder, and Kalev I. Sepp—wrote a definitive and especially useful history of Special Operations forces in Afghanistan titled *Weapon of Choice*, which I referred to throughout my writing, in particular when I was describing aspects of the K2 base camp; the inter-tribal fighting among Afghanistan's ethnic groups; and Massoud's predicament as presented in September 2001, plus the movements of soldiers in general. Thank you, Chuck et al.

ACKNOWLEDGMENTS AND SOURCES

I want to thank Major Dean Nosorog for visiting my home for a week of intensive, intellectually challenging talks and interviews, ranging from the history of the Middle East to present-day Iran, to the challenges of riding a horse into war. Dean, like many a Special Forces soldier, is a voracious observer, possessing a highly syncretic bent of mind. It was he who, among his numerous anecdotes, included a description of the refugee camps near Mazar-i-Sharif, and offered keen observations of Atta the warlord. He further dictated a fascinating account of his team's entrance into the city. I also want to thank other members of Dean's team for talking with me, including Jerry Booker, Darrin Clous, Mark House, Brad Highland, Stu Mansfield, Brian Lyle, and James Gold. Likewise, Cal Spencer, Sam Diller (now retired), John Bolduc (retired), and Lieutenant Colonel Max Bowers (retired) also opened their homes to me. Bowers loaned me his battle map of the campaign, which he carried on horseback, and upon which the post–November 2 movement of all U.S. troops in the region was planned. It was of special significance to hold in hand this dog-eared document and to study the grease pencil marks outlining the positions.

I'm also grateful for the opportunity to extensively interview Captain Mitch Nelson and his team members, including Sam Diller, Cal Spencer, Pat Essex, Ben Milo, Sonny Tatum (U.S. Air Force), and Scott Black. I want to thank Major Mark Mitchell (now colonel, commander of the Fifth Special Forces Group) for taking the time to patiently answer my questions, especially during one long stretch in the Tampa Public Library, and in countless phone calls and e-mails. All of the insights and recollections of these men, and their colleagues, were integral to reconstructing the thoughts, words, and actions of the Horse Soldiers and the battle at Qala-i-Janghi. Throughout, dialogue is drawn from primary interviews, previously published accounts, videotapes, transcripts, journals, and monographs of the events, and from people later privy to them. Along with Mitchell, I want to thank Kurt Sonntag, Pete Bach, Martin Homer, Steve Billings, Roger Palmer, Burt Docks, Malcolm Victors, Kevin Leahy, Dave Betz, and Ernest Bates. Many people welcomed me into their lives and entrusted me with their stories: I am grateful.

For an understanding of guerrilla warfare, I was greatly aided by my weeklong visit and participation in a Special Forces training exercise, called Robin Sage, in the woods of North Carolina. There, Special Forces soldiers to-be are immersed in a real-time scenario of life in a foreign guerrilla camp. The arriving Americans must ingratiate themselves with the "local population," grasp the locals' will, and fight alongside their warlord toward a desired political and social change. I'm grateful to Major Scott Stearns (retired), Major Kathleen Devine (retired), and Major General Jerry Boykin, who welcomed me at the John F. Kennedy Special Warfare Center and School at Fort Bragg and introduced me to members of the training community, among them the entertaining Brian Bolger.

I also greatly benefited from discussions about Afghanistan and Pakistan, past and present, with Greg Mortenson, author of *Three Cups of Tea*. Mortenson's work on behalf of the citizens of these countries is inspiring and groundbreaking, and his sense of selflessness is something to aspire to. Likewise, I benefited from meeting Thomas Gouttierre, director of the Center for Afghanistan Studies at the University of Nebraska, Omaha. These two educators are on the cutting edge for crafting new diplomacy in the world's troubled spots.

For a complete listing of secondary sources, see the bibliography. In particular, I want to acknowledge the following authors and their work:

For information about the last days of hijacker Mohammed Atta and his cohorts, and the attacks of September 11, 2001, and America's military response to these attacks, I referred to numerous reports by Terry McDermott of the *Los Angeles Times*. I also used "Four Corners" from the Australian Broadcasting Corporation, broadcast November 12, 2001; "Four in 9/11 Plot Are Called Tied to Qaeda in '00," by Douglas Jehl, published in the *New York Times*, August 9, 2005; "Atta's Odyssey," published October 8, 2001, in *Time*; "The Plot Comes Into Focus," by John Cloud, published October 1, 2001, also in *Time*; "The Hijackers We Let Escape," by Michael Isikoff and Daniel Klaidman, published June 10, 2002, in *Newsweek*; "They Had a Plan," published August 12, 2002, in *Time*; "The Night Before Terror," from staff reports, published October 5, 2001, in the *Portland Press Herald*; "Atta's Will Found," posted on www.abcnews.com, October 4, 2001; *The 9/11 Commission Report: Final Report of the National Commission on Terrorist Attacks Upon the United States*, published by W.W. Norton; and *The 9/11 Report: The National Commission on Terrorist Attacks Upon the United States*, published by St. Martin's Press, 2004. Technical information about the hijacked planes came from the National Institute of Standards and Technology, September 2005, and the book-length reporting of Bob Woodward, Rowan Scarborough, Norman Friedman, George Friedman, and Gerald Posner.

In addition to my primary interviews and other sources of research, for information about Ahmed Shah Massoud, including his assassination, I referred to *The Lion's Grave: Dispatches from Afghanistan*, by Jon Lee Anderson; "Slowly Stalking the Lion," by Craig Pyes and William Rempel, published in the *Los Angeles Times*, June 14, 2002; "The Lion in Winter," by Sebastian Junger, published in *National Geographic Adventure*, March/April 2001; "Good at War, Poor at Peace," by Luke Harding, published in the *Guardian*, September 12, 2001; "Afghanistan Reporter Looks Back on Two Decades of Change," by D. L. Parsell, *National Geographic News*, published November 19, 2001; and "Massoud's Last Words," a *Newsweek* Web exclusive, posted September 20, 2001. For details about Massoud's appearance after his assassination, I referred to "The Assassins," by Jon Lee Anderson, published in the *New Yorker*, June 10, 2002. I also referenced "A Gruesome Record," by Michael Griffin, published in the *Guardian*, November 16, 2001; "The Afghan Who Won the Cold War," by Robert D. Kaplan, published in the *Wall Street Journal*, May 5, 1992; and www.Afgha.com for a biography of Massoud, posted August 31, 2006.

In addition to primary interviews and other sources, for information about the surrender at Konduz and some details about the surrender at Qala-i-Janghi between General Dostum and Mullah Faisal, I referred to the following written reports: "Paper Surrender Blowing in the Wind," by Luke Harding, published November 23, 2001, in the *Guardian*; "Doomed Arab Units Prepare for Final Battle Against the Odds," by Khaled Dawoud, Julian Borger, and Nicholas Watt, published November 20, 2001, in the *Guardian*; and the following reporting of Ian Cobain, published by Times Newspapers Ltd.: "Foreign Fighters Resist Alliance," November 15,

2001; "Refugees Tell of Frenzied Killing in Besieged City," November 19, 2001; and "America Will Take No Prisoners," November 20, 2001. Also helpful for these topics were: "Alliance Says Non-Afghan Taliban Unwilling to Negotiate in Kunduz," by Sharon LaFraniere, published November 20, 2001, in the *Washington Post*; "The Rout of the Taliban, Part Two" by Peter Beaumont, Kamal Ahmed, Ed Vulliamy, Jason Burke, Chris Stephen, Tim Judah, and Paul Harris, published November 18, 2001, in the *Guardian*; and "Kunduz: Northern Stronghold Ready to Capitulate," by Luke Harding, Nicolas Watt, and Brian Whitaker, published November 22, 2001, in the *Guardian*.

For further understanding about the history of Special Forces, and war fighting past and future, I was helped by decades of thoughtful work published in dozens of books and articles about Special Forces and Afghanistan. To research the evolution of conventional military units into "culturally responsive" Special Forces teams, I consulted *The Devil's Brigade* by Robert H. Adleman and Colonel George Walton; and *From OSS to Green Berets: The Birth of Special Forces*, by (retired) Colonel Aaron Bank. Of special interest was the provocative and lucid *The New Face of War: How War Will Be Fought in the 21st Century*, by Bruce Berkowitz, and *The Transformation of War* by Martin van Creveld. These are incredibly compelling works. I also found helpful information in A *Tribute to Special Operations*, as described in "The Green Berets," by John D. Gresham; "WWII Special Operations Forces," by Dwight J. Zimmerman; and "USASOC History: From Jedburghs to Devils and Snakes," by Barbara Hall, published 2003 by Faircount LLC.

For insight into America's relationship to post–Cold War Afghanistan, Anthony Cordesman's book *The Lessons of Afghanistan: War Fighting, Intelligence and Force Transformation* was also insightful, as was Samuel B. Griffith's translation of *On Guerrilla Warfare*, by Mao Tse-tung.

I also referred to Colonel Francis J. Kelly's *U.S. Army Special Forces, 1961–1971*; *The Oxford Companion to American Military History*, edited by John Whiteclay Chambers II; *OSS: The Secret History of America's First Central Intelligence Agency*, by Richard Harris Smith; *U.S. Special Forces: A Guide to America's Special Operations Units, The World's Most Elite Fighting Force*, by Samuel

A. Southworth and Stephen Tanner; *Our Vietnam: The War 1954–1975*, by A.J. Langguth; and Shelby L. Stanton's *Green Berets at War: U.S. Army Special Forces in Southeast Asia, 1956–1975*. For a fascinating look at the traits of the Special Forces soldier, see "The Making of a Perfect Soldier," by Linda Carroll, broadcast by MSNBC on March 7, 2002; "Walking Point," by Linda Robinson, published October 18, 2004, in *U.S. News & World Report*; and "A Bulletproof Mind," by Peter Maass, published November 10, 2002, in the *New York Times Magazine*.

For information about the financial cost of the campaign in Afghanistan and the number of personnel involved, I referred to Gary Berntsen's *Jawbreaker* and Bob Woodward's *Bush at War*. For information and insight into General Atta, Afghan politics and history, and Afghan battle experiences in particular, I have relied on extensive interviews, often spanning several time periods, with the following men on Dean Nosorog's team: Stu Mansfield, Darrin Clous, Brad Highland, Jerry Booker, James Gold, Mark House, Brian Lyle, Donny Boyle (Air Force combat controller), and Brett Walden. Some of these men also provided me with journals, maps, photos, and reports, these last recorded in the field as the battle unfolded. All of these materials offered a palpable sense of the battle as seen from horseback. I also referred to "Afghan Militias 'Should Disband,'" by Jannat Jalil, broadcast by the BBC, July 19, 2003.

On Mitch Nelson's team, a number of people provided similar, candid insight into their experiences, as well as the thoughts and actions of the Afghan soldiers fighting with them. They include General Dostum, Cal Spencer, Sam Diller, Scott Black, Ben Milo, Pat Essex, and Sonny Tatum (Air Force combat controller). For biographical information about Sergeant First Class William Bennett, I drew from "Three Soldiers, Many Mourners," by Scott Pelley, from the CBS News broadcast on *60 Minutes II*, July 28, 2004. In addition to primary interviews with teammates, information about Air Force combat controller Malcolm Victors was drawn from "Mazar I Sharif," by Wil S. Hylton, published in *Esquire*, August 2002, as well as "The Liberation of Mazar-e Sharif: 5th SF Group Conducts UW In Afghanistan," by personnel of 3rd Battalion, Fifth Special Forces Group, published in *Special Warfare* magazine, June 2002, and "The Story of ODA 595," by Barbara Hall, released by the

United States Special Forces Command. See also PBS's *Frontline: Campaign Against Terror,* broadcast on August 2, 2002, in which team members described some of their adventures on horseback.

For information about the amazing feats of flight performed by the pilots and crew of the 160th SOAR, I want to thank the following members of the Nightstalker community for sharing their experiences with me at the Fort Campbell headquarters: Greg Gibson, John Garfield, Tom Dingman, Jerry Edwards, Steve Porter, Carson Millhouse, and Will Ferguson. Pilot Greg Gibson went to great effort in arranging a "flight" in a Chinook helicopter over Afghanistan. Pilot Jerry Edwards provided me with a journal of events and personal thoughts recorded as the war unfolded. These interviews were invaluable in re-creating the flight of the Special Forces teams into Afghanistan. Truly, the Nightstalkers' story in Afghanistan is an amazing one.

The following were valuable in describing the history and actions of the CIA: *Secret Armies: The Full Story of the SAS, Delta Force and Spetsnaz,* by James Adams; *Secret Warriors: Inside the Covert Military Operations of the Reagan Era,* by Steve Emerson; *The Book of Honor: The Secret Lives and Deaths of CIA Operatives,* by Ted Gup; and *The CIA at War: Inside the Secret Campaign Against Terror,* by Ronald Kessler.

For insight into the personality, thoughts, words, and actions of CIA paramilitary officer Mike Spann, I was aided by magazine articles about Spann and his wife, Shannon, and Spann's parents and extended family, as well as numerous news accounts concerning the CIA's overall presence in Afghanistan. In particular, Shannon Spann has spoken in articles and public speeches about her thoughts and feelings about Mike while he was deployed; and Spann's father, Johnny Spann, has spoken in print and on television about the ordeal of his son's death. Two books by former CIA officers, who were also contemporaries of Spann's in Afghanistan, were helpful in sorting out the physical details of the movements of Agency officers, including Spann's, as well as of the officers of the Afghan soldiers. The books helped elucidate General Atta Mohammed Noor's state of mind. These books are: *First In,* by Gary Schroen, and *Jawbreaker,* by Gary Berntsen. It was from these books that I drew the anecdotes of Schroen's encounter with Counter-

Terrorist Center Director Cofer Black; the beheading of a Taliban soldier by a Northern Alliance member after a horse charge; and the encounter between Schroen and Northern Alliance leader Fahim Khan, as well as other details about CIA officers' actions, including conversations among the CIA officers present during a horse charge. CNN's documentary *House of War: Uprising at Mazar-e Sharif*, also provided invaluable images and dialogue about Mike Spann and Dave Olson's interrogation of John Walker Lindh, as did "He's Got to Decide If He Wants to Live or Die Here," a *Newsweek* Web exclusive, by Colin Soloway, December 6, 2001, which includes a transcript of the interrogation. Insight into the actions of Dave Olson and Mike Spann and the events they participated in was also provided by my interviews with the Special Forces soldiers who traveled and worked closely with both men.

The following books, articles, and websites were instrumental in re-creating Mike Spann's journey from Winfield, Alabama, his hometown, to his final hours at Qala-i-Janghi. Of special importance was "Love in a Time of War," by Edward Klein, published August 18, 2002, in *Parade* magazine, in which Shannon Spann gave a detailed account of her life with Mike.

Other reporting by the *New York Times* was also helpful: "One for His Country, and One Against It," by Blaine Harden with Kevin Sack, published December 11, 2001; "Agent Praised as Patriot in Graveside Ceremony," by Diana Jean Schemo, published December 11, 2001; and "CIA Names Agent Killed in Fortress," by James Risen, published November 29, 2001. An article titled "Community Recalls a Native Son with Clear Goals," by Kevin Sack, published November 29, 2001, contained details of Spann's boyhood pastimes. And I also drew from CNN.com transcripts: "CIA Officer Michael Spann Buried at Arlington National Cemetery," which aired December 10, 2001; "Discussion with Widow of First American to Die in Afghan Combat," which aired September 14, 2002; "Interview with Shannon Spann," which aired July 16, 2002; "Family of Michael Spann Speak to Reporters Following Lindh Not Guilty Plea," which aired February 13, 2002; and *CNN Presents: House of War: The Uprising at Mazar-e Sharif*, which aired August 3, 2002. The following articles by Richard Serrano, published in the *Los Angeles Times*, were also helpful: "Detainees Describe CIA

Agent's Slaying," published December 8, 2004, and "Driven by a Son's Sacrifice," published April 7, 2005.

The following helpful articles about Spann and his family were written by Jeffrey McMurray and published by the Associated Press: "Father on Crusade to Prove Afghanistan Ambush Killed CIA Officer," published March 12, 2005, and "CIA Agent's Dad Probes Deadly Afghan Riot," published March 13, 2005. Several websites also offer photos and information about Mike Spann, his extended family, and his career in the U.S. Marine Corps and the CIA, including his 1999 CIA application essay. The websites are www.honor mikespann.com and the Arlington National Cemetery website, featuring "Johnny Micheal Spann, An American Hero." For information about correspondence between Shannon Spann and Mike Spann while he was in Afghanistan, I referred to the report "Shannon Spann," published on the website www.embracehisgrace.com. For information about Spann's death announcement and eulogies about his death and burial, see also various press releases and statements released by the Central Intelligence Agency; "CIA's Spann Buried at Arlington," by Mary Orndorff, published December 11, 2001, in the *Birmingham News*; and "CIA Reports Officer Killed in Prison Uprising," by Vernon Loeb and Josh White, published November 29, 2001, in the *Washington Post*. "The CIA's Secret Army," by Douglas Waller, published February 3, 2003, in *Time* was especially illuminating, as was "A Street Fight," by Evan Thomas, published April 29, 2002, in *Newsweek*. The website www.winfield city.org provided details about Spann's hometown.

Likewise, I was aided in reconstructing the thoughts, words, and actions of John Walker Lindh by consulting voluminous pages of court documents from Lindh's trial, describing Lindh's movements, and numerous newspaper and magazine accounts about Lindh and his family, as well as Mark Kukis's courageously researched *My Heart Became Attached*. Lindh's father, Frank Lindh, has written and spoken publicly about his family's ordeal, offering insights into his son's journey from California to Afghanistan. See especially the elder Lindh's speech before the Commonwealth Club of California, January 19, 2006, titled "The Human Rights Implication of 'The American Taliban' Case," as well as "The Real Story of John Walker Lindh," by Frank Lindh, on AlterNet, posted January 24, 2006; and

"Father of a U.S. Taliban Fighter Speaks Out," posted January 20, 2006. See also "Taking the Stand, The Crimeless Crime: The Prosecution of John Walker Lindh," by Frank R. Lindh, www.dcbar.org, May 2005; "He's a Really Good Boy," by Karen Breslau and Colin Soloway, a *Newsweek* Web exclusive, updated December 7, 2001; and "John Lindh Not a Traitor, Father Argues," by Kevin Fagan published by the *San Francisco Chronicle*, January 20, 2006.

For information about Lindh's classmates' perceptions of him in Yemen, I referred to "Bright Boy from the California Suburbs Who Turned Taliban Warrior," by Julian Borger, published by the *Guardian*, October 5, 2002.

For information (physical details, dialogue, and psychological insight), the following magazine articles were especially helpful: Jane Mayer's keenly observed "Lost in the Jihad," published in the *New Yorker*, March 10, 2003; "The Making of John Walker Lindh," published in *Time*, October 7, 2002; "The Long Strange Trip to the Taliban," by Evan Thomas, published in *Newsweek*, December 17, 2001; "Periscope," by Colin Soloway, published in *Newsweek*, December 31, 2001; "Tale of an American Talib," a *Newsweek* Web exclusive, by Colin Soloway, posted December 1, 2001; "In Defense of John Walker Lindh," a *Newsweek* Web exclusive, posted March 16, 2002, by Karen Breslau; and "Innocent," by Tom Junod, published in *Esquire*, July 2006. The Center for Cooperative Research has also published useful, encyclopedic-like information about Lindh. See "Are You Going to Talk to Us?" published December 17, 2001, by *Newsweek*, for a transcript of Spann's interrogation of Lindh; and "U.S. Taliban Fighter Describes Fortress Horror," by Michael Ellison, published by the *Guardian*, December 3, 2001. See www.CNN.com, "Transcript of John Walker Interview," posted July 4, 2002, for information about Lindh's state of mind and medical treatment after his capture; and "Walker's Brush with Bin Laden," by Daniel Klaidman and Michael Isikoff, published December 31, 2001, in *Newsweek*. In particular, Colin Soloway has been an omnipresent reportorial lens on the Lindh story.

For my account of Lindh's actions at the Turkish Schoolhouse in Mazar-i-Sharif, I also relied on my interviews with soldiers tasked with guarding him. In particular, Lindh's attorney James Brosnahan spoke to me about the legal issues concerning the trial and provided me, among sundry documents, with the transcript of Lindh's encounter with Special Forces medic William Bennett and a free-lance reporter, Robert Pelton, after Lindh's capture. For my account of Colin Soloway's discovery of Lindh in Afghanistan, I referred to Soloway's article in *Newsweek*, December 7, 2001, describing his odd encounter with Lindh after the Qala-i-Janghi battle, as well as a written account of the event provided to me by photographer Damien Degueldre, who accompanied Soloway during the discovery of Lindh.

For information about the Qala-i-Janghi Fortress and the battle there, I relied on my interviews with a majority of the soldiers involved, particularly Mark Mitchell, Kurt Sonntag, Dave Betz, Roger Palmer, Steve Billings, Martin Homer, Pete Bach, Jason Kubanek, Ernest Bates, and Malcolm Victors, Air Force combat controller; Burt Docks, Air Force combat controller; Don Winslow, Paul Syverson, Craig McFarland, and Kevin Leahy; and a number of U.S. Army Tenth Mountain Division soldiers.

Roger Palmer and Steve Billings recounted to me Dave Olson's account of Spann's death, including Olson's mad dash to safety. Najeeb Quarishy graciously provided invaluable information about the battle at Qala-i-Janghi, as well as a thorough account of the Americans' entrance into the town, and a portrait of Afghan life under Taliban rule. Thank you, Najeeb.

Further details were provided to me through extensive monographs written by the soldiers involved in the battle, which offer a nearly hour-by-hour account. I also consulted after-action reports written by Mitchell and Sonntag. Further documentation and photos of the Qala-i-Janghi battle were gleaned from copious U.S. Army briefings delivered by the men afterward. For Major Mark Mitchell's statement at his November 14, 2003, Distinguished Service Cross ceremony, I referenced the videotaped and transcribed record of the event, as well as the recollections of Mitchell himself.

To see the site of the battle firsthand, I spent several days with former Northern Alliance soldier Ali Sarwar, walking through Qala-i-Janghi, stepping over bullet casings, land mines, and shards of human bone left from the battle. (In a corner of the fort, we found a human skull that had resurfaced in the mud.) As we walked, Ali narrated his and his men's actions, a bold, harrowing account. Ali posed on the fort's south parapet and explained the fighting that had raged several hundred yards below him. I am grateful for this rare, close-up view of the battle.

Likewise, I want to express my admiration and gratitude to *Time* reporter Alex Perry and videographer Dodge Billingsley of Combat Films for their invaluable assistance. These two men made their way into the fort during the fighting, recording some of it in print and film.

I am grateful to Dodge for sending me a dozen hours of his raw footage, and to Alex for studiously and with good humor answering my copious questions by phone and e-mail, about both his experience in the fort and the events unfolding around the Americans, Brits, Afghans, and other reporters during the battle. Perry's prescient article for *Time* "Inside the Battle at Qala-i-Janghi," published December 1, 2001, as well as his other reporting about events before and after the war in Afghanistan make for fascinating, humane, and accurate reading about this historical moment. See especially Perry's Q & A titled "Update: American Rescued from Taliban-held Fort," published by *Time*, November 27, 2001. Heartfelt thanks to both reporters.

For information and details about German reporter Arnim Stauth's encounter with CIA officer Dave Olson, I consulted "Those Would Have Killed Us," published August 2002; and from *Transnational Broadcasting Studies*, Issue 9, "Thirteen Months After the 9/11 Attacks: Terrorism, Patriotism and Media Coverage." I also relied on the documentary *House of War: The Uprising at Mazar-e Sharif* as well as the recollections of reporter Alex Perry and videographer Dodge Billingsley.

For information about Mullah Faisal's surrender at Qala-i-Janghi, as well as Dostum's actions and state of mind after returning to the fort following the uprising, I referred to Damien McElroy's story "I'm Sick of Death, Says Dostum the Warlord," published November 29, 2001, by the *Telegraph* (London); Lieutenant Colonel Max Bowers's firsthand recollections of accompanying Dostum during the surrender; the reporting of *Guardian* reporter Luke Harding, Kurt Sonntag, and Mark Mitchell's post-uprising analysis of its genesis; the work of reporter and videographer Dodge Billingsley; and the recollections of numerous Special Forces privy to the uprising's aftermath. News of the blood money paid by Mullah Faisal to Dostum comes from Jane Mayer's "Lost in the Jihad," published in the *New Yorker*, March 10, 2003. Information about various traits and concerns of Taliban and Al Qaeda soldiers are drawn from their statements made, after their capture, to U.S. government officials contained in "Detainee Statements. Combatant Status Review Tribunals conducted at Guantanamo Bay Naval Base" (www.defense link.mil). Other information about the Taliban was drawn from "Mazar-e Taliban, R.I.P.," by Daniel Lak, published by *Outlook India*, December 16, 2001; "The Massacre in Mazar-i-Sharif," *Human Rights Watch*, November 8, 1998 (vol. 10, no. 7); "Taliban Kabul Diary," by Jason Burke, published by the *London Review of Books* (vol. 23, no. 6), March 22, 2001; "Country Reports on Human Rights Practices for 1994," submitted to the Committee on International Relations, U.S. House of Representatives, and the Committee on Foreign Relations, U.S. Senate, by the Department of State, 1995; "Country Reports on Human Rights Practices for 1995–96," submitted to the Committee on International Relations, U.S. House of Representatives and the Committee on Foreign Relations, U.S. Senate, 1996; *Taliban: Militant Islam, Oil & Fundamentalism in Central Asia*, by Ahmed Rashid; and *Reaping the Whirlwind: The Taliban Movement in Afghanistan*, by Michael Griffin.

While touring Qala-i-Janghi and Mazar-i-Sharif, I received important logistical assistance and welcome hospitality from Colonel Brian Harris, Lieutenant Daryl Hodges, Staff Sergeant Jeffrey Ewing, Sergeant Christopher Carpenter, Sergeant Kasey Phillips, Specialist Damien Miller, and Specialist Oliver Jackson. Major Eric Bloom was a gracious and helpful press officer at Bagram Air Base, Afghanistan; and while I was there, Sergeant Major Stan Parker helped get me onto an instructive ride on a Chinook helicopter. Thanks also to Brigadier General James Champion who, as deputy commanding general (Operations) of Combined Joint Task Force-76, met with me while I was at Bagram. Thanks as well to photographer Jonas Dovydenas for his hospitality in Kabul. Too numerous to thank individually are all of the other Special Forces and regular Army soldiers at Bagram and Camp Tillman, near Mazar-i-Sharif, who aided my travel to key research sites. A special thanks to Jesse Ooten for loaning his hootch at Bagram to this traveling writer.

I am indebted to Lieutenant Colonel Pablo Hernandez for his guidance as I interviewed Afghan government officials and former mujahideen fighters at the presidential headquarters in Kabul. Chuck Ricks, of Indiana, was also an invaluable ally in Kabul. When we met, Chuck was working in the Office of Parliamentary Relations and Public Affairs, Afghanistan Ministry of Defense, and he immediately helped arrange interviews with General Abdul Rashid Dostum and General Atta Mohammed Noor, as well as with key former members of the Northern Alliance then serving in the government. To meet these men, many of whom had fought against the Soviets, was an honor.

I want to thank the following Afghan generals and soldiers for their generosity in meeting with me in Kabul and Mazar-i-Sharif, where some of them provided important insights and details about the overall battle: General Atta Mohammad Noor, General Abdul Rashid Dostum, Deputy Defense Minister A. Yusuf Nuristani, Muhamad Tamimi Huma, Matin Sharifi, General Taj Mohammed, General Atta Yama, Deputy Defense Minister Mohammad Humayun Fawzi, General Baz Mohammad Jowhari, Deputy Defense Minister of Policy and Strategy Major General Muhebbulah, Major General Taj Mohammad, General Azimi, and subcommander Ali Sarwar, of General Dostum's army. Some information about the events of Ahmed Shah Massoud's last day come from my interview with "Colonel Paima," who was with the Tajik leader the day of his assassination.

Other information about General Dostum came from: "Mujaheddin Write Their Name in Blood," by Jon Swain, published by the *Sunday Times* (London), November 11, 2001; "Makeover for a Warlord," by Anthony Davis, published in *Time*, June 3, 2002; "Profile: General Rashid Dostum," on BBC World News, September 25, 2001; and "Rashid Dostum, The Treacherous General," by Patrick Cockburn, published December 1, 2001, by the *Independent* (UK).

Other important interviews, as well as assistance with translation and logistical support, graciously came from, among many, Mohibullah Quarishy, Najeeb Quarishy, Nadir Shihab, "Rocky"

ACKNOWLEDGMENTS AND SOURCES

Bahari, Yama Bassam, Nadir Ali, and, last, Abdul Matin, who accompanied me on key visits with Dostum and Atta in Kabul and Mazar-i-Sharif. I apologize in advance for any others whose names I have inadvertently missed.

For information about Afghan wildlife and woodcraft, I consulted Crosslines Essential Field Guides *Afghanistan,* second edition, 2004, Media Action International. Information about Afghan villages and the devastation inflicted by Taliban soldiers came from reports prepared by the United Nations High Commissioner for Refugees. For fascinating information about the country, see *Afghanistan,* by Louis Dupree, a book that has been forgotten but should be read by everyone wishing to grasp the complexity of Afghan life.

I want to thank the various U.S. Air Force pilots who shared with me their combat experiences over Afghanistan, in particular the nighttime attack of the Spectre gunship over Qala-i-Janghi Fortress.

I am also indebted to the brave soldiers of the Tenth Mountain Division for sharing their recollections about their rescue of the Special Forces soldiers injured in the errant bombing: Major General Franklin "Buster" Hagenback, Colonel Robert Caslen Jr., Lieutenant Colonel Bryan Hilferty, Sergeant Major Dennis Carey, Command Sergeant Major Frank Grippe, Staff Sergeant Thomas Abbott, Private First Class Eric Andreason, Private First Class Thomas Beers, Sergeant Douglas Covell, Sergeant Jerry Higley, Specialist David Hine, Private First Class Michael Hoke, Specialist Roland Miskimon, and Private First Class Thomas Short.

For some details about these soldiers at the errant JDAM bombing, I referred to "U.S. Soldiers Recount Smart Bomb Blunder," by Vernon Loeb, published February 2, 2002, in the *Washington Post*; "Troops of 10th Recount Mayhem at Mazar-e Sharif," by Paul Hornak, published by the *Watertown Daily Times*, April 4, 2002. Recollections by Special Forces soldiers who witnessed the Tenth Mountain troops in action also provided important details.

I also want to thank editor and author Heather Shaw, author of the wonderfully reviewed literary novel *Smallfish Clover,* for her invaluable research and editorial assistance in organizing the material early in the process, and for her help sketching early versions of some of the book's maps. Thank you, Heather. Thanks also to good neighbor John William for his help with map creation, and to Terrie Taylor for her interest. Last, we lost a dear friend and fellow author, Lori Hall Steele, September 1, 1964–November 19, 2008. Thanks to her family and young son, Jack: *prayers and memories*.

I'd like to express my thanks and appreciation to the publishing and editorial staff at Scribner, the best place a writer could hope to find himself or herself working. I'd like to thank Susan Moldow, Roz Lippel, Brian Belfiglio, Katie Monaghan, Jessica Manners, Katie Rizzo, and Elisa Rivlin, for making the wheels turn and for making the writing and publishing of 12 *Strong* such a rewarding experience.

I also want to acknowledge the dedicated work of *Washington Post* researcher and fact-checker Julie Tate, whose professionalism, thoroughness, and eye for facts is simply amazing. Thank you, Julie.

Last, my editor, Colin Harrison, is the kind of editor a writer dreams of working with. There is not enough hyperbole to convey the pleasure of this working relationship. Colin, because he is also a novelist, gets "it." Likewise, my agent, Sloan Harris, is the man you want in your corner. He has been a champion of this book from its beginning, a tireless fighter, and an insightful reader of its drafts. Thank you, Sloan. As well, I am extremely grateful to Ron Bernstein at ICM and to Jann Wenner, Will Dana, and Brad Wieners at *Men's Journal* for their important support.

Vic and Amy Reynolds and the staff of the superb Horizon Books in Traverse City, have long supported writers and their books, as have *The Northern Express* and the *Traverse City Record-Eagle*. All three of these entities have resisted calls for censorship. Thank you for being the best at what you do.

For welcome support, thanks go to: the dedicated staff, faculty, and parents of the Pathfinder School in Traverse City, Michigan; Bart Lewis, Janet Leahy, and Anne Cooper; writer, actor, and Iraq veteran Ben Busch; historian Dr. Tracy Busch; Sid Van Slyke, the best banker a writer could hope to meet; and Tod Williams and Kip Williams for sharing their enthusiasm for this story and for reading an early version and making welcome comments. Thanks to Betsy Beers for her long-standing interest and support, to Kima Cramer for her good cheer, and to Dr. Steve Andriese for his interest in this story. To Rob and Jen Hughes, thanks for the porch, beer, and friendship. Much appreciation to Barb and Jan Doran, Bob and Randi Sloan, Joe Mielke and Jodee Taylor, Bruce and Susan Makie, Tim Nielsen and Emily Mitchell, Tim and Terry Bazzett, Jan Richardson, and to the ever-gracious Peter Phinny, plus Ken and Joan Richmond and William Hosner. Hats off to: the crew at Cuppa Joe for keeping the cup full; the Thursday night club at Stella's: Dave Lint, Chris Smith, Ken Gum, and Grant Parsons; Jerry and Teresa Gertiser, and realtor Rick Stein, for their generosity; novelist and counterterrorism expert Chuck Pfarrer; humanitarians and box-ing coaches Bill and Robin Bustance; and friends Mike and Stephanie Long, and Nancy Flowers and Bob Butz. To Ronda and Dave Barth and the extended Stanton, Earnest, Edwards, and Ger-tiser families: thank you. To my parents, much love and apprecia-tion. To Grant and Paulette Parsons, admiration and gratitude. To my family, Anne, John, Kate, and Will, who lived this book with me: *I promise to be home. I love you all.*

This book required long absences and travel to distant and some-times dangerous places. At the beginning of the project, meeting Special Forces soldiers to interview was easier said than done. Armed with a homemade press kit I'd shipped ahead to my room at a Country Inn and Suites near Fort Campbell, I drove through post security bearing letters of introduction. I believed arranging inter-views would be a snap.

This was after the men had come home from Afghanistan. At that point they were training to deploy to Iraq (though most of America didn't know this at the time). Finding someone to inter-view was going to be tough.

In addition, there wasn't a press officer at Fifth Group to handle my requests for interviews. Special Forces soldiers, it turned out, were not in the habit of cooperating with writers. They were indeed "the Quiet Professionals." The idea that I could interview any of these personnel, even if they were available, seemed to amuse many of the people I met at Fort Campbell.

However, after several trips, I met soldiers who knew soldiers who'd fought alongside Dostum and Atta and at Qala-i-Janghi. And after almost a year, I had completed more than a few interviews. I knew I'd turned a corner when a staff officer asked *me* if I knew where a particular Fifth Group soldier was located. It turned out, I did know. I said the guy was in Arizona at a training exercise.

But this familiarity had not come easily. One day early on, I walked into a team room filled with muddy gear, weapons, radios, and maps, and asked if a soldier named Mark House happened to be there. His name had been given to me as someone who might be willing to meet with me.

One of the soldiers in the room stepped forward and asked what I wanted. He looked at me suspiciously.

"I'm working on a book," I said.

Blank stare.

Then I threw a Hail Mary: I told him that I wanted to know what it was like to wake in the predawn hours on a tree-lined street in the middle of America and leave for war. . . . Children's toys fill the cracked driveways of the neighbors' houses up and down the street. . . .

A man steps outside, walks to his car, and turns for a last look. He may not see this place again.

This was the face I wanted to see, I said to the soldier—the face of that man, in those private hours.

He held out his hand. "I'm Mark House," he said.

He smiled. "You found him."

BIBLIOGRAPHY

BOOKS

Adams, James. *Secret Armies: The Full Story of the SAS, Delta Force and Spetsnaz.* New York: Pan Books Limited,1988.

Adleman, Robert H., and Col. George Walton. *The Devil's Brigade.* Philadelphia: Chilton Press, 1966.

Alexander, Col. John B., U.S. Army (ret.). *Winning the War: Advanced Weapons, Strategies, and Concepts for the Post-9/11 World.* New York: Thomas Dunne Books, 2003.

Ali, Tariq. *The Clash of Fundamentalisms: Crusades, Jihads and Modernity.* New York: Verso, 2002.

Anderson, Jon Lee. *The Lion's Grave: Dispatches from Afghanistan.* New York: Grove Press, 2002.

Archer, Chalmers Jr. *Green Berets in the Vanguard.* Annapolis, MD: U.S. Naval Institute Press, 2001.

Bank, Col. Aaron (ret.). *From OSS to Green Berets: The Birth of Special Forces.* California: Pocket Books, 1986.

Barker, Jonathan. *The No-Nonsense Guide to Terrorism.* Oxford, UK, and London: New Internationalist Publications in Association with Verso, 2003.

Beckwith, Col. Charlie A. (ret.), and Donald Knox. *Delta Force: The Army's Elite Counterterrorist Unit.* New York: Avon, 1983.

Berkowitz, Bruce. *The New Face of War: How War Will Be Fought in the 21st Century.* New York: Free Press, 2003.

Berntsen, Gary, and Ralph Pezzullo. *Jawbreaker: The Attack on Bin Laden and Al-Qaeda: A Personal Account by the CIA's Key Field Commander.* New York: Crown Publishers, 2005.

Bowman, John S. *Columbia Chronologies of Asian History and Culture.* New York: Columbia University Press, 2000.

Brisard, Jean-Charles, and Guillaume Dasquié. *Forbidden Truth: U.S.-Taliban Secret Oil Diplomacy and the Failed Hunt for Bin Laden.* New York: Thunder's Mouth Press/Nation Books, 2002.

Briscoe, Charles H., Richard L. Kiper, James A. Schroder, and Kalev I. Sepp. *Weapon of Choice: U.S. Army Special Operations Forces in Afghanistan.* Fort Leavenworth: Combat Studies Institute Press, 2005.

Burke, Jason. *Al-Qaeda: Casting a Shadow of Terror.* New York: I.B. Tauris, 2004.

BIBLIOGRAPHY

Buruma, Ian, and Avishai Margalit. *Occidentalism: The West in the Eyes of Its Enemies.* New York: Penguin, 2004.

Chambers II, John Whiteclay (editor). *The Oxford Companion to American Military History.* United Kingdom: Oxford University Press, 1999.

Coll, Steve. *Ghost Wars: The Secret History of the CIA, Afghanistan, and Bin Laden, from the Soviet Invasion to September 10, 2001.* New York: Penguin, 2004.

Cordesman, Anthony H. *The Lessons of Afghanistan: War Fighting, Intelligence, and Force Transformation.* Washington, D.C.: Center for Strategic and International Studies, 2002.

Crile, George. *Charlie Wilson's War: The Extraordinary Story of the Largest Covert Operation in History.* New York: Atlantic Monthly Press, 2003.

Dupree, Louis. *Afghanistan.* Princeton, NJ: Princeton University Press, 1980.

Emerson, Steven. *Secret Warriors: Inside the Covert Military Operations of the Reagan Era.* New York: Putnam Adult, 1988.

Ewans, Martin. *Afghanistan: A New History.* United Kingdom: Curzon Press, 2002.

Friedman, George. *America's Secret War: Inside the Hidden Worldwide Struggle Between America and Its Enemies.* New York: Doubleday, 2004.

Friedman, Norman. *Terrorism, Afghanistan, and America's New Way of War.* Annapolis, MD: U.S. Naval Institute Press, 2003.

Fuller, J.F.C. *The Generalship of Alexander the Great.* Piscataway, NJ: Rutgers University Press, 1960.

Giardet, Edward, and Jonathan Walter. *Afghanistan.* Geneva: Crosslines Publications, 2004.

Grau, Lester W. (editor). *The Bear Went Over the Mountain: Soviet Combat Tactics in Afghanistan.* London and Portland, OR: Frank Cass Publishers, 2001.

Griffin, Michael. *Reaping the Whirlwind: The Taliban Movement in Afghanistan.* Sterling, VA: Pluto Press, 2001.

Gunaratna, Rohan. *Inside Al Qaeda: Global Network of Terror.* New York: Columbia University Press, 2002.

Gup, Ted. *The Book of Honor: The Secret Lives and Deaths of CIA Operatives.* New York: Random House, 2000.

Holmes, Tony. *American Eagles.* United Kingdom: Classic Publications, 2001.

Hopkirk, Peter. *The Great Game: The Struggle for Empire in Central Asia.* New York: Kodansha International, 1994.

Jalali, Ali Ahmad, and Lester W. Grau. *Afghan Guerrilla Warfare.* St. Paul, MN: MBI Publishing Company, 2001.

Johnson, Chalmers. *The Sorrows of Empire: Militarism, Secrecy, and the End of the Republic.* New York: Henry Holt and Co., 2004.

Kaplan, Robert D. *The Coming Anarchy: Shattering the Dreams of the Post Cold War.* New York: Vintage Books, 2001.

———. *Soldiers of God: With Islamic Warriors in Afghanistan and Pakistan.* New York: Vintage Books, 2001.

Kelly, Col. Francis J. *U.S. Army Special Forces, 1961–1971.* Washington, D.C.: CMH Publications, Department of the Army, 1989.

Kepel, Gilles. *Bad Moon Rising: A Chronicle of the Middle East Today.* London: Saqi Books, 2003.

Kessler, Ronald. *Inside the CIA: Revealing the Secrets of the World's Most Powerful Spy Agency.* New York: Pocket Books, 1992.

BIBLIOGRAPHY

———. *The CIA at War: Inside the Secret Campaign Against Terror*. New York: St. Martin's Press, 2003.

Kirkbride, Wayne A. *The Capture of Che Guevara, Special Forces: The First Fifty Years*. Tampa: Faircourt LLC for the Special Forces Association, 2002.

Kukis, Mark. *My Heart Became Attached: The Strange Odyssey of John Walker Lindh*. Dulles, VA: Brassey's, Inc., 2003.

Landau, Alan M., Frieda W. Landau, Terry Griswold, D. M. Giangreco, and Hans Halberstadt. *U.S. Special Forces: Airborne Rangers, Delta & U.S. Navy Seals*. Osceola, WI: MBI Publishing Company, 1992.

Langguth, A. J. *Our Vietnam: The War 1954–1975*. New York: Simon & Schuster, 2000.

Lawrence, Bruce. *Messages to the World: The Statements of Osama Bin Laden*. New York: Verso, 2005.

Leeming, David. *From Olympus to Camelot: The World of European Mythology*. United Kingdom: Oxford University Press, 2003.

Mayer, Jane. *The Dark Side: The Inside Story of How the War on Terror Turned into a War on American Ideals*. New York: Doubleday, 2008.

Meyer, Karl E. *The Dust of Empire: The Race for Mastery in the Asian Heartland*. New York: Century Foundation, 2003.

Meyer, Karl E., and Shareen Blair Brysac. *Tournament of Shadows: The Great Game and the Race for Empire in Central Asia*. Washington, D.C.: Counterpoint, 1999.

Miller, John, and Michael Stone, with Chris Mitchell. *The Cell: Inside the 9/11 Plot, and Why the FBI and CIA Failed to Stop It*. New York: Hyperion, 2002.

Moore, Robin. *The Hunt for Bin Laden: On the Ground with the Special Forces in Afghanistan*. New York: Random House, 2003.

National Commission on Terrorist Attacks. *The 9/11 Commission Report*. New York: W.W. Norton & Company, 2004.

Neillands, Robin. *In the Combat Zone*. New York: New York University Press, 1998.

The 9/11 Report: The National Commission on Terrorist Attacks Upon the United States. New York: St. Martin's Press, 2004.

Peters, Ralph. *Fighting for the Future: Will America Triumph?* Pennsylvania: Stackpole Books, 1999.

Posner, Gerald L. *Why America Slept: The Failure to Prevent 9/11*. New York: Ballantine Books, 2003.

Rashid, Ahmed. *Taliban: Militant Islam, Oil & Fundamentalism in Central Asia*. New Haven, CT: Yale University Press, 2001.

Ruthven, Malise. *Islam: A Very Short Introduction*. United Kingdom: Oxford University Press, 1997.

Scarborough, Rowan. *Rumsfeld's War: The Untold Story of America's Anti-Terrorist Commander*. Washington, D.C.: Regnery Publishing, Inc., 2004.

Schroen, Gary C. *First In: An Insider's Account of How the CIA Spearheaded the War on Terror in Afghanistan*. New York: Presidio Press, 2005.

Smith, Richard Harris. *OSS: The Secret History of America's First Central Intelligence Agency*. Berkeley: University of California Press, 1972.

Southworth, Samuel A., and Stephen Tanner. *U.S. Special Forces: A Guide to America's Special Operations Units, The World's Most Elite Fighting Force*. Cambridge, MA: Da Capo Press, 2002.

BIBLIOGRAPHY

Stanton, Shelby L. *Green Berets at War: U.S. Army Special Forces in Southeast Asia, 1956–1975.* New York: Ballantine Books, 1999.

Sulima and Hala. *Behind the Burqa: Our Life in Afghanistan and How We Escaped to Freedom* as told to Batya Swift Yasgur (excerpted in the *Guardian*). New York: Wiley, 2002.

Tanner, Stephen. *Afghanistan: A Military History from Alexander the Great to the Fall of the Taliban.* New York: Da Capo Press, 2002.

Tenet, George. *At the Center of the Storm: My Years at the CIA.* New York: Harper-Collins, 2007.

Townshend, Charles. *Terrorism: A Very Short Introduction.* United Kingdom: Oxford University Press, 2002.

Tse-tung, Mao. *On Guerrilla Warfare.* Translation by Samuel B. Griffith. Urbana and Chicago: University of Illinois Press, 2000.

Tzu, Sun. *The Illustrated Art of War.* Translation by Samuel B. Griffith. United Kingdom: Oxford University Press, 1969.

Van Creveld, Martin. *The Transformation of War: The Most Radical Reinterpretation of Armed Conflict Since Clausewitz.* New York: Free Press, 1991.

Weaver, Mary Anne. *Pakistan: In the Shadow of Jihad and Afghanistan.* New York: Farrar, Straus and Giroux, 2003.

Weinberger, Caspar W., and Wynton C. Hall. *Home of the Brave, Honoring the Unsung Heroes in the War on Terror.* New York: Tom Doherty Associates, LLC, 2006.

Whitfield, Susan. *Life Along the Silk Road.* Berkeley: University of California Press,1999.

Wise, James E. Jr., and Scott Baron. *The Navy Cross: Extraordinary Heroism in Iraq, Afghanistan, and Other Conflicts.* Maryland: Naval Institute Press, 2007.

Woodward, Bob. *Bush at War.* New York: Simon & Schuster, 2002.

Yousef, Mohammad, and Mark Adkin. *Afghanistan: The Bear Trap.* Pennsylvania: Casemate, 2001.

GOVERNMENT DOCUMENTS

Attorney General Transcript, John Walker Lindh Press Conference, DOJ Conference Center, January 15, 2002.

"Brief Synopsis of 10th Mountain Division's Major Operations in Afghanistan, 2001–2002." Published by the U.S. Army, Tenth Mountain Division.

CIA World Fact Book, "Yemen," 1998.

"Colonel John F. Mulholland, Jr." Biography published by Fifth Special Forces Group (Airborne), United States Army, July 2001.

"Country Reports on Human Rights Practices for 1994," submitted to the Committee on International Relations, U.S. House of Representatives, and the Committee on Foreign Relations, U.S. Senate, by the Department of State, 1995.

"Country Reports on Human Rights Practices for 1995–1996."

Department of Defense document concerning John Walker Lindh: DOI: (U) 20011201.

Detainee Statements. Combatant Status Review Tribunals conducted at Guantánamo Bay Naval Base (www.defenselink.mil).

BIBLIOGRAPHY

Hall, Barbara POC (final release authority, Colonel Manuel A. Diemer). *The Story of ODA 595*. Undated.

Harlow, Bill. "Public Acknowledgment of CIA Officer Killed in Line of Duty," www.cia.gov, December 3, 2001.

"Honoring Johnny Michael Spann, First American Killed in Combat in War Against Terrorism in Afghanistan, and Pledging Continued Support for Members of Armed Forces." Congressional Record, December 11, 2001 (House), page H9149-H9152.

Letter from John Walker Lindh's Defense Attorney, James Brosnahan, to Ashcroft, Rumsfeld, Tenet, and Powell.

Spann, Michael. CIA Application Essay, 1999.

Spann, Michael. CIA Statement, November 28, 2001.

Tenet, George J. Statement on the Death of a CIA Officer in Afghanistan, November 28, 2001.

United States of America v. *John Philip Walker Lindh, a/k/a/ "Suleyman al-Faris," a/k/a "Abdul Hamid."* Affidavit in Support of a Criminal Complaint and an Arrest Warrant, in the United States District Court for the Eastern District of Virginia, Alexandria Division, January 15, 2002.

U.S. District Court website, text extracted by Jim Scanlon.

ELECTRONIC MEDIA

House of War: Uprising at Mazar-e Sharif. CNN, August 3, 2002.

Pelley, Scott. "Three Soldiers, Many Mourners." *60 Minutes II*, July 28, 2004.

"Profile: General Rashid Dostum." BBC News, September 25, 2001.

Timeline of Terrorists, Through 9/11/2001. Four Corners, Australian Broadcasting Corporation.

Vinci, Alessio. "CNN: Reporter's Notebook: A Scene of Human Carnage and Rubble." CNN, November 26, 2001.

NEWSPAPERS

Ashan, Syed Badrul. "The Afghanistan Story." *Independent*, June 21, 2002.

Beaumont, Peter, Kamal Ahmed, Ed Vulliamy, Jason Burke, Chris Stephen, Tim Judah, and Paul Harris. "The Route of the Taliban: Part Two." *Observer*, November 18, 2001.

Benson, Ross. "Fall of Kabul: Chilling Truth about the Butchers Who Routed the Taliban." *Daily Mail*, November 14, 2001.

Bernton, Hal. "Americans Were Spotted in Terror Training Camps." *Seattle Times*, December 5, 2001.

Borger, Julian. "Bright Boy from the California Suburbs Who Turned Taliban Warrior." *Guardian*, October 5, 2002.

Borger, Julian, and Nicholas Watt. "Doomed Arabs Prepare for Final Battle." *Guardian Weekly*, November 22, 2001.

Burke, Jason. "The Making of the World's Most Wanted Man: Part 1." *Observer*, October 29, 2001.

Cobain, Ian. "Foreign Fighters Resist Alliance; War on Terror." Times Newspapers Limited, November 15, 2001.

———. "Refugees Tell of Frenzied Killing in Besieged City: Battle for Konduz; War on Terror." Times Newspapers Limited, November 19, 2001.

Cobain, Ian, and Damian Whitworth. "U.S. Scorns Deal to Free al-Qaeda's Trapped Mercenaries." Times Newspapers Limited, November 20, 2001.

Cockburn, Patrick. "Rashid Dostum: The Treacherous General." *Independent* (UK), December 1, 2001.

Connerty-Marin, David, and Josie Huang. "The Night Before Terror." *Portland Press Herald/Maine Sunday Telegram*, October 5, 2001.

Crary, David. "Many in U.S. Fought for Foreign Armies." Associated Press, December 24, 2001.

———. "Broad Effort Launched After '98 Attacks." *Washington Post*, December 19, 2001.

Cullison, Alan, and Robert S. Greenberger. "Airstrikes Are Criticized by Northern Alliance, as U.S. Looks for Broader Post-Taliban Plans." *Wall Street Journal*, October 15, 2001.

Dawoud, Khaled, Julian Borger, and Nicholas Watt. "Doomed Arab Units Prepare for Final Battle Against the Odds." *Guardian*, November 20, 2001.

Ellison, Michael. "U.S. Taliban Fighter Describes Fortress Horror." *Guardian*, December 3, 2001.

Estes, Tracy. "Winfield War Hero Honored by CBS News on Memorial Day." *Marion County Journal Record*, May 2002.

"Father of a U.S. Taliban Fighter Speaks Out." Associated Press, January 20, 2006.

Filkins, Dexter. "U.S. Warns of Bounties Posing Threat to Westerners." *New York Times*, April 6, 2002.

Gellman, Barton. "A Strategy's Cautious Evolution Before Sept. 11; the Bush Anti-Terror Effort Was Mostly Ambition." *Washington Post*, January 20, 2002.

Griffin, Michael. "A Gruesome Record: The Northern Alliance May Be Trying to Rebrand Themselves, but the People of Kabul Are Unlikely to Forget Their Past Atrocities." *Guardian*, November 16, 2001.

Harden, Blaine, with Kevin Sack, Douglas Frantz, and Carlotta Gall. Evelyn Nieves also contributed. "A Nation Challenged: The Meeting; One for His Country, and One Against It." *New York Times*, December 11, 2001.

Harding, Luke. "Good at War, Poor at Peace." *Guardian*, September 12, 2001.

———. "Paper Surrender Blowing in the Wind." *Guardian*, November 23, 2001.

Harding, Luke, Nicholas Watt, and Brian Whitaker. "Kunduz: Northern Stronghold Ready to Capitulate." *Guardian*, November 22, 2001.

Hornak, Paul. "Troops of 10th Recount Mayhem at Mazar-e-Sharif." *Watertown Daily Times*, April 4, 2002.

Jehl, Douglas. "Four in 9/11 Plot Are Called Tied to Qaeda in '00." *New York Times*, August 9, 2005.

Kaplan, Robert. "The Afghan Who Won the Cold War." *Wall Street Journal*, May 5, 1992.

Kurtz, Howard. "Interview Sheds Light on Bin Laden." *Washington Post*, February 7, 2002.

LaFraniere, Sharon. "Bombing of Enclave Intensifies; Several Thousand Hold Out in Taliban's Northern Stronghold." *Washington Post*, November 19, 2001.

Lindh, John Walker. "Excerpts from Statement by John Walker Lindh in Court." *New York Times*, October 5, 2002.

Litvinenko, Alexendar. "Watch Your Alliances." *Washington Post*, August 31, 2002.

Loeb, Vernon. "U.S. Soldiers Recount Smart Bomb Blunder." *Washington Post*, February 2, 2002.

Loeb, Vernon, and Josh White. "CIA Reports Officer Killed in Prison Uprising." *Washington Post*, November 29, 2001.

McDermott, Terry. "Sunday Report; A Perfect Soldier." *Los Angeles Times*, January 27, 2002.

———. "Sunday Report; The Plot; How Terrorists Hatched a Simple Plan to Use Planes as Bombs." *Los Angeles Times*, September 1, 2002.

McDermott, Terry, and Dirk Laabs. "The World; Column One; Prelude to 9/11: A Hijacker's Love, Lies." *Los Angeles Times*, January 27, 2003.

McDermott, Terry, Josh Meyer, and Patrick J. McDonnell. "Sunday Report; The Plots and Designs of Al Qaeda's Engineer." *Los Angeles Times*, December 22, 2002.

McMurray, Jeffrey. "CIA Agent's Dad Probes Deadly Afghan Riot." Associated Press, March 13, 2005.

———. "Father on Crusade to Prove Afghanistan Ambush Killed CIA Officer." Associated Press, March 12, 2005.

Oliver, Mark, and Derek Brown. "Mullah Faizal." *Guardian*, December 3, 2001.

Oppel, Richard A., Abdul Waheed Wafa, and Sangar Rahimi. "20 Dead as Taliban Attackers Storm Kabul Offices." *New York Times*, February 11, 2009.

Orndorff, Mary. "CIA's Spann Buried at Arlington." *Birmingham News*, December 11, 2001.

Parsell, D. L. "Afghanistan Reporter Looks Back on Two Decades of Change." *National Geographic News*, November 19, 2001.

Risen, James. "A Nation Challenged: The Casualty; C.I.A. Names Agent Killed in Fortress." *New York Times*, November 29, 2001.

Rohde, David, and C. J. Chivers. "A Nation Challenged: Qaeda's Grocery Lists and Manuals of Killing." *New York Times*, March 17, 2002.

Russell, Rosalind. "Afghan Warlord Dostum Says He Needs Arms to Advance." Reuters, October 31, 2001.

Sack, Kevin. "A Nation Challenged: A Town Remembers; Community Recalls a Native Son with Clear Goals." *New York Times*, November 29, 2001.

Schemo, Diana Jean. "A Nation Challenged: The Burial; Agent Praised as Patriot in Graveside Ceremony." *New York Times*, December 11, 2001.

Serrano, Richard A. "Detainees Describe CIA Agent's Slaying." *Los Angeles Times*, December 8, 2004.

———. "Driven by a Son's Sacrifice." *Los Angeles Times*, April 7, 2005.

Swain, Jon. "The Northern Alliance's Cruel History: Mujaheddin Write Their Name in Blood." *Sunday Times* (London), November 11, 2001.

Torchia, Christopher. "Afghan Commander Gets Indoor Pool." Associated Press, September 8, 2002.

Weinstein, Joshua. "Link to Portland Grows." *Portland Press Herald/Maine Sunday Telegram*, September 14, 2001.

Wilford, John Noble. "How Catapults Married Science, Politics, and War." *New York Times*, February 24, 2004.

BIBLIOGRAPHY

MAGAZINES AND JOURNALS

Anderson, Jon Lee. "Letter from Kabul: The Assassins." *New Yorker*, June 10, 2002.

Anderson, Martin Edwin. "How Col. John Boyd Beat the Generals." *Insight* magazine, August 2002.

Barry, John, and Michael Hirsh. "Commandos: The Real Tip of the Spear." *Newsweek*, October 8, 2003.

Burke, Jason. "Taliban Kabul Diary." *London Review of Books*, March 22, 2001.

Cloud, John. "Atta's Odyssey." *Time*, October 8, 2001.

———. "The Plot Comes Into Focus." *Time*, October 1, 2001.

Davis, Anthony. "Makeover for a Warlord." *Time*, June 3, 2002.

Dorr, Robert F. "Air Force Special Operations Command." A *Tribute to Special Operations*, January 2003.

Dunlap, Col. Charles J. Jr., USAF. "Special Operations Forces after Kosovo." *Joint Force Quarterly*, Spring/Summer 2001.

Gannon, Kathy. "Letter from Afghanistan: Road Rage." *New Yorker*, March 22, 2004.

Gray, Colin S. "Why Strategy Is Difficult." *Joint Force Quarterly*, Summer 1999.

Gresham, John D. "Rangers, Nightstalkers and Green Berets: The U.S. Army Special Operations Command." A *Tribute to Special Operations*, January 2003.

———. "Snake Eaters Ball: Operation Enduring Freedom." A *Tribute to Special Operations*, January 2003.

Grizwold, Eliza. "Where the Taliban Roam." *Harper's Magazine*, September 2003.

Gutman, Roy, and John Barry. "Periscope: Digging Up the Truth." *Newsweek*, September 30, 2002.

Hall, Barbara. "USASOC History: From Jedburghs to Devils and Snakes." A *Tribute to Special Operations*, January 2003.

———. "The Liberation of Mazar-e Sharif: 5th SF Group Conducts UW in Afghanistan." *Special Warfare*, June 2002.

Hammonds, Keith H. "The Strategy of the Fighter Pilot." *Fast Company*, June 2002.

Hylton, Wil S. "Mazar I Sharif." *Esquire*, August 2002.

Isikoff, Michael, and Daniel Klaidman. "The Hijackers We Let Escape." *Newsweek*, June 10, 2002.

Junger, Sebastian. "The Lion in Winter." *National Geographic Adventure*, March/April 2001.

Junod, Tom. "Innocent: Can America and Islam Coexist?" *Esquire*, July 2006.

Kaplan, David. "Made in the U.S.A." *U.S. News & World Report*, June 10, 2002.

Klaidman, Daniel, and Michael Isikoff. "Walker's Brush with Bin Laden." *Newsweek*, December 31, 2001.

Klein, Edward. "Love in a Time of War." *Parade* magazine, August 18, 2002.

Maass, Peter. "A Bulletproof Mind." *New York Times Magazine*, November 10, 2002.

Mayer, Jane. "Lost in the Jihad." *New Yorker*, March 10, 2003.

McCaffrey, General Barry R., USA (ret.). "Lessons of Desert Storm." *Joint Force Quarterly*, Winter 2000–2001.

Mishra, Pankaj. "The Afghan Tragedy." *New York Review of Books*, January 17, 2002.

———. "The Real Afghanistan." *New York Review of Books*, March 10, 2005.

Pelton, Robert. "The Legend of Heavy D & the Boys (or How the Green Berets Learned to Stop Worrying and Love the Warlord)." *National Geographic Adventure*, March 2002.

Perry, Alex. "Friday Night in Tashkent." *Time*, October 29, 2001.

———. "Reporters' Notebook: Alex Perry in Uzbekistan. Halloween at the End of the Earth." *Time Asia*, November 2, 2001.

———. "Eyewitness: The Taliban Undone." *Time*, November 14, 2001.

———. "Sexual Liberation." *Time*, November 17, 2001.

———. "Lying to Refugees." *Time*, November 23, 2001.

———. "Mass Slaughter of the Taliban's Foreign Jihadists." *Time*, November 26, 2001.

Personnel of the Third Battalion, Fifth Special Forces Group. "The Liberation of Mazar-e Sharif: 5th SF Group Conducts UW in Afghanistan." *Special Warfare*, June 2002.

Pyes, Craig, and William C. Rempel. "Slowly Stalking an Afghan Lion." *Los Angeles Times*, June 14, 2002.

Ratnesar, Romesh. "The Afghan Way of War." *Time*, November 19, 2001.

Robinson, Linda. "Walking Point." *U.S. News & World Report*, October 18, 2004.

Roche, Timothy, Brian Bennett, Anne Berryman, Hilary Hylton, Siobhan Morrissey, and Amany Radwan. "The Making of John Walker Lindh." *Time*, October 7, 2002.

Soloway, Colin. "Periscope: Reporters' Notebooks: Our Eyewitnesses to History." *Newsweek*, December 31, 2001.

Stein, Harry. "How the Father Figures." *City Journal*, January 28, 2002.

Thomas, Evan. "A Street Fight." *Newsweek*, April 29, 2002.

———. "The Story of September 11." *Newsweek*, December 31, 2001.

Waller, Douglas. "Inside the CIA's Covert Forces." *Time*, December 10, 2001.

———. "The CIA's Secret Army." *Time*, February 3, 2003.

Zimmerman, Dwight J. "WWII Special Operations Forces." *A Tribute to Special Operations*, January 2003.

INTERNET SITES

"American Taliban Not Linked to CIA Agent's Death." www.afgha.com, April 2, 2002.

Atta, Mohammad. "Suicide Note." www.abcnews.com, September 28, 2001.

Barela, Timothy P., Tech. Sgt. "Let Your Constellation Be Your Guide." www.af.mil/news/airman/0696/cons.htm.

"Bin Laden Granted Afghan Nationality." www.paknews.org, November 9, 2001.

"Biography: Ahmad Shah Massoud." www.afgha.com, August 31, 2006.

"BMP-3 Russian infantry fighting vehicle." www.wikipedia.com, November 24, 2006.

Breslau, Karen, and Colin Soloway. "'He's a Really Good Boy.' The Parents of the American Talib Describe Their Son." *Newsweek* Web exclusive, December 7, 2001.

Brown, Janelle. "The Taliban's Bravest Opponents." www.salon.com, October 2, 2001.

BIBLIOGRAPHY

Carroll, Linda. "The Making of a Perfect Soldier." www.MSNBC.com, March 7, 2003.

"CIA Officer Michael Spann Buried at Arlington National Cemetery." www.CNN.com/Transcripts, aired December 10, 2001.

Dana, Peter H. Global Positioning System Overview, Department of Geography, University of Texas at Austin, 1994, www.colorado.edu/geography/gcraft/notes/gps/gps_f.html.

"Discussion with Widow of First American to Die in Afghan Combat." www.CNN.com/Transcripts, aired September 14, 2002.

Fagan, Kevin. "John Lindh Not a Traitor, Father Argues. He Says He Didn't Fight U.S., but Was Rescued." www.SFGate.com, January 20, 2006.

"Family of Michael Spann Speak to Reporters Following Lindh Not Guilty Plea." www.CNN.com/Transcripts, aired February 13, 2002.

Farzana. "Massoud." www.afgha.com, August 31, 2006.

"Father of Slain CIA Agent Blames Lindh." www.CBSnews.com, July 23, 2002.

"Frontline: Campaign Against Terror." www.pbs.org/wgbh/pages/frontline/shows/campaign/interviews/595.html, undated.

House of War: The Uprising at Mazar-e Sharif. www.CNN.com/Transcripts, aired August 3, 2002.

"Interview with Joseph Biden; Stories of Relatives of Heroes of United Flight 93." www.CNN.com/Transcripts, aired December 7, 2001.

"Interview with Shannon Spann." www.CNN.com/Transcripts, aired July 16, 2002.

King, Larry. Interview with Mark Miller and Colin Soloway, transcript from www.CNN.com, December 7, 2001.

Koval, Gretel C. "Walker: The Road Ahead." *Newsweek* Web exclusive, December 12, 2001.

Lak, Daniel. "Mazar-e-Taliban, R.I.P." www.afgha.com, December 16, 2001.

Lindh, Frank R. "The Real Story of John Walker Lindh." www.alternet, January 24, 2006.

———. "Taking the Stand: The Crimeless Crime: The Prosecution of John Walker Lindh." www.dcbar.com, May 2005.

Madison, David. "American Taliban." www.indyweek.com, January 30, 2002.

Marlowe, Ann. "'Warlords' and 'Leaders.'" *National Review* Online, Febuary 19, 2002.

"Massoud's Last Words." *Newsweek* Web exclusive, September 20, 2001.

McElroy, Damien. "I'm Sick of Death, Says Dostum the Warlord." www.telegraph.co.uk, November 29, 2001.

Perry, Alex. Reports by satellite. "Update: American Rescued from Taliban-held Fort." *Time* Online Edition, November 25, 2001.

———. "Inside the Battle at Qala-i-Janghi." *Time* Online Edition, December 1, 2001.

"Profile: John Walker Lindh." (Collection of media reports, page snapshot, November 12, 2008), www.historycommons.org.

Qadri, Dr. Hafiz Haggani. "Muslims Are Not Ignoramuses." www.binoria.org.

Sills, Sam. "The Abraham Lincoln Brigade of the Spanish Civil War." www.english.upenn.edu.

Soloway, Colin. "He's Got to Decide if He Wants to Live or Die Here." *Newsweek* Web exclusive, December 6, 2001.

BIBLIOGRAPHY

"Spann Described as a Hero." www.CNN.com, posted November 28, 2001.
"Spann's Father Wants Stiff Sentence for Lindh." www.CNN.com, July 22, 2002.
Spann, Johnny. "False and Misleading Statements by Mr. Frank Lindh Omit Many Known Facts." www.honormikespann.com, February 2006.
———. "Johnny Spann's Letter on Recent Trip to Afghanistan." www.honor mikespann.org, updated January 2003.
Spann, Shannon. "Embrace His Grace." www.embracehisgrace.com/Inspirational .htm.
"Spirit Fest Helps Spann Widow Deal with Loss of Event Honoree." www.arlington cemetery.com/jmspann.htm, press report, June 23, 2002.
"Translation of Hijacker's Note." www.ABCNews.com, September 28, 2001.
www.af.mil; www.af.mil/news/airman/0690/space2.htm (U.S. Air Force)
www.afghan-network.net
www.bannu.itgo.com
www.colorado.edu/geography/gcraft/notes/gps/gps_f.html
www.defenselink.mil
www.globalsecurity.org
www.jlcint.com (Yemen Language Center)
www.nasm.si.edu/galleries/gps (Smithsonian Institute, National Air and Space Museum), 1998
www.pbs.org/newshour/terrorism/international/fatwa_1996.html
www.winfieldcity.org
www.wpafb.af.mil (USAF Museum)

OTHER SOURCES

"Afghanistan: The Massacre in Mazar-i-Sharif." Human Rights Watch, November 1998.
"Are You Going to Talk to Us?" Videotape of Mike Spann's interrogation of John Walker Lindh, transcript published by Newsweek, December 17, 2001.
Billingsley, Dodge. Unedited video of fort siege. November 26–December 1, 2001.
Federation of American Scientists
Logistics of World Trade Center Explosion, National Institute of Standards and Technology, September 2005.
"Parents' Grief for American Taleban." Transcript of BBC program on www.bbc.co.uk (November 15, 2002).
Transcript, "Distinguished Service Cross Award Ceremony for Mark Mitchell." MacDill Air Force Base, Tampa, FL, November 14, 2003.
World Book Multimedia Encyclopedia, 2002.

ABOUT THE AUTHOR

DOUG STANTON is the author of *The Odyssey of Echo Company: The 1968 Tet Offensive and the Epic Battle to Survive the Vietnam War* and the *New York Times* bestsellers *In Harm's Way: The Sinking of the USS* Indianapolis *and the Extraordinary Story of Its Survivors* and *Horse Soldiers: The Extraordinary Story of a Band of U.S. Soldiers Who Rode to Victory in Afghanistan*, which is the basis for the major motion picture *12 Strong*. Stanton is a founder of the National Writers Series, a year-round book festival, and lives in his hometown of Traverse City, Michigan, with his wife, Anne Stanton, and their three children, John, Katherine, and Will. His writing has appeared in *Esquire*, the *New York Times*, the *New York Times Book Review*, *Time*, the *Washington Post*, *Men's Journal*, *Outside*, *The Daily Beast*, and *Newsweek*.

Turn the page for an excerpt
from Doug Stanton's powerful new book

THE ODYSSEY OF ECHO COMPANY

THE 1968 TET OFFENSIVE AND THE EPIC BATTLE TO SURVIVE THE VIETNAM WAR

"A book for all Americans to read."
—Tom Brokaw

"Stanton's searing tale of war and homecoming
will soon find its place on a rarefied shelf alongside
Matterhorn, *A Bright Shining Lie*, *The Things They Carried*,
and *We Were Soldiers Once . . . and Young*, which is to say,
among the classics of the Vietnam War."
—Hampton Sides

"We are finally ready to learn about Vietnam,
and no book tells the story better than this one."
—*Library Journal* (starred review)

INVOCATION

January 31, 1968

Landing Zone Jane, Northern I Corps near Hai Lang, South Vietnam

It's 4:00 a.m. when they attack.

Stan lifts himself up from the puddle, whiskered face dripping wet. He's been sleeping in cold water, trying to pull an imaginary blanket over his aching shoulders, water filling his nose, his ears . . . He hears whistles. Shrill whistles, like a referee's, the same sound he'd heard in high school when he was about to pin a wrestling opponent. But now soldiers are running past him. What's happening?

He feels a blow to the head and he's knocked dizzy.

He tries standing, but he can't. He can't move. His cold bed made of muddy water has left him numb. He's scared. He's so utterly embarrassed and ashamed about this. Who's watching? *he wonders.* Who will know I can't move? *He sees an enemy soldier running in his direction. A sweaty face. Lit by flashes as it gallops toward him, coming into focus. Stan freezes. He sees the soldier's silver bayonet riding toward him, the sharp point swinging back and forth in the damp air, hunting him, when the soldier pitches forward and the bayonet falls past Stan's face, just missing. The enemy soldier crashes down on Stan.*

He looks up and sees another soldier, an American, an older seasoned trooper, jerking his own bayonet from the attacking soldier's back. The trooper yells, urging him to get up.

The dead soldier is dressed in khaki pants and shirt, the uniform of the North Vietnamese Army. He's young, maybe Stan's age, nineteen or twenty. Stan's never been this close to or touched a dead person before, except when he said good-bye to his mother at her funeral a few months back. He pushes the body away and jumps up.

And he starts running.

He runs toward the concertina wire at the rim of the hill where the landing zone (LZ Jane) is situated. Stan knows the 101st Airborne soldiers have to hold the wire—but how? Artillery is blasting now, long shells crunching over Stan's head, sailing into the dark. The ground is shaking. Illumination flares throw gray shadows against the trees, making enormous apparitions that crawl through the branches, limb to limb.

Stan looks down the hill and sees hundreds of NVA soldiers pouring up the scrubby draw at him. He's afraid of dying. He drops to one knee and starts firing his M-16. And then he sees something even more amazing.

The enemy soldiers start pole-vaulting over the wire.

They run up the hill, bamboo poles bouncing on their shoulders, plant the poles, swing up into the night, illuminated by the flares, hang there, captured at this apex as if in a photograph . . . and fall back to earth on the other side of the wire. Others who leap across the sky disappear in red mist or an expanding cloud of bone as the machine gunners pour fire into them. The heavy rounds eat the men right out of the air.

The surviving pole vaulters run past Stan, headed toward the center of the LZ, clutching canvas pouches.

Stan hears someone yell, "Sappers!"

In the pouches are explosives. The sappers are headed on suicide missions to the command bunker.

Everyone in the platoon fires at these running men. When hit, the men detonate with a mighty force. Stan looks up—it's begun to rain. He tastes blood.

It's raining men, exploded men.

And then something flashes in him, some loom is unfolded in him, and the loom's shuttle commences back and forth, across the soft treasure of who he is. Shuttle, whisper, weave; shuttle. Back in high school, he heard stories of nights like these, of U.S. soldiers overrun in their camp by the North Vietnamese and the Viet Cong soldiers. Around him, the long barrels of machine guns start to glow red. An NVA soldier, ten feet away, raises his rifle to fire— Stan shoots him. He lunges with his bayonet into another man as he rushes past him.

Stan catches a quick movement—something lands on his shoulder, light as a bird.

He turns and sees two detached fingers perched on his uniform, trembling as he breathes. Pointed upward, they grow still. He lifts his eyes.

The sky, the sky. Who is watching this? Who?

PART I

THE GIRL WITH THE PEACHES

May 27, 2005

Kabul, Afghanistan

I first met Stan Parker in May 2005 when I was climbing into a Chinook helicopter on the tarmac at Bagram Air Base in Afghanistan. I was trying to get to a U.S. Army Special Forces camp in Khost, on the Pakistani border. Wearing wraparound shades and a keffiyeh, Stan was older by at least twenty years than many of the soldiers I saw walking around the airstrip. He was in charge of air operations at Bagram Air Base, and this meant that he had charge of me. Unfortunately, the weather was not cooperating, and we weren't flying anywhere. But fortunately for me, Stan Parker started telling stories while we waited.

Short, sandy-haired, broad-chested, Stan had an easy smile and talked a lot like Robert Duvall in *Lonesome Dove*. He was wearing body armor, with an M-4 carbine slung across his chest. I wouldn't want to mess with him. After thirty-five years of military service, he told me, he was finally thinking of "getting ready for retirement."

His had been quite a career. In 1993, he'd been in the Battle of Mogadishu in Somalia. He'd been part of the Special Forces operation in Honduras, the Philippines, Korea, and Eritrea, Africa. He'd been in gun battles in Afghanistan. When I asked Stan how many firefights he'd been in, he could not come up with an answer—hundreds, perhaps. He had achieved the rank of sergeant major and had been assigned to U.S. Special Operations Command, at MacDill Air Force Base, in Tampa, Florida. He was one of the Army's senior elite soldiers, and, on top of this, he was deeply wired into America's counterterrorism fight across the globe. He knew things. He'd seen things, he told me, "beyond a civilized person's comprehension."

Yet when I'd asked him to name the scariest part of his military career, he said, "Coming home from Vietnam."

And on the day we met, that's what Stan Parker really wanted to talk about: what had happened to him thirty-seven years earlier when he was twenty, during the 1968 Tet Offensive.

Vietnam. 1968. January.

Stan had read my book *In Harm's Way*, and this had given him an idea. He'd seen that story about World War II, and the sinking of a Navy ship and the ordeal of its crew, as a survival story.

Stan wanted to know if I would ever write about how he and his buddies had survived the Tet Offensive, when his forty-six-man Reconnaissance Platoon attacked and was attacked by well-trained North Vietnamese Army and Viet Cong fighters. Stan's odyssey had lasted ninety days, until he was wounded for a third time and forced to leave the intense brotherhood of his unit. A soldier awarded three Purple Hearts had no choice but to be shipped home. In order to stay in Vietnam, he'd refused the third award. (He was subsequently assigned to another unit.)

I took Stan's suggestion about writing this story and filed it under "Maybe." I didn't think America was ready to hear that story while in the midst of the wars in Iraq and Afghanistan. Maybe the country never would be ready. Part of this had to do with the age of the men who'd fought in Vietnam. Now in their late fifties and early sixties, they weren't old enough to want to talk, not just yet.

In September 2012, seven years later, I gave a lecture to cadets and command staff at the U.S. Air Force Academy in Colorado Springs, and I remembered that Stan Parker, now retired, lived nearby. We hadn't spoken in several years, and I called him.

"I've been waiting for you," he said, surprising me. "I've told my buddies about you. We're ready. We want to talk about Vietnam."

The truth was that I'd never forgotten meeting Sergeant Major Stan Parker on that helicopter in Afghanistan. I still remembered the sunlight coming in through the green helo's rear ramp as this old soldier asked me to write about something that had happened to him and his Recon buddies years earlier, something that had affected them deeply and that they didn't understand.

Tom Brokaw, he'd said then, was the only civilian who'd ever looked at his Combat Infantryman Badge with star and recognized that he'd served in Vietnam. Stan had met Brokaw a few months earlier on a helicopter in Afghanistan when he was reporting an NBC News special. Stan had been tasked as his bodyguard. He proudly showed me a photo of their meeting. Back in the late 1990s, Brokaw had discovered the willingness of World War II veterans to unpack their secrets about their war as they turned seventy and felt perhaps that it was time to unburden themselves.

Maybe, just maybe, I wondered, the same might now be true of the more than 2.7 million Americans who had deployed in Vietnam, 1.6 million of whom experienced combat or the threat of attack. Their average ages now ranged from sixty-five to sixty-nine.

That's a lot of people, I thought, *a lot of untold stories*. It seemed time for them to come home.

• • •

Even before he retired, Stan wanted to track down his former Recon 1/501 platoon-mates and ask them what they remembered. Some of them he found on the Internet didn't want to be reminded of the past and hung up when we called. He spent a surreal dinner with a former Recon member who made the nervous admission that he'd never told his wife that he'd served in Vietnam. He pulled Stan aside and begged him not to tell his family. Stan made up a fib that they once worked together many years earlier.

This pattern of evasion troubled him. He wondered what it meant when a person couldn't admit that a major chapter of his life, perhaps its most potent and transformative moment, had ever taken place.

And so Stan and I started to talk at great length over a span of several years. As with so many other stories about Vietnam, Stan's would take a long time to tell. As he and I spoke, I occasionally saw a shadow in his eyes—call it memory, call it flashback. Something was there, traveling back and forth through his consciousness, unresolved. *So many things, Doug. If I talk, I might be whole, I'll be unburdened, I'll be heard. I won't be a better person, I know that, but I might be the person I am. Does that make sense?*

It does.

I want to go on.

Go on.